Methods in Molecular Biology

Series Editor
John M. Walker
School of Life and Medical Sciences
University of Hertfordshire
Hatfield, Hertfordshire, UK

For further volumes:
http://www.springer.com/series/7651

For over 35 years, biological scientists have come to rely on the research protocols and methodologies in the critically acclaimed *Methods in Molecular Biology* series. The series was the first to introduce the step-by-step protocols approach that has become the standard in all biomedical protocol publishing. Each protocol is provided in readily-reproducible step-by-step fashion, opening with an introductory overview, a list of the materials and reagents needed to complete the experiment, and followed by a detailed procedure that is supported with a helpful notes section offering tips and tricks of the trade as well as troubleshooting advice. These hallmark features were introduced by series editor Dr. John Walker and constitute the key ingredient in each and every volume of the *Methods in Molecular Biology* series. Tested and trusted, comprehensive and reliable, all protocols from the series are indexed in PubMed.

Alternative Splicing

Methods and Protocols

Edited by

Peter Scheiffele and Oriane Mauger

Biozentrum, University of Basel, Basel, Switzerland

Humana Press

Editors
Peter Scheiffele
Biozentrum, University of Basel
Basel, Switzerland

Oriane Mauger
Biozentrum, University of Basel
Basel, Switzerland

ISSN 1064-3745 ISSN 1940-6029 (electronic)
Methods in Molecular Biology
ISBN 978-1-0716-2520-0 ISBN 978-1-0716-2521-7 (eBook)
https://doi.org/10.1007/978-1-0716-2521-7

This Humana imprint is published by the registered company Springer Science+Business Media, LLC, part of Springer Nature.
The registered company address is: 1 New York Plaza, New York, NY 10004, U.S.A.

Preface

Alternative splicing is a fundamental, evolutionarily conserved process that vastly expands the coding power of genomes. Alternative splicing regulation is a key step in gene regulation. Maybe best recognized as the major source of proteome diversification, alternative splicing also controls transcript dynamics including their localization, stability, and translation. Consequently, alternative splicing underlies fundamental cellular processes including cell growth, differentiation, and death, and its misregulation is associated with severe pathologies.

At the time of its discovery in the late 1970s, alternative splicing was thought to be rare and affect only a few genes. Over the past two decades and with the breakthrough of high-throughput methods, alternative splicing has been recognized to impact almost every human gene. Given the high prevalence of alternative splicing and its essential functions, many laboratories whose primary focus is not on RNA have begun investigating the regulation and relevance of alternative splicing in their biological model. Furthermore, while historically studies on alternative splicing mostly relied on biochemistry and molecular biology approaches, the field has significantly evolved over the past years and involved a vast array of methods, including high-resolution microscopy and bioinformatics. Thus, we thought that having a compendium of chapters describing key methods on alternative splicing would be helpful for both RNA aficionados that want to implement novel approaches in their labs and novices undertaking alternative splicing projects.

In this book, we gathered chapters written by experts in their fields that have implemented commonly used and cutting-edge methods to analyze alternative splicing. The first part of the book comprises two introductory chapters laying down a foundation for the alternative splicing mechanism and its targeting for therapeutic strategies. In the second part of the book, we cover a large spectrum of experimental and analytical approaches with four major goals: (1) analyzing alternative splicing profiles in complex biological systems, (2) visualizing and localizing alternative spliced transcripts with cellular and subcellular resolution, (3) probing regulators of alternative splicing, and (4) assessing the functional consequences of alternative splicing. Each chapter contains a brief introduction section to define the scope of the method and detailed protocols with extensive tips to succeed in implementing novel approaches in the lab.

We would like to sincerely thank all authors for their effort and engagement. Through their valuable contributions, we are confident this book will be useful for scientists who want to enter or dig deeper into the fascinating splicing world.

Basel, Switzerland

Peter Scheiffele
Oriane Mauger

Contents

Contributors

KAUR ALASOO • *Institute of Computer Science, University of Tartu, Tartu, Estonia*
FRÉDÉRIC H. -T. ALLAIN • *Department of Biology, Institute of Biochemistry, Swiss Federal Institute of Technology, Zurich, Switzerland*
ERIC ALLEMAND • *Laboratory of Cellular and Molecular Mechanisms of Hematological Disorders and Therapeutic Implications, Institut IMAGINE, INSERM, Paris, France*
ARTURO ANDRADE • *Department of Biological Sciences, College of Life Sciences and Agriculture, University of New Hampshire, Durham, NH, USA; Robert J. and Nancy D. Carney Institute for Brain Science, Brown University, Providence, RI, USA*
FABRICE ANGO • *INM, University of Montpellier, INSERM, Montpellier, France*
PAULINA BRÖNNIMANN • *Department of Chemistry and Applied Biosciences, Institute of Pharmaceutical Sciences, ETH Zürich, Zürich, Switzerland; Translational Research Unit, Institute of Pathology, University of Bern, Bern, Switzerland*
ALEXANDRA BUNDA • *Department of Biological Sciences, College of Life Sciences and Agriculture, University of New Hampshire, Durham, NH, USA*
JOHN A. CALARCO • *Department of Cell and Systems Biology, University of Toronto, Toronto, ON, Canada*
SÉBASTIEN CAMPAGNE • *Department of Biology, Institute of Biochemistry, Swiss Federal Institute of Technology, Zurich, Switzerland; ARNA Laboratory, INSERM U1212, CNRS 5320, University of Bordeaux, Bordeaux, France*
JESSICA L. CENTA • *Center for Genetic Diseases, Chicago Medical School, Rosalind Franklin University of Medicine and Science, North Chicago, IL, USA*
JEFFREY A. CHAO • *Friedrich Miescher Institute for Biomedical Research, Basel, Switzerland*
NICHOLAS K. CONRAD • *Department of Microbiology, UT Southwestern Medical Center, Dallas, TX, USA*
TEBBE DE VRIES • *Department of Biology, Institute of Biochemistry, Swiss Federal Institute of Technology, Zurich, Switzerland*
GIULIA DI BARTOLOMEI • *Biozentrum of the University of Basel, Basel, Switzerland*
ANDREA B. EBERLE • *Department of Chemistry, Biochemistry and Pharmaceutical Sciences, University of Bern, Bern, Switzerland*
BASTIAN TH. EICHENBERGER • *Friedrich Miescher Institute for Biomedical Research, Basel, Switzerland; University of Basel, Basel, Switzerland*
GORDON FISHELL • *Stanley Center for Psychiatric Research, The Broad Institute, Cambridge, MA, USA*
LUCA GIORGETTI • *Friedrich Miescher Institute for Biomedical Research, Basel, Switzerland*
ANDRÉ GOHR • *Centre for Genomic Regulation, Barcelona Institute of Science and Technology, Barcelona, Spain*
ASHWIN GOVINDAN • *Department of Microbiology, UT Southwestern Medical Center, Dallas, TX, USA*
JONATHAN HALL • *Department of Chemistry and Applied Biosciences, Institute of Pharmaceutical Sciences, ETH Zürich, Zürich, Switzerland*
FRANÇOIS HALLOY • *Department of Chemistry and Applied Biosciences, Institute of Pharmaceutical Sciences, ETH Zürich, Zürich, Switzerland; Department of Paediatrics, Medical Sciences Division, University of Oxford, Oxford, UK*

FURSHAM HAMID • *Centre for Developmental Neurobiology, King's College London, London, UK*

MICHELLE L. HASTINGS • *Center for Genetic Diseases, Chicago Medical School, Rosalind Franklin University of Medicine and Science, North Chicago, IL, USA*

STEPHANIE HEINRICH • *Institute of Biochemistry, ETH Zurich, Zurich, Switzerland*

MARTIN HEMBERG • *Wellcome Sanger Institute, Cambridge, UK; Evergrande Center for Immunologic Diseases, Harvard Medical School and Brigham and Women's Hospital, Boston, MA, USA*

ANTONIO HERMOSO-PULIDO • *Centre for Genomic Regulation, Barcelona Institute of Science and Technology, Barcelona, Spain*

MARIA HONDELE • *Biozentrum, University of Basel, Basel, Switzerland*

MANUEL IRIMIA • *Centre for Genomic Regulation, Barcelona Institute of Science and Technology, Barcelona, Spain; Universitat Pompeu Fabra, Barcelona, Spain; ICREA, Barcelona, Spain*

DANIEL JUTZI • *United Kingdom Dementia Research Institute Centre, Institute of Psychiatry, Psychology and Neuroscience, King's College London, Maurice Wohl Clinical Neuroscience Institute, London, UK*

BOXING LI • *Department of Physiology, Guangdong Provincial Key Laboratory of Brain Function and Disease, Zhongshan School of Medicine, Sun Yat-sen University, Guangzhou, China*

JINGHONG LI • *Department of Chemistry, Key Laboratory of Bioorganic Phosphorus Chemistry and Chemical Biology, Tsinghua University, Beijing, China*

ZHELIN JEFF LI • *Division of Biomedical Sciences, University of California, Riverside, CA, USA*

EUGENE MAKEYEV • *Centre for Developmental Neurobiology, King's College London, London, UK*

FEDERICA MANTICA • *Centre for Genomic Regulation, Barcelona Institute of Science and Technology, Barcelona, Spain*

YAMILE MÁRQUEZ • *Centre for Genomic Regulation, Barcelona Institute of Science and Technology, Barcelona, Spain*

OLIVER MÜHLEMANN • *Department of Chemistry, Biochemistry and Pharmaceutical Sciences, University of Bern, Bern, Switzerland*

GUILLERMO E. PARADA • *Wellcome Sanger Institute, Cambridge, UK; Donnelly Centre for Cellular and Biomolecular Research and Department of Molecular Genetics, University of Toronto, Toronto, ON, Canada*

PALLAVI P. PILAKA-AKELLA • *Department of Cell and Systems Biology, University of Toronto, Toronto, ON, Canada*

EWA PISKADLO • *Friedrich Miescher Institute for Biomedical Research, Basel, Switzerland*

YUXIN QIN • *Department of Physiology, Guangdong Provincial Key Laboratory of Brain Function and Disease, Zhongshan School of Medicine, Sun Yat-sen University, Guangzhou, China*

XIAOJUN REN • *Department of Chemistry, Key Laboratory of Bioorganic Phosphorus Chemistry and Chemical Biology, Tsinghua University, Beijing, China; Department of Chemistry and Biology, Faculty of Environment and Life Science, Beijing University of Technology, Beijing, China*

MARC-DAVID RUEPP • *United Kingdom Dementia Research Institute Centre, Institute of Psychiatry, Psychology and Neuroscience, King's College London, Maurice Wohl Clinical Neuroscience Institute, London, UK*

ANNA M. SCARBOROUGH • *Department of Microbiology, UT Southwestern Medical Center, Dallas, TX, USA*
PETER SCHEIFFELE • *Biozentrum, University of Basel, Basel, Switzerland*
ALEXANDER SCHMIDT • *Biozentrum of the University of Basel, Basel, Switzerland*
DIETMAR SCHREINER • *Biozentrum of the University of Basel, Basel, Switzerland*
DANIEL SCHÜMPERLI • *Department of Chemistry and Applied Biosciences, Institute of Pharmaceutical Sciences, ETH Zürich, Zürich, Switzerland*
JAVIER TAPIAL • *Centre for Genomic Regulation, Barcelona Institute of Science and Technology, Barcelona, Spain*
JAAK VILO • *Institute of Computer Science, University of Tartu, Tartu, Estonia*
ZICHENG WEI • *Department of Physiology, Guangdong Provincial Key Laboratory of Brain Function and Disease, Zhongshan School of Medicine, Sun Yat-sen University, Guangzhou, China*
SIKA ZHENG • *Division of Biomedical Sciences, University of California, Riverside, CA, USA*

Chapter 1

Alternative Splicing in Human Biology and Disease

Daniel Jutzi and Marc-David Ruepp

Abstract

Alternative pre-mRNA splicing allows for the production of multiple mRNAs from an individual gene, which not only expands the protein-coding potential of the genome but also enables complex mechanisms for the post-transcriptional control of gene expression. Regulation of alternative splicing entails a combinatorial interplay between an abundance of *trans*-acting splicing factors, *cis*-acting regulatory sequence elements and their concerted effects on the core splicing machinery. Given the extent and biological significance of alternative splicing in humans, it is not surprising that aberrant splicing patterns can cause or contribute to a wide range of diseases. In this introductory chapter, we outline the mechanisms that govern alternative pre-mRNA splicing and its regulation and discuss how dysregulated splicing contributes to human diseases affecting the motor system and the brain.

Key words Alternative splicing, snRNPs, *Trans*-acting splicing factors, *Cis*-acting regulatory elements, Myotonic dystrophy type 1, Spinal muscular atrophy, Amyotrophic lateral sclerosis, Frontotemporal dementia with parkinsonism linked to chromosome 17

1 Introduction

The vast majority of human genes are interrupted by multiple non-coding introns, which must be excised from precursor messenger RNAs (pre-mRNAs) to yield mature mRNAs that can serve as template for protein synthesis [1]. This modular gene architecture allows for alternative splicing, such as the inclusion or skipping of cassette exons, intron retention, alternative 5′ and 3′ splice site usage and more sophisticated splicing patterns, to generate different mRNA isoforms from the same gene [2, 3]. Alternative splicing is particularly prevalent in humans, where more than 95% of all genes undergo alternative splicing [4, 5]. While many alternative splicing events are either rare, under relaxed evolutionary selection pressure or only subject to minor variation between tissues or conditions, other alternative splicing events are common, evolutionarily conserved and stringently regulated to only occur in specific cell types, during a specific phase of development or in

Peter Scheiffele and Oriane Mauger (eds.), *Alternative Splicing: Methods and Protocols*,
Methods in Molecular Biology, vol. 2537, https://doi.org/10.1007/978-1-0716-2521-7_1,

response to external stimuli [6, 7]. The resulting mRNA isoforms can differ in their protein coding sequence, translation efficiency, stability and subcellular localization. Hence, alternative splicing not only increases proteome complexity but also provides an additional layer of post-transcriptional gene regulation [8, 9].

2 Mechanism of Pre-mRNA Splicing

The pre-mRNA splicing reaction is orchestrated by a dynamic multimegadalton RNA/protein machine called the spliceosome, which consists of stable building blocks termed small nuclear ribonucleoproteins (snRNPs, U1, U2, U4/U6, and U5), and numerous accessory proteins [10, 11]. Each snRNP is composed of a U snRNA (or two in the U4/U6 di-snRNP) that is complemented by a variable number of snRNP-specific proteins and a shared set of seven unique proteins (Sm or LSm proteins) that wrap around the snRNA as a doughnut-shaped ring [12, 13]. Rather than being a preformed enzymatically active complex, the spliceosome de novo assembles on its substrate pre-mRNA in a stepwise manner and undergoes a series of compositional and structural rearrangements to gradually form a catalytic center to remove the non-coding introns and ligate their adjacent exons in two consecutive *trans*-esterification reactions [14]. Several landmark intermediates of this assembly process have been purified and their atomic structures have recently been solved by Cryo-EM [15–17]. Transitions between these intermediates are reversible and either occur spontaneously or are driven by ATP-dependent RNA helicases [18, 19]. The current model of spliceosome assembly is depicted in Fig. 1.

Spliceosome assembly is initiated by the formation of base-pairing interactions between the U1 snRNA and the 5′-splice site (5′-ss) of the pre-mRNA, while splicing factor 1 (SF1) and the U2 auxiliary factor (U2AF) heterodimer cooperatively bind to the branchpoint sequence (BPS), polypyrimidine tract (PPT), and 3-′-splice site (3′-ss) [20, 21]. Together, these molecular interactions play a central role in the initial definition of 5′-ss and 3′-ss and are referred to as E complex. In the subsequent ATP-dependent step, the RNA helicase Prp5 displaces SF1 from the pre-mRNA and allows for base pairing between the U2 snRNA and the branchpoint to form the A complex [22]. Following recruitment of a preformed U4/U6.U5 tri-snRNP, the resulting pre-B complex is remodelled by the action of the RNA helicase Prp28, which releases the 5′-ss from the U1 snRNA and transfers it to the U6 snRNA, resulting in the ejection of the U1 snRNP and the formation of the B complex [23]. During catalytic activation, the RNA helicase Brr2 then unwinds the extensively base-paired U4 and U6 snRNAs, which ejects the U4 snRNP and frees the U6 snRNA to form the catalytic

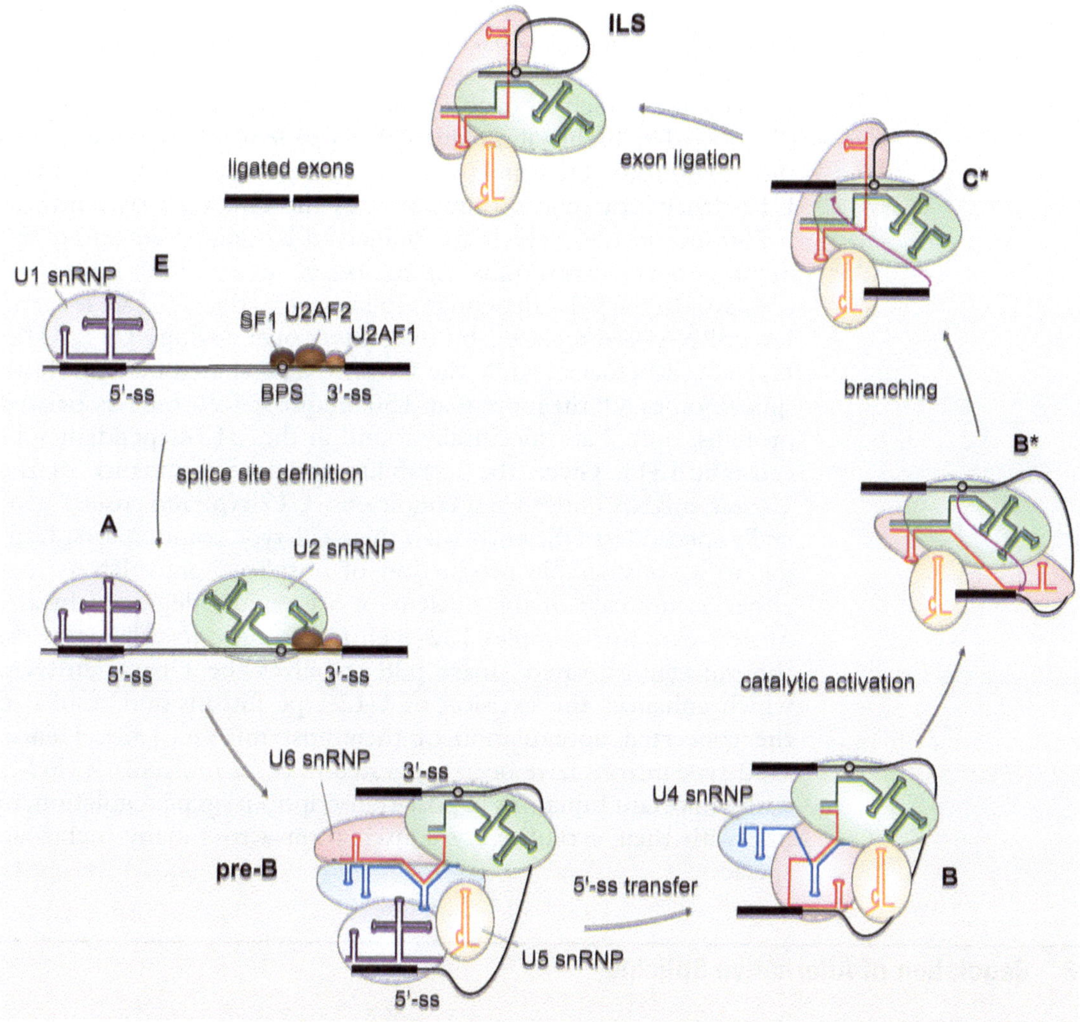

Fig. 1 The pre-mRNA splicing cycle. Schematic depiction of the stepwise assembly and disassembly of the spliceosome during the pre-mRNA splicing reaction. In short, the U1 and U2 snRNPs mark intron/exon boundaries and recruit the preformed U4/U6.U5 tri-snRNP. The resulting pre-spliceosome is then catalytically activated by the action of RNA helicases which extensively remodel the RNA-RNA interaction network and eject the U1 and U4 snRNPs from the complex. Following the two-step splicing reaction which ligates the adjacent exons and excises the intron as a lariat, the spliceosome is released and recycled for the next round of the splicing cycle. See main text for details

center with the U2 snRNA, which in turn directly contacts the branchpoint [24]. In combination with other structural and compositional rearrangements, this yields the catalytically active B* complex, where the first trans-esterification reaction (branching) occurs. The resulting step I spliceosome (C complex) then matures into the step II catalytically activated C* complex where the second *trans*-esterification (exon ligation) completes the pre-mRNA splicing reaction [25]. The excised intron lariat and the spliceosome then form the ILS complex, which is finally recycled for the next round of the splicing [26].

Humans, like most metazoans, contain two distinct types of spliceosomes which co-evolved and splice separate classes of introns [27, 28]. The highly abundant U2-dependent spliceosome excises the ubiquitous U2-type introns, whereas the lowly abundant U12-dependent spliceosome removes the rare U12-type introns (~700 in humans), which are delineated by unique *cis*-acting elements and whose positions within host genes are highly conserved [29, 30]. In the U12-dependent spliceosome, the U1, U2, U4, and U6 snRNAs are replaced by their functional analogs U11, U12, U4atac, and U6atac, while the U5 snRNA is shared between both spliceosomes. Of the more than 150 unique spliceosome-associated proteins, only 7 are specifically found in the U12-dependent spliceosome [31]. Given the instability and low abundance of the U6atac snRNA under basal conditions, U12-type introns are generally spliced less efficiently than their U2-type counterparts, leading to a considerable production of unspliced pre-mRNAs that either accumulate in the nucleus or are rapidly degraded by the nuclear exosome complex [32–34]. In response to cellular stress, the mitogen-activated kinase p38 stabilizes the U6atac snRNA, which enhances the excision of U12-type introns and results in the concerted upregulation of their host mRNAs [33]. Hence, U12-type introns have been suggested to act as molecular switches with important functions in post-transcriptional gene regulation, in line with their evolutionary conservation across many metazoan species.

3 Regulation of Alternative Splicing

In order to allow for alternative splicing to occur and be regulated, the recognition of splice sites by the core splicing machinery must exhibit a certain amount of flexibility. In higher eukaryotes, this flexibility is provided by relatively loosely conserved splice site sequences, which have a weak affinity for their spliceosomal interaction partners and themselves are not sufficient to unambiguously define an exon. Therefore, *cis*-acting sequence elements are required to guide faithful spliceosome assembly and separate bona fide splice sites from cryptic splice sites, which are commonly found in long introns [35]. Auxiliary elements that favor splice site recognition are called exonic or intronic splicing enhancers, while their repressive counterparts are referred to as exonic or intronic splicing silencers (Fig. 2a) [36, 37]. Besides ensuring proper splice site recognition, this antagonistic system forms the basis for the regulation of alternative splicing. In most cases, *cis*-acting elements serve as recruitment platforms for *trans*-acting splicing factors that either stabilize or prevent the binding of core spliceosomal components to

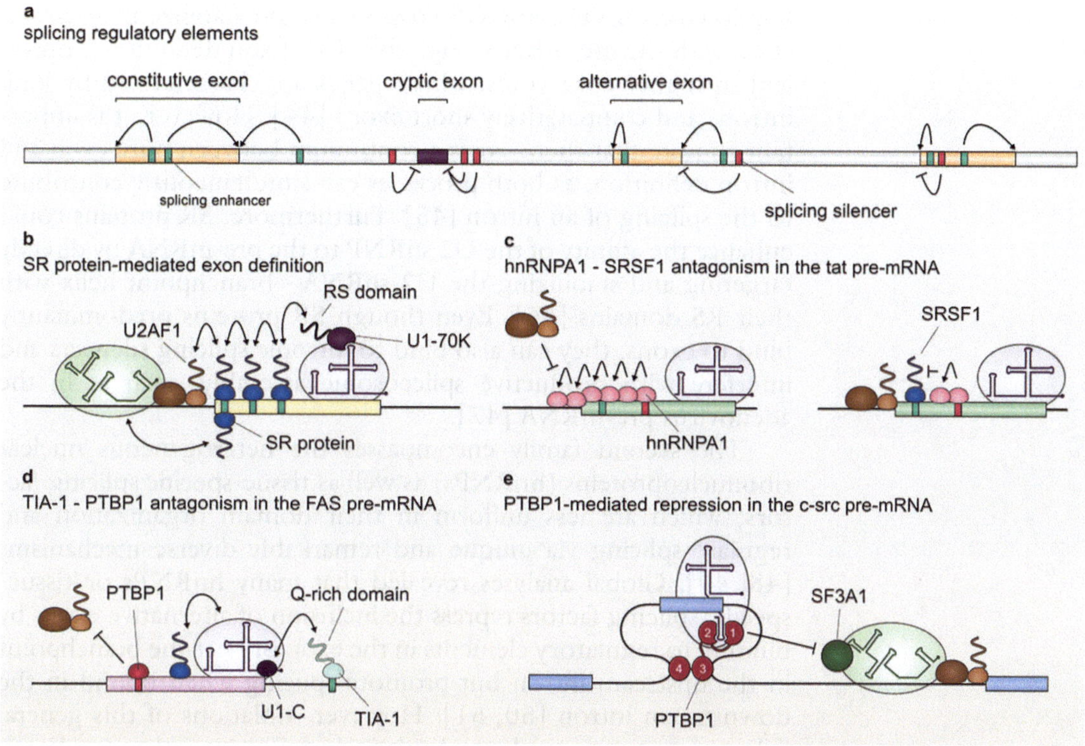

Fig. 2 Mechanisms of pre-mRNA splicing regulation. (**a**) *Cis*-acting splicing regulatory elements modulate the usage of splice sites. Genomic regions harboring constitutive exons are rich in splicing enhancers, whereas cryptic exons are repressed by splicing silencers. At alternative exons, enhancers and silencers dynamically antagonize each other to provide the flexibility required for splicing regulation. (**b**) SR proteins employ their intrinsically unstructured RS domains to build exon-definition complexes that connect U2AF1 at the 3′-ss to U1-70K at the 5′-ss. (**c**) In the tat pre-mRNA, oligomerization of hnRNPA1 prevents access of the U2AF heterodimer to the 3′-ss. Binding of SRSF1 to an upstream splicing enhancer blocks the propagation of hnRNP A1 and hence allows for efficient 3′-ss recognition. (**d**) In the FAS pre-mRNA, PTBP1 represses exon inclusion by interfering with U2AF binding at the 3′-ss, whereas TIA-1 interacts with U1-C to promote 5′-ss recognition. (**e**) In the c-src pre-mRNA, PTBP1 employs RRM3/4 to loop out the alternative exon, while RRM1/2 interact with stem-loop 4 of the U1 snRNA to block the formation of cross-intron contacts with SF3A1

the pre-mRNA, but they may also function by affecting the secondary structure of the transcript [38, 39].

An interesting feature of many splicing factors is that they can both repress and promote splicing, often depending on where they are bound on the pre-mRNA [40, 41]. Based on their position-dependent effects, splicing factors are commonly divided into two families with largely antagonistic functions. The first family consists of the serine/arginine-rich proteins (SR proteins), which are defined by one or two N-terminal RNA recognition motifs and a C-terminal RS domain rich in serine and arginine residues [42]. More often than not, SR proteins bind to exonic splicing enhancers and employ their RS domains to engage in exon definition interactions which physically link the U2AF heterodimer to the downstream U1 snRNP across exons and stabilize their association with the pre-mRNA (Fig. 2b) [43]. Exon definition is prevalent in higher eukaryotes, whose genes are characterized by long introns and comparatively short exons [44]. However, it is important to note that there exists a continuum between pure exon and intron definition, as both processes can simultaneously contribute to the splicing of an intron [45]. Furthermore, SR proteins could enhance the affinity of the U2 snRNP to the pre-mRNA by directly targeting and stabilizing the U2 snRNA—branchpoint helix with their RS domains [46]. Even though SR proteins predominantly bind to exons, they can also bind to intronic splicing silencers and interfere with productive spliceosome assembly, such as in the adenovirus pre-mRNA [47].

The second family encompasses the heterogeneous nuclear ribonucleoproteins (hnRNPs) as well as tissue-specific splicing factors, which are less uniform in their domain organization and regulate splicing via unique and remarkably diverse mechanisms [48, 49]. Global analyses revealed that many hnRNPs or tissue-specific splicing factors repress the inclusion of alternative exons by binding to regulatory elements in the exon or near the branchpoint in the upstream intron but promote splicing when bound in the downstream intron [50, 51]. However, violations of this general rule are common and mechanistic insights into the regulatory mechanisms of hnRNPs and tissue-specific splicing factors primarily stem from detailed studies of individual splicing events. A particularly well studied example is hnRNP A1, which specifically binds to an exonic splicing silencer in the HIV tat mRNA and subsequently oligomerizes across the pre-mRNA in the 3′ to 5′ direction by a series of non-specific binding events, thus preventing access of U2AF to the 3′-ss. This repression is counteracted by SRSF1 binding to an exonic splicing enhancer, which blocks the propagation of hnRNP A1 along the pre-mRNA (Fig. 2c) [52, 53]. When simultaneously bound to intronic splicing silencers in both introns flanking a cassette exon, the hnRNPs A1 and C as well as the polypyrimidine tract-binding protein PTBP1 can loop out the intervening exon and thus prevent its incorporation into mature mRNA [54–56]. Conversely, T-cell restricted intracellular antigen-1 (TIA-1) promotes splicing of the FAS pre-mRNA by stabilizing the U1 snRNP on the 5′-ss via a direct interaction with the U1 snRNP-specific protein U1-C [57]. This is counteracted by PTBP1, which binds to an exonic splicing silencer and prevents the association of U2AF to the 3′-ss by interfering with exon definition interactions (Fig. 2d) [58]. Intriguingly, some hnRNPs also contact the splicing machinery directly via snRNAs. Examples include PTBP1, which contacts stem-loop 4 of the U1 snRNA after the initial splice site definition of the c-src N1 exon to prevent its interaction with the U2 snRNP associated factor SF3A1 (Fig. 2e) [59, 60], and Fused in Sarcoma (FUS) which specifically recognizes

U1 snRNA stem-loop 3 [61]. Another splicing factor that regulates alternative splicing only in the later stages of the splicing cycle is RNA-binding motif protein 5 (RBM5), which directly interacts with the U4/U6.U5 tri-snRNP to prevent splice site pairing around exon 6 of the FAS pre-mRNA [62].

The ultimate decision if a potential splice site is used or not results from the combinatorial interplay between multiple splicing factors and their concerted effects on the core splicing machinery. After all, it is only by cooperation and competition that splicing factors, which individually bind RNA with low affinity and specificity, can form the highly specific and stable regulatory complexes required to accurately define splice sites. Thus, it is the repertoire of available splicing factors and their relative expression levels that enable cell-type and developmental phase-specific patterns of alternative splicing and integration of external signals.

4 Alternative Splicing in Human Disease

Given the extent and biological importance of alternative splicing, it is not surprising that widespread alterations in splicing patterns have been observed in numerous human diseases, ranging from neurodegenerative and neuromuscular diseases to cancer and developmental disorders [31, 63, 64]. However, it is important to consider that such alterations may arise as a secondary consequence of the underlying disease process or merely be a co-phenomenon of tissue degeneration, inflammation, regeneration or transformation [65]. Therefore, specific splicing alterations that are causally linked to the development or progression of a disease must be identified. The most compelling evidence for a role of alternative splicing in disease mechanisms is provided by genetic mutations which can either disrupt *cis*-acting elements, and thus alter the splicing of a single transcript, or affect *trans*-acting factors to have more widespread effects on gene expression. It has been estimated that at least 15%, possibly even 50%, of human genetic diseases arise from mutations that affect pre-mRNA splicing [66, 67]. Below, we discuss a few well-studied examples of diseases linked to the dysregulation of alternative splicing, with a focus on diseases linked to the motor system and the brain.

5 Myotonic Dystrophy Type I (DM1)

Myotonic dystrophy type 1 (DM1) is a multisystem neuromuscular disease, whose clinical symptoms typically include myotonia and muscle atrophy, cardiac arrhythmia, diabetes, and cognitive defects [68]. The disease is caused by a heterozygous CTG repeat expansion in the 3′-UTR of the DM protein kinase (DMPK) gene [69–

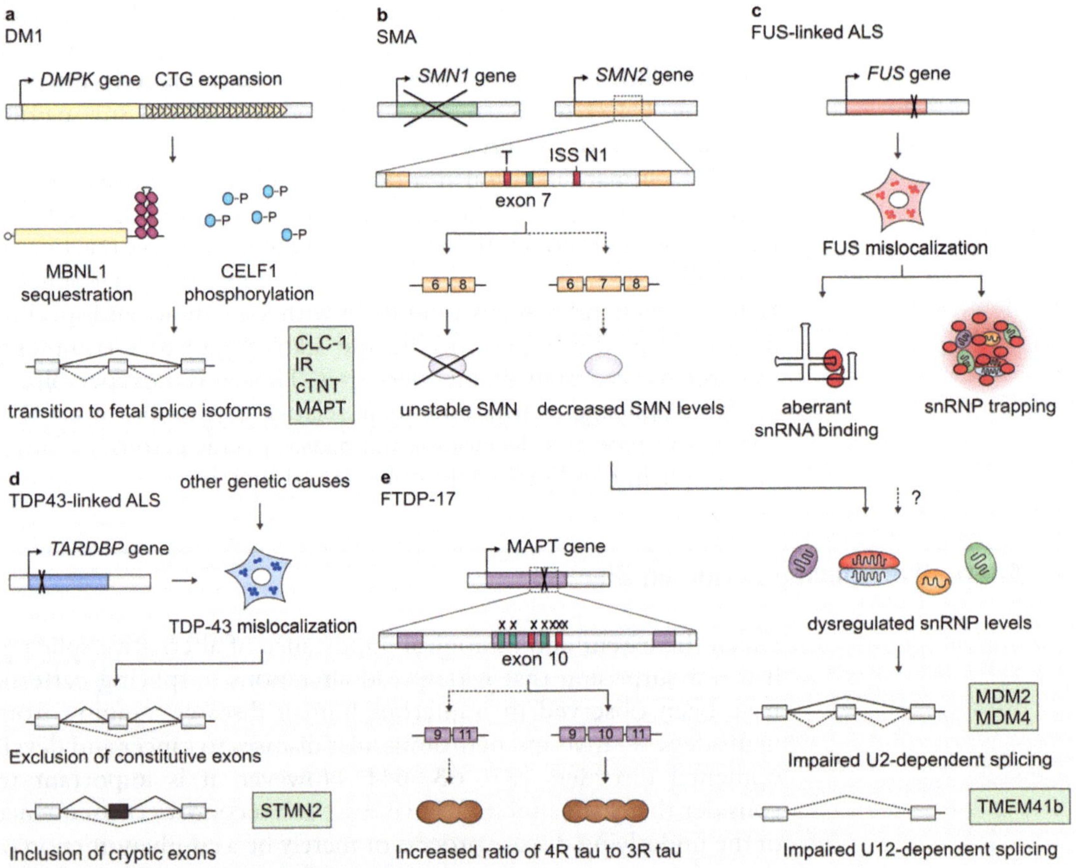

Fig. 3 Dysregulated alternative splicing in human disease. (**a**) A CTG expansion in the 3′-UTR of the DMPK gene leads to MBNL1 sequestration and CELF1 phosphorylation and thereby induces a shift toward fetal splicing patterns in DM1. (**b**) SMA patients lack a functional *SMN1* gene and rely on the small amount of full-length SMN protein produced from the *SMN2* gene, where a C to T transition in exon 7 favors exon skipping. Reduced SMN levels impair snRNP homeostasis and lead to global alterations in alternative splicing. (**c**) Mutations in the *FUS* gene typically induce a mislocalization of FUS to the cytoplasm, where it aberrantly interacts with the U1 snRNA and traps snRNP biogenesis intermediates in cytoplasmic granules. (**d**) Along with most genetic causes of ALS, mutations in the *TARDBP* gene lead to cytoplasmic redistribution of TDP-43 accompanied by widespread mis-splicing of pre-mRNAs. (**e**) A subset of FTDP-17 linked mutations in the MAPT gene cluster in and around exon 10 and promote its inclusion to increase the ratio of 4R to 3R tau

71], which leads to the expression of CUG repeat-containing RNAs that are retained in nuclear RNA foci [72] and sequester splicing factors of the muscleblind-like (MBNL) family [73], thereby depleting the nuclear pool that is available for splicing regulation (Fig. 3a). Moreover, the repeat RNA foci induce the hyperphosphorylation and stabilization of the CUGBP/ELAV-like family member CELF1 via protein kinase C (PKC) [74]. A physiological role of both MBNL and CELF proteins is to coordinately promote the switch from fetal to adult splicing patterns during development of the heart, skeletal muscles and brain [75–77]. Consistently, their dysregulation in DM1 induces the aberrant expression of neonatal mRNA isoforms, some of which have been linked to clinical features of the disease. In the muscle, mis-splicing of the pre-mRNA encoding the muscle-specific chloride channel 1 (ClC-1) produces aberrant mRNA isoforms containing premature termination codons, which is sufficient to cause myotonia [78, 79], whereas an isoform switch of the insulin receptor (IR) mRNA contributes to insulin resistance [80]. In the heart, alternative splicing of the cardiac troponin T (cTNT) pre-mRNA is associated with cardiac dysfunctions [81], whereas the expression of neonatal mRNA isoforms of the microtubule-associated protein tau (MAPT) is linked to the formation of neurofibrillary tangles and cognitive defects in the brain of DM1 patients [82].

6 Spinal Muscular Atrophy (SMA)

Spinal muscular atrophy (SMA) is the most common motor neuron disease in children and is characterized by the selective degeneration of lower motor neurons in the spinal cord and brainstem, leading to muscle atrophy [83]. The disease is caused by insufficient levels of the SMN protein, due to homozygous loss or mutation of the *SMN1* gene, which is responsible for the bulk production of SMN [84] (Fig. 3b). As a complete loss of SMN is incompatible with life [85], SMA patients depend on the small amount of SMN that is expressed from the paralogous *SMN2* gene, where a C to T transition in exon 7 transforms an exonic splicing enhancer, bound by SRSF1, into an exonic splicing silencer, bound by hnRNP A1 [86, 87]. Together with a strong intronic silencer (ISS-N1) in intron 7, this favors to the production of an mRNA isoform lacking exon 7 which encodes an unstable SMN protein that is rapidly degraded [88, 89]. Nevertheless, about 10% of the SMN2 mRNAs include exon 7 and hence can be translated into functional protein.

SMN executes its biological functions as part of the macromolecular SMN complex composed of SMN, Gemins 2–8, and Unrip. The best-characterized function of this complex is to chaperone the assembly of Sm proteins onto snRNAs, an essential step in the biogenesis of Sm-class snRNPs that occurs in the cytoplasm [90–92]. As a consequence, snRNP levels are dysregulated in spinal cords of SMA models, and particularly in motor neurons, leading to widespread splicing defects [93, 94]. Intriguingly, U12-dependent snRNPs are preferentially affected by reduced SMN activity [95, 96], which has led to the hypothesis that impaired splicing of U12-type introns could contribute to motor neuron degeneration [97]. This idea has recently been reinforced by the finding that AAV9-mediated delivery of U12-dependent snRNA genes into the spinal cord of severe SMA mice prolonged survival by a few days [98]. A specific U12-type intron-containing transcript linked to the SMA phenotype is the Stasimon (also known as TMEM41B) mRNA which encodes an ER-resident transmembrane protein involved in autophagy [99, 100]. However, altered U2-type splicing events have also been connected to the SMA disease mechanism. For example, reduced inclusion of U2-dependent cassette exons in the Mdm2 and Mdm4 mRNAs lead to the activation of p53, a process previously linked to motor neuron degeneration in SMA [101, 102]. Although it is an appealing hypothesis that mis-splicing of a small set of pre-mRNAs drives the disease cascade in SMA, we cannot exclude that a global and possibly mild reduction in splicing activity affects the processing of hundreds of pre-mRNAs, which together compromise neuronal survival. It is also important to note that the SMN protein has been linked to other cellular processes including mRNA transport and translation [103, 104], and although these proposed functions currently await rigorous biochemical validation, their possible contribution to SMA should not be neglected.

7 Amyotrophic Lateral Sclerosis Linked to TDP-43 and FUS

Amyotrophic lateral sclerosis (ALS) is the most common motor neuron disease in adults and is characterized by the loss of both upper motor neurons in the cortex as well as lower motor neurons in the brainstem and spinal cord [105, 106]. Mutations in more than 30 genes have been linked to the familial form of the disease and typically encode proteins with roles in protein homeostasis, cellular trafficking or RNA metabolism [107, 108]. However, almost all genetic backgrounds ultimately lead to the mislocalization and deposition of insoluble TAR DNA-binding protein 43 (TDP-43) in cytoplasmic aggregates in affected neurons and glia cells, the pathological hallmark of 97% of ALS cases [109] (Fig. 3d).

TDP-43 is an hnRNP-type RNA-binding protein that binds to GU-rich repeat sequences and regulates several aspects of RNA processing, including alternative splicing. Transcriptome-wide analyses suggest that TDP-43 represses splicing when bound in proximity of the 3′-splice site or more than 150 nucleotides downstream of the 5′-splice site, whereas binding of TDP-43 close to the 5-′-splice site promotes splicing [110]. Besides cassette exons, TDP-43 is increasingly recognized to suppress the inclusion of non-conserved cryptic exons in model systems ranging from HeLa cells to mouse brains [111–113]. A loss of this repressive splicing activity due to mutation or nuclear clearance of TDP-43 in ALS prompts mis-splicing of relevant mRNAs such as Stahmin-2, one of the most abundant transcripts in motor neurons [114, 115]. An essential protein in mammals, Stathmin-2 is a regulator of microtubule dynamics and allows motor axons to regenerate in response to injury [116]. However, TDP-43-linked ALS can also be modelled in mice, which lack the cryptic exon found in the human *STMN2* gene, arguing that a combination of multiple splicing alterations, rather than a single splicing event, in combination with other aberrant processes initiates and propagates a complex disease cascade [117, 118]. Furthermore, some artificial TDP-43 mutants—which are not found in ALS patients—are hyperactive in their splicing function to the extent that they suppress the inclusion of constitutive exons, the so-called skiptic exons. Importantly, such skiptic splicing events were observed in the brain of an ALS mouse model and ALS-patient derived fibroblasts [119].

A rare, but very aggressive type of ALS that lacks TDP-43 positive aggregates is caused by mutations in the *FUS* gene encoding the RNA-binding protein Fused in Sarcoma (FUS) [120, 121]. These mutations typically destroy the nuclear localization signal of FUS, leading to cytoplasmic mislocalization and aggregation of the protein [122] (Fig. 3c). As a highly abundant hnRNP-type protein, FUS preferentially binds to a bipartite motif consisting of a YNY-containing stem-loop next to a GGU trinucleotide and regulates various steps of RNA metabolism, including alternative splicing [123–125]. However, as recent mouse models strongly indicate that FUS causes motor neuron loss primarily through a cytoplasmic gain-of-function [126, 127], it is unlikely that the splicing regulatory function, or any other nuclear function, of FUS contributes to the disease. Nevertheless, a growing body of evidence suggests that ALS-associated FUS interferes with snRNP biogenesis in the cytoplasm and could therefore affect the splicing of mRNAs that are sensitive to reduced snRNP levels, thus establishing a molecular link between the motor neuron diseases ALS and SMA [61, 128–130].

8 Frontotemporal Dementia with Parkinsonism Linked to Chromosome 17 (FTDP-17)

Frontotemporal dementia (FTD) is the most common form of dementia in people aged <60 years and is characterized by the loss of cortical neurons in the frontal and temporal lobes, leading to language difficulties and behavioral alterations [131, 132]. Given that about 15% of individuals with ALS also develop symptoms of FTD, pure ALS and FTD are now commonly thought of as extreme presentations of a continuous disease spectrum, a view that is supported by substantial overlap of their genetic causes [109].

Mutations in the *MAPT* gene on chromosome 17 encoding the microtubule-associated protein tau cause pure FTD that is sometimes accompanied by symptoms of parkinsonism such as bradykinesia and rigidity in the absence of resting tremor [133, 134]. The deposition of insoluble phosphorylated tau in filamentous structures is the pathological hallmark of FTDP-17 [135]. In the adult nervous system, alternative splicing of MAPT exons 2 and 3 (which encode N-terminal inserts) and exon 10 (which encodes one of four C-terminal microtubule-binding repeats) result in the production of six tau protein isoforms with either zero, one or two N-terminal inserts (0N, 1N and 2N) and three or four microtubule-binding repeats (3R and 4R) [136] (Fig. 3e). In healthy individuals, the ratio of 3R to 4R isoforms is close to 1 and a disruption of this ratio in either direction is associated with neurodegeneration, although the underlying mechanisms remain to be elucidated [137].

To date, almost 40 point mutations in MAPT have been linked to FTDP-17, with a mutational hotspot in exon 10 and its flanking intronic region at the 5′-splice site. With only minor exceptions, virtually all of these mutations promote exon 10 inclusion and thus favor the production of 4R tau, thereby shifting the 4R-to-3R ratio to 2.5–3 in FTDP-17 patients [138]. Several mechanisms have been proposed to explain the increase in exon 10 splicing [139]. For example, intronic mutations could destabilize a putative stem-loop structure that sequesters the 5′-splice site and hence prevents its recognition by the U1 snRNP [133, 140]. Alternatively, mutations at the 5′-splice site could directly enhance the affinity to the U1 snRNP by increasing base pair complementarity with the U1 snRNA [141] or by affecting intronic *cis*-acting elements that govern the recruitment of *trans*-acting splicing factors like RBM4 [142], various CELF proteins [143], SRSF7 [144] or the RNA helicase p68 [145]. Conversely, a point mutation in exon 10 strengthens an exonic splicing enhancer that is bound by Tra2ß [146].

Even though many mutations cluster around exon 10 in the MAPT gene and affect pre-mRNA processing and thus the 4R-to-3R tau ratio, other FTDP-17 linked missense mutations are scattered throughout the open reading frame and can alter biochemical properties of the tau protein, such as microtubule binding or seed-induced aggregation [147]. Regardless of the direct consequences of the individual mutations, the downstream processes that lead to neurodegeneration are likely to be shared between them and remain to be uncovered.

9 Conclusion and Outlook

Alternative pre-mRNA splicing is a key mechanism that diversifies the transcriptional output of the human genome and thus provides opportunities for the post-transcriptional control of gene expression. The regulation of alternative splicing is facilitated by *trans*-acting splicing factors which bind to *cis*-acting sequence elements in the pre-mRNA and modulate the usage of nearby splice sites via remarkably diverse mechanisms. While many splicing factors do so by interacting with core spliceosomal proteins, recent findings suggest that they can also directly contact spliceosomal snRNAs. Taking into account that splicing factors typically contain multiple RNA-binding domains, this could enable novel mechanisms for splicing regulation.

Dysregulated alternative splicing is frequently observed in human diseases and can affect the processing of a single transcript or hundreds of transcripts, depending on the nature of the mutation. In the latter scenario, the identification of individual splicing events that are causally linked to a disease is challenging but represents a key step to uncover relevant targets for therapeutic intervention, for example by using splice switching antisense oligonucleotides (ASOs) or small molecule splicing modulators.

Acknowledgments

This work was supported by the UK Dementia Research Institute which receives its funding from UK DRI Ltd., funded by the Medical Research Council, Alzheimer's Society, and Alzheimer's Research UK.

References

1. Sharp PA (1994) Split genes and RNA splicing. Cell 77:805–815. https://doi.org/10.1016/0092-8674(94)90130-9
2. Black DL (2003) Mechanisms of alternative pre-messenger RNA splicing. Annu Rev Biochem 72:291–336
3. Lee Y, Rio DC (2015) Mechanisms and regulation of alternative pre-mRNA splicing. Annu Rev Biochem 84:291–323
4. Barbosa-Morais NL, Irimia M, Pan Q et al (2012) The evolutionary landscape of alternative splicing in vertebrate species. Science 338:1587–1593. https://doi.org/10.1126/science.1230612
5. Merkin J, Russell C, Chen P, Burge CB (2012) Evolutionary dynamics of gene and isoform regulation in mammalian tissues. Science 338:1593–1599. https://doi.org/10.1126/science.1228186
6. Shin C, Manley JL (2004) Cell signalling and the control of pre-mRNA splicing. Nat Rev Mol Cell Biol 5:727–738
7. Xing Y, Lee C (2006) Alternative splicing and RNA selection pressure—evolutionary consequences for eukaryotic genomes. Nat Rev Genet 7:499–509
8. Nilsen TW, Graveley BR (2010) Expansion of the eukaryotic proteome by alternative splicing. Nature 463:457–463
9. Blencowe BJ (2006) Alternative splicing: new insights from global analyses. Cell 126:37–47
10. Will CL, Lührmann R (2011) Spliceosome structure and function. Cold Spring Harb Perspect Biol 3:1–2. https://doi.org/10.1101/cshperspect.a003707
11. Jurica MS, Moore MJ (2003) Pre-mRNA splicing: awash in a sea of proteins. Mol Cell 12:5–14
12. Stark H, Dube P, Luührmann R, Kastner B (2001) Arrangement of RNA and proteins in the spliceosomal U1 small nuclear ribonucleoprotein particle. Nature 409:539–542. https://doi.org/10.1038/35054102
13. Urlaub H, Raker VA, Kostka S, Lührmann R (2001) Sm protein-Sm site RNA interactions within the inner ring of the spliceosomal snRNP core structure. EMBO J 20:187–

196. https://doi.org/10.1093/emboj/20.1.187

14. Wahl MC, Will CL, Lührmann R (2009) The spliceosome: design principles of a dynamic RNP machine. Cell 136:701–718
15. Kastner B, Will CL, Stark H, Lührmann R (2019) Structural insights into nuclear pre-mRNA splicing in higher eukaryotes. Cold Spring Harb Perspect Biol 11: a032417. https://doi.org/10.1101/cshperspect.a032417
16. Wilkinson ME, Charenton C, Nagai K (2020) RNA splicing by the spliceosome. Annu Rev Biochem 89:359–388
17. Yan C, Wan R, Shi Y (2019) Molecular mechanisms of pre-mRNA splicing through structural biology of the spliceosome. Cold Spring Harb Perspect Biol 11:a032409. https://doi.org/10.1101/cshperspect.a032409
18. Cordin O, Beggs JD (2013) RNA helicases in splicing. RNA Biol 10:83–95
19. Hoskins AA, Friedman LJ, Gallagher SS et al (2011) Ordered and dynamic assembly of single spliceosomes. Science 331:1289–1295. https://doi.org/10.1126/science.1198830
20. Kondo Y, Oubridge C, van Roon AMM, Nagai K (2015) Crystal structure of human U1 snRNP, a small nuclear ribonucleoprotein particle, reveals the mechanism of 5′ splice site recognition. Elife 4:e04986. https://doi.org/10.7554/eLife.04986
21. Berglund JA, Abovich N, Rosbash M (1998) A cooperative interaction between U2AF65 and mBBP/SF1 facilitates branchpoint region recognition. Genes Dev 12:858–867. https://doi.org/10.1101/gad.12.6.858
22. Plaschka C, Lin PC, Charenton C, Nagai K (2018) Prespliceosome structure provides insights into spliceosome assembly and regulation. Nature 559:419–422. https://doi.org/10.1038/s41586-018-0323-8
23. Charenton C, Wilkinson ME, Nagai K (2019) Mechanism of 5′ splice site transfer for human spliceosome activation. Science 364:362–367. https://doi.org/10.1126/science.aax3289
24. Haselbach D, Komarov I, Agafonov DE et al (2018) Structure and conformational dynamics of the human Spliceosomal Bact complex. Cell 172:454–464.e11. https://doi.org/10.1016/j.cell.2018.01.010
25. Zhang X, Yan C, Hang J et al (2017) An atomic structure of the human spliceosome. Cell 169:918–929.e14. https://doi.org/10.1016/j.cell.2017.04.033
26. Zhang X, Zhan X, Yan C et al (2019) Structures of the human spliceosomes before and after release of the ligated exon. Cell Res 29: 274–285. https://doi.org/10.1038/s41422-019-0143-x
27. Patel AA, Steitz JA (2003) Splicing double: insights from the second spliceosome. Nat Rev Mol Cell Biol 4:960–970
28. Will CL, Lührmann R (2005) Splicing of a rare class of introns by the U12-dependent spliceosome. Biol Chem 386:713–724
29. Turunen JJ, Niemelä EH, Verma B, Frilander MJ (2013) The significant other: splicing by the minor spliceosome. Wiley Interdiscip Rev RNA 4:61–76
30. Jutzi D, Akinyi MV, Mechtersheimer J et al (2018) The emerging role of minor intron splicing in neurological disorders. Cell Stress 2:40–54
31. Verma B, Akinyi MV, Norppa AJ, Frilander MJ (2018) Minor spliceosome and disease. Semin Cell Dev Biol 79:103–112
32. Niemelä EH, Frilander MJ (2014) Regulation of gene expression through inefficient splicing of U12-type introns. RNA Biol 11:1325–1329
33. Younis I, Dittmar K, Wang W et al (2013) Minor introns are embedded molecular switches regulated by highly unstable U6atac snRNA. Elife 2:e00780. https://doi.org/10.7554/eLife.00780
34. Patel AA, McCarthy M, Steitz JA (2002) The splicing of U12-type introns can be a rate-limiting step in gene expression. EMBO J 21:3804–3815. https://doi.org/10.1093/emboj/cdf297
35. Wang Z, Rolish ME, Yeo G et al (2004) Systematic identification and analysis of exonic splicing silencers. Cell 119:831–845. https://doi.org/10.1016/j.cell.2004.11.010
36. Matlin AJ, Clark F, Smith CWJ (2005) Understanding alternative splicing: towards a cellular code. Nat Rev Mol Cell Biol 6:386–398
37. Wang Z, Burge CB (2008) Splicing regulation: from a parts list of regulatory elements to an integrated splicing code. RNA 14:802–813
38. Smith CWJ, Valcárcel J (2000) Alternative pre-mRNA splicing: the logic of combinatorial control. Trends Biochem Sci 25:381–388
39. Shepard PJ, Hertel KJ (2008) Conserved RNA secondary structures promote alternative splicing. RNA 14:1463–1469. https://doi.org/10.1261/rna.1069408

40. Chen M, Manley JL (2009) Mechanisms of alternative splicing regulation: insights from molecular and genomics approaches. Nat Rev Mol Cell Biol 10:741–754
41. Fu XD, Ares M (2014) Context-dependent control of alternative splicing by RNA-binding proteins. Nat Rev Genet 15: 689–701
42. Shepard PJ, Hertel KJ (2009) The SR protein family. Genome Biol 10:242
43. Long JC, Caceres JF (2009) The SR protein family of splicing factors: master regulators of gene expression. Biochem J 417:15–27
44. Berget SM (1995) Exon recognition in vertebrate splicing. J Biol Chem 270:2411–2414
45. Li X, Liu S, Zhang L et al (2019) A unified mechanism for intron and exon definition and back-splicing. Nature 573:375–380. https://doi.org/10.1038/s41586-019-1523-6
46. Shen H, Kan JLC, Green MR (2004) Arginine-serine-rich domains bound at splicing enhancers contact the branchpoint to promote Prespliceosome assembly. Mol Cell 13: 367–376. https://doi.org/10.1016/S1097-2765(04)00025-5
47. Kanopka A, Muhlemann O, Akusjarvi G (1996) Inhibition by SR proteins splicing of a regulated adenovirus pre-mRNA. Nature 381:535–538. https://doi.org/10.1038/381535a0
48. Dreyfuss G, Matunis MJ, Piñol-Roma S, Burd CG (1993) hnRNP proteins and the biogenesis of mRNA. Annu Rev Biochem 62:289–321
49. Geuens T, Bouhy D, Timmerman V (2016) The hnRNP family: insights into their role in health and disease. Hum Genet 135:851–867
50. Witten JT, Ule J (2011) Understanding splicing regulation through RNA splicing maps. Trends Genet 27:89–97
51. Darnell RB (2010) HITS-CLIP: panoramic views of protein-RNA regulation in living cells. Wiley Interdiscip Rev RNA 1:266–286. https://doi.org/10.1002/wrna.31
52. Zhu J, Mayeda A, Krainer AR (2001) Exon identity established through differential antagonism between exonic splicing silencer-bound hnRNP A1 and enhancer-bound SR proteins. Mol Cell 8:1351–1361. https://doi.org/10.1016/S1097-2765(01)00409-9
53. Mayeda A, Krainer AR (1992) Regulation of alternative pre-mRNA splicing by hnRNP A1 and splicing factor SF2. Cell 68:365–375. https://doi.org/10.1016/0092-8674(92)90477-T
54. König J, Zarnack K, Rot G et al (2010) ICLIP reveals the function of hnRNP particles in splicing at individual nucleotide resolution. Nat Struct Mol Biol 17:909–915. https://doi.org/10.1038/nsmb.1838
55. Oberstrass FC, Auwetor SD, Erat M et al (2005) Structural biology - structure of PTB bound to RNA: specific binding and implications for splicing regulation. Science 309: 2054–2057. https://doi.org/10.1126/science.1114066
56. Blanchette M, Chabot B (1999) Modulation of exon skipping by high-affinity hnRNP A1-binding sites and by intron elements that repress splice site utilization. EMBO J 18: 1939–1952. https://doi.org/10.1093/emboj/18.7.1939
57. Förch P, Puig O, Martínez C et al (2002) The splicing regulator TIA-1 interacts with U1-C to promote U1 snRNP recruitment to 5′ splice sites. EMBO J 21:6882–6892. https://doi.org/10.1093/emboj/cdf668
58. Izquierdo JM, Majós N, Bonnal S et al (2005) Regulation of fas alternative splicing by antagonistic effects of TIA-1 and PTB on exon definition. Mol Cell 19:475–484. https://doi.org/10.1016/j.molcel.2005.06.015
59. Sharma S, Maris C, Allain FHT, Black DL (2011) U1 snRNA directly interacts with polypyrimidine tract-binding protein during splicing repression. Mol Cell 41:579–588. https://doi.org/10.1016/j.molcel.2011.02.012
60. Martelly W, Fellows B, Senior K et al (2019) Identification of a noncanonical RNA binding domain in the U2 snRNP protein SF3A1. RNA 25:1509–1521. https://doi.org/10.1261/rna.072256.119
61. Jutzi D, Campagne S, Schmidt R et al (2020) Aberrant interaction of FUS with the U1 snRNA provides a molecular mechanism of FUS induced amyotrophic lateral sclerosis. Nat Commun 11:6341. https://doi.org/10.1038/s41467-020-20191-3
62. Bonnal S, Martínez C, Förch P et al (2008) RBM5/Luca-15/H37 regulates Fas alternative splice site pairing after exon definition. Mol Cell 32:81–95. https://doi.org/10.1016/j.molcel.2008.08.008
63. Nik S, Bowman TV (2019) Splicing and neurodegeneration: insights and mechanisms. Wiley Interdiscip Rev RNA 10:e1532
64. Escobar-Hoyos L, Knorr K, Abdel-Wahab O (2019) Aberrant RNA splicing in cancer. Annu Rev Cancer Biol 3:167–185
65. Orengo JP, Ward AJ, Cooper TA (2011) Alternative splicing dysregulation secondary to skeletal muscle regeneration. Ann Neurol

69:681–690. https://doi.org/10.1002/ana.22278

66. Faustino NA, Cooper TA (2003) Pre-mRNA splicing and human disease. Genes Dev 17:419–437
67. Pagani F, Baralle FE (2004) Genomic variants in exons and introns: identifying the splicing spoilers. Nat Rev Genet 5:389–396
68. Meola G, Cardani R (2015) Myotonic dystrophies: an update on clinical aspects, genetic, pathology, and molecular pathomechanisms. Biochim Biophys Acta Mol basis Dis 1852:594–606
69. Brook JD, McCurrach ME, Harley HG et al (1992) Molecular basis of myotonic dystrophy: expansion of a trinucleotide (CTG) repeat at the 3′ end of a transcript encoding a protein kinase family member. Cell 68:799–808. https://doi.org/10.1016/0092-8674(92)90154-5
70. Fu YH, Pizzuti A, Fenwick RG et al (1992) An unstable triplet repeat in a gene related to myotonic muscular dystrophy. Science 255:1256–1258. https://doi.org/10.1126/science.1546326
71. Mahadevan M, Tsilfidis C, Sabourin L et al (1992) Myotonic dystrophy mutation: an unstable CTG repeat in the 3′ untranslated region of the gene. Science 255:1253–1255. https://doi.org/10.1126/science.1546325
72. Davis BM, Mccurrach ME, Taneja KL et al (1997) Expansion of a CUG trinucleotide repeat in the 3′ untranslated region of myotonic dystrophy protein kinase transcripts results in nuclear retention of transcripts. Proc Natl Acad Sci U S A 94:7388–7393. https://doi.org/10.1073/pnas.94.14.7388
73. Miller JW, Urbinati CR, Teng-Umnuay P et al (2000) Recruitment of human muscleblind proteins to (CUG)(n) expansions associated with myotonic dystrophy. EMBO J 19:4439–4448. https://doi.org/10.1093/emboj/19.17.4439
74. Kuyumcu-Martinez NM, Wang GS, Cooper TA (2007) Increased steady-state levels of CUGBP1 in myotonic dystrophy 1 are due to PKC-mediated hyperphosphorylation. Mol Cell 28:68–78. https://doi.org/10.1016/j.molcel.2007.07.027
75. Kalsotra A, Xiao X, Ward AJ et al (2008) A postnatal switch of CELF and MBNL proteins reprograms alternative splicing in the developing heart. Proc Natl Acad Sci U S A 105:20333–20338. https://doi.org/10.1073/pnas.0809045105
76. Ladd AN, Charlet-B N, Cooper TA (2001) The CELF family of RNA binding proteins is implicated in cell-specific and developmentally regulated alternative splicing. Mol Cell Biol 21:1285–1296. https://doi.org/10.1128/mcb.21.4.1285-1296.2001
77. Charizanis K, Lee KY, Batra R et al (2012) Muscleblind-like 2-mediated alternative splicing in the developing brain and dysregulation in myotonic dystrophy. Neuron 75:437–450. https://doi.org/10.1016/j.neuron.2012.05.029
78. Charlet BN, Savkur RS, Singh G et al (2002) Loss of the muscle-specific chloride channel in type 1 myotonic dystrophy due to misregulated alternative splicing. Mol Cell 10:45–53. https://doi.org/10.1016/S1097-2765(02)00572-5
79. Mankodi A, Takahashi MP, Jiang H et al (2002) Expanded CUG repeats trigger aberrant splicing of ClC-1 chloride channel pre-mRNA and hyperexcitability of skeletal muscle in myotonic dystrophy. Mol Cell 10:35–44. https://doi.org/10.1016/S1097-2765(02)00563-4
80. Savkur RS, Philips AV, Cooper TA (2001) Aberrant regulation of insulin receptor alternative splicing is associated with insulin resistance in myotonic dystrophy. Nat Genet 29:40–47. https://doi.org/10.1038/ng704
81. Philips AV, Timchenko LT, Cooper TA (1998) Disruption of splicing regulated by a CUG-binding protein in myotonic dystrophy. Science 280:737–741. https://doi.org/10.1126/science.280.5364.737
82. Sergeant N, Sablonnière B, Schraen-Maschke S et al (2001) Dysregulation of human brain microtubule-associated tau mRNA maturation in myotonic dystrophy type 1. Hum Mol Genet 10:2143–2155. https://doi.org/10.1093/hmg/10.19.2143
83. Lunn MR, Wang CH (2008) Spinal muscular atrophy. Lancet 371:2120–2133
84. Lefebvre S, Bürglen L, Reboullet S et al (1995) Identification and characterization of a spinal muscular atrophy-determining gene. Cell 80:155–165. https://doi.org/10.1016/0092-8674(95)90460-3
85. Schrank B, Götz R, Gunnersen JM et al (1997) Inactivation of the survival motor neuron gene, a candidate gene for human spinal muscular atrophy, leads to massive cell death in early mouse embryos. Proc Natl Acad Sci U S A 94:9920–9925. https://doi.org/10.1073/pnas.94.18.9920
86. Cartegni L, Krainer AR (2002) Disruption of an SF2/ASF-dependent exonic splicing enhancer in SMN2 causes spinal muscular atrophy in the absence of SMN. Nat Genet

30:377–384. https://doi.org/10.1038/ng854

87. Kashima T, Manley JL (2003) A negative element in SMN2 exon 7 inhibits splicing in spinal muscular atrophy. Nat Genet 34:460–463. https://doi.org/10.1038/ng1207
88. Singh NN, Androphy EJ, Singh RN (2004) In vivo selection reveals combinatorial controls that define a critical exon in the spinal muscular atrophy genes. RNA 10:1291–1305. https://doi.org/10.1261/rna.7580704
89. Burnett BG, Muñoz E, Tandon A et al (2009) Regulation of SMN protein stability. Mol Cell Biol 29:1107–1115. https://doi.org/10.1128/mcb.01262-08
90. Meister G, Bühler D, Pillai R et al (2001) A multiprotein complex mediates the ATP-dependent assembly of spliceosomal U snRNPs. Nat Cell Biol 3:945–949. https://doi.org/10.1038/ncb1101-945
91. Pellizzoni L, Yong J, Dreyfuss G (2002) Essential role for the SMN complex in the specificity of snRNP assembly. Science 298: 1775–1779. https://doi.org/10.1126/science.1074962
92. Neuenkirchen N, Englbrecht C, Ohmer J et al (2015) Reconstitution of the human U sn RNP assembly machinery reveals stepwise Sm protein organization. EMBO J 34:1925–1941. https://doi.org/10.15252/embj.201490350
93. Ruggiu M, McGovern VL, Lotti F et al (2012) A role for SMN exon 7 splicing in the selective vulnerability of motor neurons in spinal muscular atrophy. Mol Cell Biol 32: 126–138. https://doi.org/10.1128/mcb.06077-11
94. Zhang Z, Lotti F, Dittmar K et al (2008) SMN deficiency causes tissue-specific perturbations in the repertoire of snRNAs and widespread defects in splicing. Cell 133:585–600. https://doi.org/10.1016/j.cell.2008.03.031
95. Workman E, Saieva L, Carrel TL et al (2009) A SMN missense mutation complements SMN2 restoring snRNPs and rescuing SMA mice. Hum Mol Genet 18:2215–2229. https://doi.org/10.1093/hmg/ddp157
96. Gabanella F, Butchbach MER, Saieva L et al (2007) Ribonucleoprotein assembly defects correlate with spinal muscular atrophy severity and preferentially affect a subset of spliceosomal snRNPs. PLoS One 2:e921. https://doi.org/10.1371/journal.pone.0000921
97. Doktor TK, Hua Y, Andersen HS et al (2017) RNA-sequencing of a mouse-model of spinal muscular atrophy reveals tissue-wide changes in splicing of U12-dependent introns. Nucleic Acids Res 45:395–416. https://doi.org/10.1093/nar/gkw731
98. Osman EY, van Alstyne M, Yen PF et al (2020) Minor snRNA gene delivery improves the loss of proprioceptive synapses on SMA motor neurons. JCI Insight 5:e130574. https://doi.org/10.1172/jci.insight.130574
99. Lotti F, Imlach WL, Saieva L et al (2012) An SMN-dependent U12 splicing event essential for motor circuit function. Cell 151:440–454. https://doi.org/10.1016/j.cell.2012.09.012
100. Simon CM, Van Alstyne M, Lotti F et al (2019) Stasimon contributes to the loss of sensory synapses and motor neuron death in a mouse model of spinal muscular atrophy. Cell Rep 29:3885–3901.e5. https://doi.org/10.1016/j.celrep.2019.11.058
101. Simon CM, Dai Y, Van Alstyne M et al (2017) Converging mechanisms of p53 activation drive motor neuron degeneration in spinal muscular atrophy. Cell Rep 21:3767–3780. https://doi.org/10.1016/j.celrep.2017.12.003
102. Van Alstyne M, Simon CM, Sardi SP et al (2018) Dysregulation of Mdm2 and Mdm4 alternative splicing underlies motor neuron death in spinal muscular atrophy. Genes Dev 32:1045–1059. https://doi.org/10.1101/gad.316059.118
103. Pagliardini S, Giavazzi A, Setola V et al (2000) Subcellular localization and axonal transport of the survival motor neuron (SMN) protein in the developing rat spinal cord. Hum Mol Genet 9:47–56. https://doi.org/10.1093/hmg/9.1.47
104. Kye MJ, Niederst ED, Wertz MH et al (2014) SMN regulates axonal local translation via miR-183/mTOR pathway. Hum Mol Genet 23:6318–6331. https://doi.org/10.1093/hmg/ddu350
105. Kiernan MC, Vucic S, Cheah BC et al (2011) Amyotrophic lateral sclerosis. Lancet 377: 942–955
106. Rowland LP, Shneider NA (2001) Medical Progress. Amyotrophic lateral sclerosis. N Engl J Med 344:1688–1700
107. Mathis S, Goizet C, Soulages A et al (2019) Genetics of amyotrophic lateral sclerosis: a review. J Neurol Sci 399:217–226
108. Renton AE, Chiò A, Traynor BJ (2014) State of play in amyotrophic lateral sclerosis genetics. Nat Neurosci 17:17–23
109. Ling SC, Polymenidou M, Cleveland DW (2013) Converging mechanisms in als

and FTD: disrupted RNA and protein homeostasis. Neuron 79:416–438

110. Tollervey JR, Curk T, Rogelj B et al (2011) Characterizing the RNA targets and position-dependent splicing regulation by TDP-43. Nat Neurosci 14:452–458. https://doi.org/10.1038/nn.2778
111. Humphrey J, Emmett W, Fratta P et al (2017) Quantitative analysis of cryptic splicing associated with TDP-43 depletion. BMC Med Genet 10:38. https://doi.org/10.1186/s12920-017-0274-1
112. Ling JP, Pletnikova O, Troncoso JC, Wong PC (2015) TDP-43 repression of nonconserved cryptic exons is compromised in ALS-FTD. Science 349:650–655. https://doi.org/10.1126/science.aab0983
113. Tan Q, Yalamanchili HK, Park J et al (2016) Extensive cryptic splicing upon loss of RBM17 and TDP43 in neurodegeneration models. Hum Mol Genet 25:5083–5093. https://doi.org/10.1093/hmg/ddw337
114. Klim JR, Williams LA, Limone F et al (2019) ALS-implicated protein TDP-43 sustains levels of STMN2, a mediator of motor neuron growth and repair. Nat Neurosci 22:167–179. https://doi.org/10.1038/s41593-018-0300-4
115. Melamed Z, López-Erauskin J, Baughn MW et al (2019) Premature polyadenylation-mediated loss of stathmin-2 is a hallmark of TDP-43-dependent neurodegeneration. Nat Neurosci 22:180–190. https://doi.org/10.1038/s41593-018-0293-z
116. Chauvin S, Sobel A (2015) Neuronal stathmins: a family of phosphoproteins cooperating for neuronal development, plasticity and regeneration. Prog Neurobiol 126:1–18
117. Ebstein SY, Yagudayeva I, Shneider NA (2019) Mutant TDP-43 causes early-stage dose-dependent motor neuron degeneration in a TARDBP Knockin mouse model of ALS. Cell Rep 26:364–373.e4. https://doi.org/10.1016/j.celrep.2018.12.045
118. Mitchell JC, Constable R, So E et al (2015) Wild type human TDP-43 potentiates ALS-linked mutant TDP-43 driven progressive motor and cortical neuron degeneration with pathological features of ALS. Acta Neuropathol Commun 3:36. https://doi.org/10.1186/s40478-015-0212-4
119. Fratta P, Sivakumar P, Humphrey J et al (2018) Mice with endogenous TDP -43 mutations exhibit gain of splicing function and characteristics of amyotrophic lateral sclerosis. EMBO J 37:e98684. https://doi.org/10.15252/embj.201798684
120. Vance C, Rogelj B, Hortobágyi T et al (2009) Mutations in FUS, an RNA processing protein, cause familial amyotrophic lateral sclerosis type 6. Science 323:1208–1211. https://doi.org/10.1126/science.1165942
121. Kwiatkowski TJ, Bosco DA, LeClerc AL et al (2009) Mutations in the FUS/TLS gene on chromosome 16 cause familial amyotrophic lateral sclerosis. Science 323:1205–1208. https://doi.org/10.1126/science.1166066
122. Dormann D, Rodde R, Edbauer D et al (2010) ALS-associated fused in sarcoma (FUS) mutations disrupt transportin-mediated nuclear import. EMBO J 29:2841–2857. https://doi.org/10.1038/emboj.2010.143
123. Loughlin FE, Lukavsky PJ, Kazeeva T et al (2019) The solution structure of FUS bound to RNA reveals a bipartite mode of RNA recognition with both sequence and shape specificity. Mol Cell 73:490–504.e6. https://doi.org/10.1016/j.molcel.2018.11.012
124. Rogelj B, Easton LE, Bogu GK et al (2012) Widespread binding of FUS along nascent RNA regulates alternative splicing in the brain. Sci Rep 2:603. https://doi.org/10.1038/srep00603
125. Lagier-Tourenne C, Polymenidou M, Hutt KR et al (2012) Divergent roles of ALS-linked proteins FUS/TLS and TDP-43 intersect in processing long pre-mRNAs. Nat Neurosci 15:1488–1497. https://doi.org/10.1038/nn.3230
126. Devoy A, Kalmar B, Stewart M et al (2017) Humanized mutant FUS drives progressive motor neuron degeneration without aggregation in "FUSDelta14" knockin mice. Brain 140:2797–2805. https://doi.org/10.1093/brain/awx248
127. Scekic-Zahirovic J, Sendscheid O, El Oussini H et al (2016) Toxic gain of function from mutant FUS protein is crucial to trigger cell autonomous motor neuron loss. EMBO J 35:1077–1097. https://doi.org/10.15252/embj.201592559
128. Gerbino V, Carrì MT, Cozzolino M, Achsel T (2013) Mislocalised FUS mutants stall spliceosomal snRNPs in the cytoplasm. Neurobiol Dis 55:120–128. https://doi.org/10.1016/j.nbd.2013.03.003
129. Sun S, Ling SC, Qiu J et al (2015) ALS-causative mutations in FUS/TLS confer gain and loss of function by altered association with SMN and U1-snRNP. Nat Commun 6:6171. https://doi.org/10.1038/ncomms7171

130. Yu Y, Chi B, Xia W et al (2015) U1 snRNP is mislocalized in ALS patient fibroblasts bearing NLS mutations in FUS and is required for motor neuron outgrowth in zebrafish. Nucleic Acids Res 43:3208–3218. https://doi.org/10.1093/nar/gkv157
131. Panza F, Lozupone M, Seripa D et al (2020) Development of disease-modifying drugs for frontotemporal dementia spectrum disorders. Nat Rev Neurol 16:213–228
132. Hodges JR, Piguet O (2018) Progress and challenges in frontotemporal dementia research: a 20-year review. J Alzheimers Dis 62:1467–1480
133. Hutton M, Lendon CL, Rizzu P et al (1998) Association of missense and 5′-splice-site mutations in tau with the inherited dementia FTDP-17. Nature 393:702–704. https://doi.org/10.1038/31508
134. Wszolek ZK, Tsuboi Y, Ghetti B et al (2006) Frontotemporal dementia and parkinsonism linked to chromosome 17 (FTDP-17). Orphanet J Rare Dis 1:30
135. Ehrlich M, Hallmann AL, Reinhardt P et al (2015) Distinct neurodegenerative changes in an induced pluripotent stem cell model of frontotemporal dementia linked to mutant TAU protein. Stem Cell Rep 5:83–96. https://doi.org/10.1016/j.stemcr.2015.06.001
136. D'Souza I, Schellenberg GD (2005) Regulation of tau isoform expression and dementia. Biochim Biophys Acta Mol basis Dis 1739: 104–115
137. Guo T, Noble W, Hanger DP (2017) Roles of tau protein in health and disease. Acta Neuropathol 133:665–704
138. Niblock M, Gallo JM (2012) Tau alternative splicing in familial and sporadic tauopathies. Biochem Soc Trans 40:677–680
139. Wolfe MS (2012) The role of tau in neurodegenerative diseases and its potential as a therapeutic target. Scientifica (Cairo) 2012:1–20. https://doi.org/10.6064/2012/796024
140. Grover A, Houlden H, Baker M et al (1999) 5′ splice site mutations in tau associated with the inherited dementia FTDP-17 affect a stem-loop structure that regulates alternative splicing of exon 10. J Biol Chem 274:15134–15143. https://doi.org/10.1074/jbc.274.21.15134
141. Andreadis A (2012) Tau splicing and the intricacies of dementia. J Cell Physiol 227: 1220–1225
142. Kar A, Havlioglu N, Tarn WY, Wu JY (2006) RBM4 interacts with an intronic element and stimulates tau exon 10 inclusion. J Biol Chem 281:24479–24488. https://doi.org/10.1074/jbc.M603971200
143. Chapple PJ, Anthony K, Martin TR et al (2007) Expression, localization and tau exon 10 splicing activity of the brain RNA-binding protein TNRC4. Hum Mol Genet 16:2760–2769. https://doi.org/10.1093/hmg/ddm233
144. Gao L, Wang J, Wang Y, Andreadis A (2007) SR protein 9G8 modulates splicing of tau exon 10 via its proximal downstream intron, a clustering region for frontotemporal dementia mutations. Mol Cell Neurosci 34: 48–58. https://doi.org/10.1016/j.mcn.2006.10.004
145. Kar A, Fushimi K, Zhou X et al (2011) RNA helicase p68 (DDX5) regulates tau exon 10 splicing by modulating a stem-loop structure at the 5' splice site. Mol Cell Biol 31: 1812–1821. https://doi.org/10.1128/mcb.01149-10
146. Kondo S, Yamamoto N, Murakami T et al (2004) Tra2β, SF2/ASF and SRp30c modulate the function of an exonic splicing enhancer in exon 10 of tau pre-mRNA. Genes Cells 9:121–130. https://doi.org/10.1111/j.1356-9597.2004.00709.x
147. Strang KH, Golde TE, Giasson BI (2019) MAPT mutations, tauopathy, and mechanisms of neurodegeneration. Lab Investig 99: 912–928

Chapter 2

Targeting Alternative Splicing for Therapeutic Interventions

Jessica L. Centa and Michelle L. Hastings

Abstract

Targeting of pre-mRNA splicing has yielded a rich variety of strategies for altering gene expression as a treatment for disease. The search for therapeutics that can modulate splicing has been dominated by antisense oligonucleotides (ASOs) and small molecule compounds, with each platform achieving remarkably effective results in the clinic. The success of RNA-targeting drugs has led to the exploration of new strategies to expand the repertoire of this type of therapeutic. Here, we discuss some of the more common causes of faulty gene expression and provide examples of approaches that have been developed to target and correct these defects for therapeutic value.

Key words Alternative splicing, Therapeutics, Antisense oligonucleotides, Small molecules, RNA base editing, Exon skipping, Cryptic splice sites, Mutation

1 Introduction

RNA has taken center stage in the therapeutics arena in recent years, proving to be a powerful drug targeting paradigm. There are currently six drugs on the market that treat disease by targeting splicing [1]. Five of these are steric blocking antisense oligonucleotides (ASOs) and one is a small molecule modulator of splicing (Table 1). All of these have been approved in the last six years. These advances have laid a foundation for the rapid progression of research and development in this area. There are many approaches in the pipeline and it is likely that this is only the beginning of countless possible splice-targeting strategies that can be applied on a mutation and target-specific manner to treat pathological conditions [2, 3].

RNA biogenesis is a complex process involving many steps, all of which are regulated in ways that, in many cases, are only beginning to be understood [4]. Approaches to study this complexity have grown in step with the greater appreciation of this complexity and with the growing knowledge that intervening in the different steps of RNA biogenesis is an effective therapeutic target. Many of

Peter Scheiffele and Oriane Mauger (eds.), *Alternative Splicing: Methods and Protocols*,
Methods in Molecular Biology, vol. 2537, https://doi.org/10.1007/978-1-0716-2521-7_2,

Table 1
Commercially available splice-altering drugs and year approved for market

Drug	Mechanism	Disease	Year
Nusinersen (Spinraza™)	Steric blocking ASO	Spinal muscular atrophy	2016
Eteplirsen (Exondys 51™)	Steric blocking ASO	Duchenne muscular dystrophy	2016
Golodirsen (Vyondys 53™)	Steric blocking ASO	Duchenne muscular dystrophy	2019
Viltolarsen (Viltepso™)	Steric blocking ASO	Duchenne muscular dystrophy	2020
Risdiplam (Evrysdi™)	Small molecule	Spinal muscular atrophy	2020
Casimersen (Amondys 45™)	Steric blocking ASO	Duchenne muscular dystrophy	2021

the approaches to analyze and target RNA are detailed in the following chapters. They include descriptions of techniques to better collect and assemble RNA sequencing data patterns to account for novel transcripts and single-cell sequencing analysis in addition to the incorporation of long-read RNA sequencing that eliminates ambiguity of splicing, all of which are shedding light on the extent of alternative splicing and the regulation of alternative splicing events. Analysis of ribosome-associated mRNAs and other approaches suggest that many alternatively spliced mRNAs are translated into alternative protein isoforms or even novel proteins, offering important insights on genomic and proteomic complexity. Techniques such as Basescope™, RNA-FISH, and two-color reporter systems also allow for the sensitive detection of RNA and RNA isoforms in cells. Much of the advance in RNA detection and analysis has been driven by the discoveries of new mechanisms and proteins involved in splicing and regulation of alternative splicing. Likewise, the realization that alternative splicing regulation is more complex than previously appreciated, with alternative protein coding sequences, poison exons and intron retention to control RNA stability and localization and translation, has led to new mechanistic insights on the regulation of gene expression by splicing and the regulation of splicing factors.

Hand-in-hand with the growth in our understanding of the complexity of RNA biogenesis and post-transcriptional regulation of gene expression, has come corresponding revelations that many mutations causing monogenic diseases affect alternative splicing or may be treated by modulating alternative splicing [5–7]. Likewise, mechanisms involving splicing have been found to be associated with other human conditions such as cancer, aging, and idiopathic disease. Fortunately, there has been a rapid acceleration and several successes in therapeutic approaches to manipulate splicing for treatment of disease. Here, we focus on the process of pre-mRNA splicing and alternative splicing and discuss the ways that this process can be disrupted in disease and consider approaches that can be taken to correct or modulate splicing to treat disease. This introduction to therapeutics aimed at modulating splicing serves as an entry point to the proceeding techniques presented in the following chapters that are facilitating and accelerating disease and therapeutic discovery.

2 Pre-mRNA Splicing

2.1 Constitutive Splicing

Most protein-coding RNAs undergo splicing to remove introns, non–protein-coding sequences, which intervene protein-coding sequences, called exons. Functional noncoding RNAs can also have introns and undergo splicing. The removal, or splicing out, of introns and ligation of exons gives rise to the final RNA product. This splicing reaction has two transesterification steps involving the 5′ splice site (5'ss) or donor site at the 5′ end of the intron, the 3′ splice site (3'ss) or acceptor site, at the 3′ end of the intron, and a branchpoint adenosine upstream of the 3′ splice site which is typically preceded by a pyrimidine-rich tract [8] (Fig. 1a). In addition to the conserved splice site sequences directly involved in the catalytic steps of splicing, there are many other sequences involved in splicing. These sequence elements can be located in the exon or intron and typically are recognized by proteins that, when bound to the RNA, can either promote or inhibit splicing, and hence are referred to as enhancer or silencer sequences.

Splicing at the correct splice sites involves recognition of the sites by the spliceosome, a large complex comprised of protein and small nuclear RNA (snRNA) components [9, 10]. The snRNAs form base-pairing interactions with the splice sites, ensuring proper splice site selection. The auxiliary splicing proteins, which number in the hundreds, assist in the process. There are many sequences within a pre-mRNA that resemble consensus splice sites but are not spliced by the spliceosome. The mechanisms by which the authentic splice sites are recognized is not completely understood, though many advances in recent years have improved predictability. Much work has gone into understanding the splicing code, which is the information, be it sequence, context or structural, that determines where and when splicing occurs.

2.2 Alternative Splicing

Splicing can give rise to myriad different mRNAs that arise from natural alternative splicing events. In fact, most RNAs have alternatively spliced isoforms wherein multiple RNA processing signals compete [11, 12]. Alternative splicing of a gene product can have many different consequences on gene expression. An alternatively spliced RNA may have sequences included or spliced out that control stability, translatability, coding capacity, and other qualities

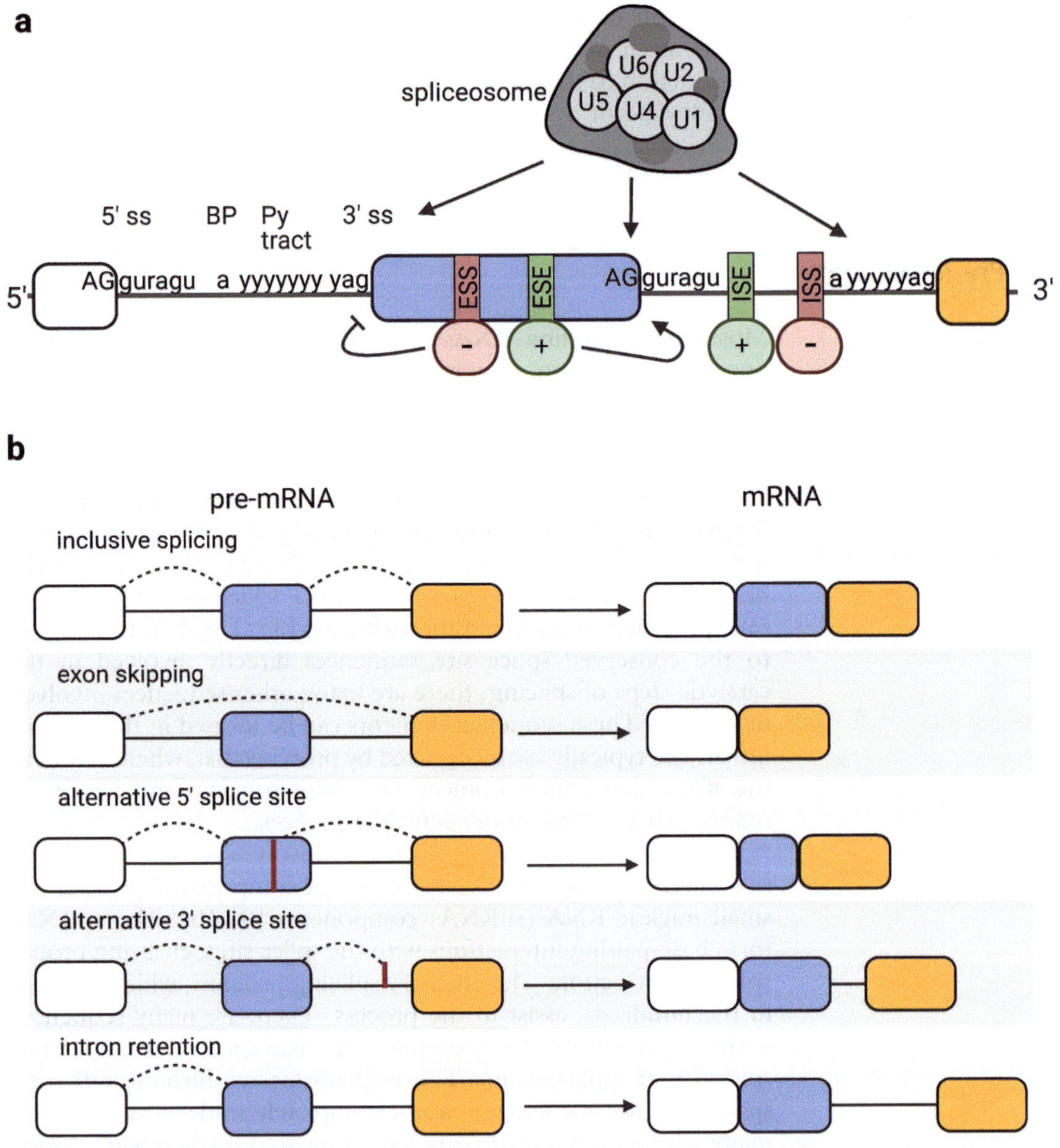

Fig. 1 Splicing sequences and patterns of alternative splicing. (**a**) Key elements involved in pre-mRNA splicing. Splicing is catalyzed by the spliceosome, a large complex of RNA and proteins including small nuclear ribonucleoprotein (snRNPs), designated as U1, U2, U4, U5, and U6. Exons are shown as boxes, separated by introns shown as horizonal lines. Conserved consensus splice site sequences at the 3′ (3′ss) and 5′ (5′ss) ends of the introns, the branch point sequence (BS), and the polypyrimidine tract (Py tract) are labeled. Splicing regulatory sequence enhancers and silencer elements in exons and introns (ESE, exonic splicing enhancer; ISE, intronic splicing enhancer; ESS, exonic splicing silencer; ISS, intronic splicing silencer) influence splicing through the recruitment of trans-acting splicing factor proteins, that promote splicing (+) or suppress splicing (−). (**b**) Common types of alternative splicing. Exons are indicated by boxes, introns by horizontal lines, and splicing events depicted with dashed lines. Inclusive splicing depicts all introns are spliced out and exons ligated together. Exon skipping involving the splicing out (exclusion) of a cassette exon. Alternative 5′ and 3′ splice sites compete with another nearby splice site and change the boundary of the exons. Intron retention, wherein an intron remains in the mature mRNA

of the RNA that may alter its ultimate functional capacity as either a message for a translated protein or untranslated RNA.

Numerous different mRNA isoforms can be produced from a single gene locus. Splicing out of an entire exon, also referred to as exon skipping, is the most common form of alternative splicing. Competing alternative 5′ or 3′ splice sites, intron retention, and alternative polyadenylation that either competes with splicing or another polyadenylation site are other common forms of alternative splicing (Fig. 1b).

Alternative splicing is a regulated process that can be controlled in a manner based on the specific tissue and/or cell-type, developmental stage, stimuli, or conditions. Complex alternative splicing patterns and regulation of gene output at the mRNA level is likely a crucial part of organismal diversity. Variations in genomic sequences that affect splicing are wide-spread and contribute to complex traits. Variants near splice sites as well as missense and nonsense variants have been found to have the strongest effect on splicing [7]. Likewise, disruptions in alternative splicing can result in pathological outcomes.

3 Strategies for Therapeutically Manipulating Splicing

Gene variants that affect splicing and are associated with human disease or disease risk are common [5–7, 13]. As aberrant splicing increasingly becomes recognized as a disease mechanism as well as an intervention point for influencing gene expression, technologies to manipulate the process in a precise manner have come to the forefront of drug development. Leading the way in the clinic are antisense oligonucleotides (referred to as ASOs or AONs) and small molecules that modify splicing for therapeutically beneficial outcomes (Table 1). Success with these modalities have sparked a surge in the study of ASOs and small molecules and other emerging technologies targeted to specific splicing events for the treatment of disease. In general, therapeutic approaches can be grouped into two categories, strategies to inhibit a specific splicing event or to activate splicing at a particular site (Fig. 2).

3.1 Antisense Oligonucleotides

Antisense technology has seen a surge in success in clinical development in recent years [1]. Five antisense drugs approved for commercial use are steric blocking ASOs that target pre-mRNA splicing. This type of ASO works by base-pairing to RNA and exerting an occupancy-only effect by obstructing natural interactions of the RNA with proteins or other RNAs [14]. Splicing relies on the recognition of single-stranded RNA sequences by RNA binding proteins and/or by RNA-RNA base-pairing interactions with small nuclear RNA (snRNA) components of the spliceosome as well as RNA secondary structures. Therefore, targeting an ASO

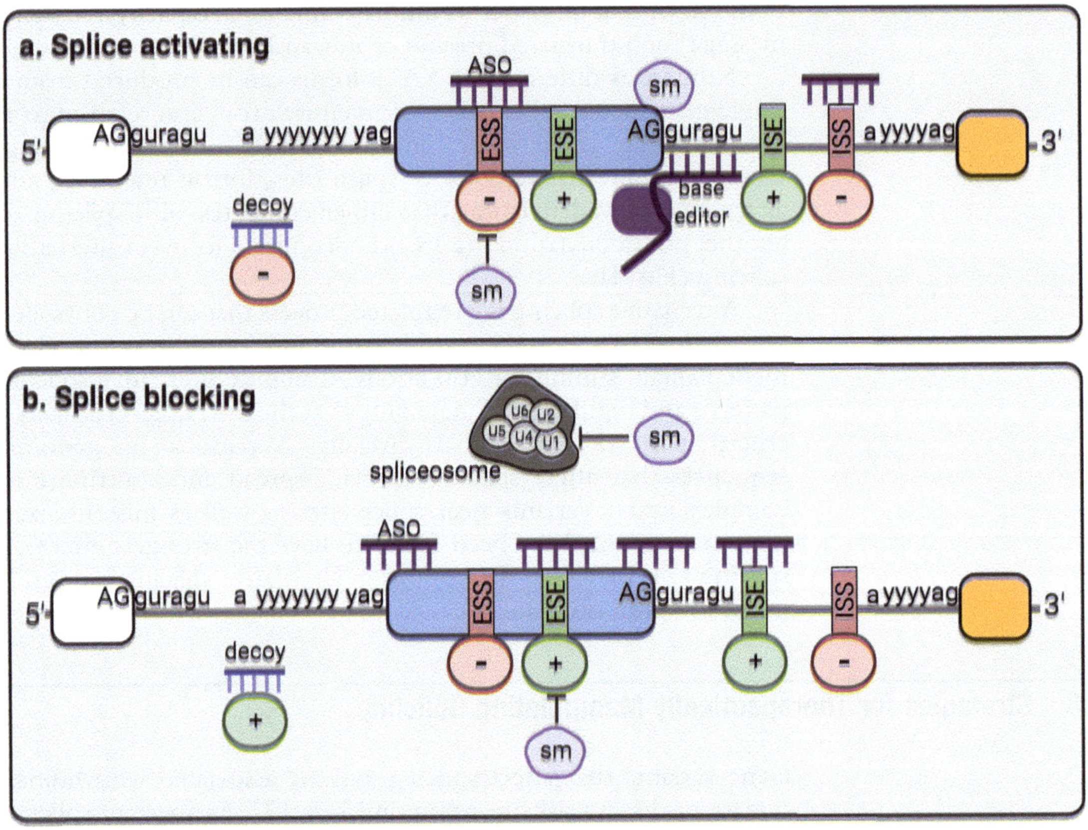

Fig. 2 Targets and tools for modulating splicing. (**a**) Splicing can be activated by stabilizing the interaction between the spliceosome and pre-mRNA using small molecules. Additionally, ASOs targeting ESS or ISS sequences and decoy oligonucleotides targeting splicing inhibitory proteins can cause exon inclusion. (**b**) Splicing can be blocked by using ASOs that base-pair at the 5′ss or 3′ss regions or at ESE or ISE sequences. Spliceosome components and auxiliary splicing factors can be inhibited using small molecules (sm) or decoy oligonucleotides. Splicing mutations, in particular those at the +1, +2, and −1 positions of the intron, can be corrected using base editing

to base-pair with these sequences creates a double-stranded RNA and masks recognition of splicing signals, thereby altering splicing in a manner that is dependent on the type of splicing signal that is targeted. The requirement for intact and accessible nucleotides of the 5′ and 3′ splice site for efficient splicing also makes the splicing reaction vulnerable to disruption when this outcome is desirable for a therapeutic outcome (Fig. 2).

3.1.1 Exon Skipping

One way that steric blocking ASOs can modify splicing is by inducing skipping of exons (Fig. 3). Exon skipping may be a desirable outcome for a number of reasons. For example, skipping of an exon that has a mutation that creates an open-reading frameshift or creates a nonsense mutation can result in restoration of the open reading frame. There are currently four FDA-approved ASO drugs

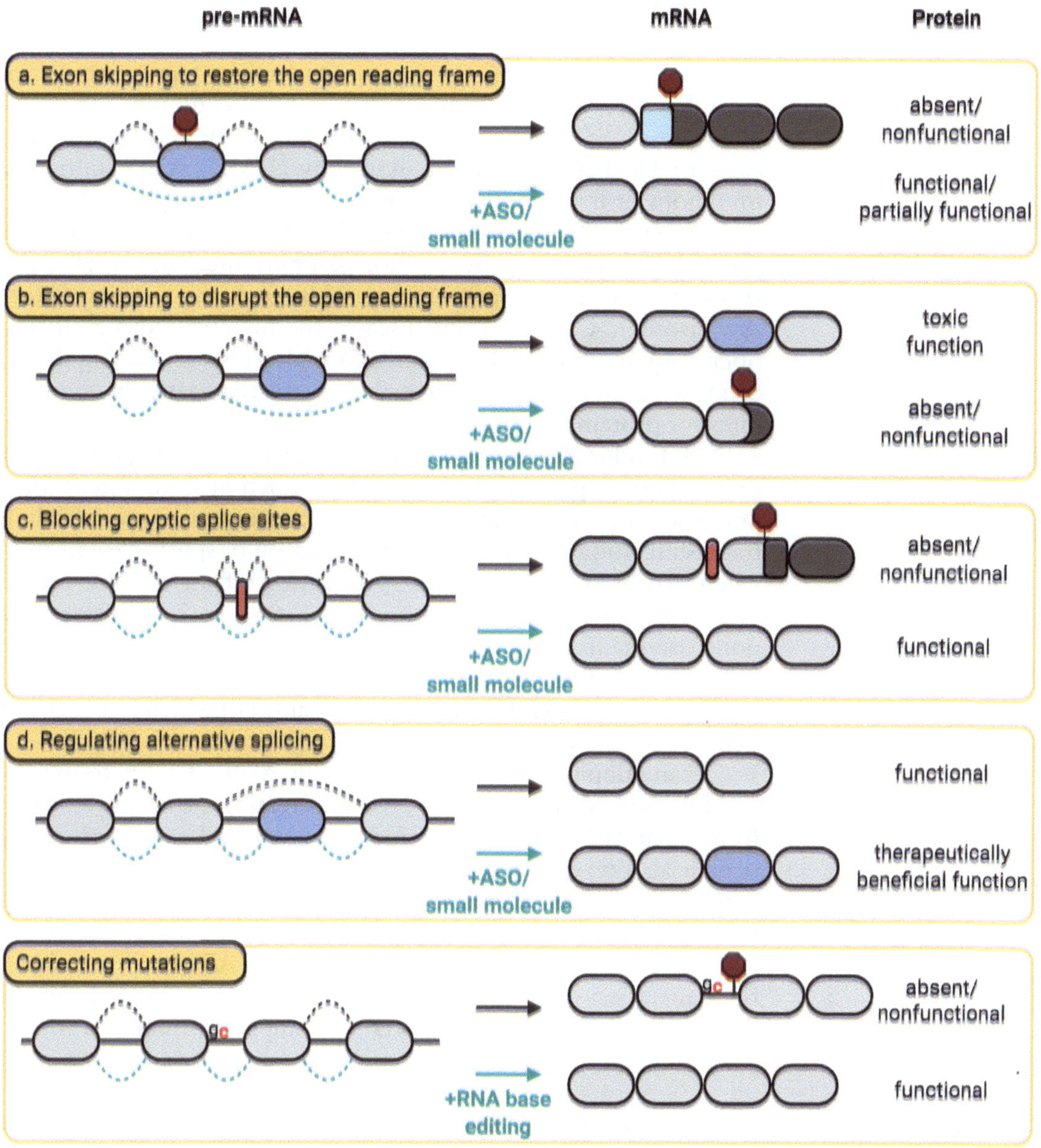

Fig. 3 Strategies for therapeutically manipulating splicing. Exons are depicted in gray and the horizonal lines represent introns. The dashed lines indicate splicing events with black being disease-causing and cyan designating the therapeutic approach (**a**) Targeted exons can be skipped to restore the open reading frame. Deletions or nonsense mutations creating an open reading frameshift can cause a premature termination codon (depicted by the octagon). This mRNA transcript may undergo NMD resulting in no protein production or be translated to a nonfunctional protein. The open reading frame can be restored by skipping an exon using ASOs or small molecules to block splicing machinery or splicing regulatory elements. This strategy can create a partially functional protein with therapeutic benefit. (**b**) In diseases characterized by the expression of toxic proteins, exon skipping using ASOs or small molecules to disrupt the reading frame and create a premature termination codon (PTC), resulting in NMD or a nonfunctional protein, may provide therapeutic benefit. (**c**) Mutations causing cryptic splice sites can be targeted by ASOs or small molecules to mask the incorrect splice site and redirect splicing to the natural site. (**d**) Natural alternative splicing events can be manipulated using ASOs or small molecules to produce a particular spliced isoform. (**e**) RNA base editing is a potential therapeutic option for splice site mutations

that act by inducing exon skipping for reading frame correction: Exondys 51, Vyondys 53, Viltepso, and Amondys 45 [15–17]. These ASOs restore the open reading frame of dystrophin in cases of Duchenne muscular dystrophy (DMD) where nonsense mutations or frameshifts occur in exons that can be skipped out to produce an mRNA encoding a partially functional protein. Another promising ASO acting by an exon skipping mechanism has been shown to have therapeutic efficacy in the pediatric lysosomal storage disorder, CLN3 Batten disease. In this case, correcting the open reading frame of *CLN3* mRNA with a disease-causing multi-exon deletion, produces a protein isoform that partially compensates for protein function [18].

ASO-induced exon skipping can also be used to generally reduce gene expression and protein production from a gene via a knockdown strategy that involves the induction of an open reading frameshift. Inducing the skipping of an exon or splicing event that, when activated, results in the production of an mRNA with a premature termination codon (PTC) that is a substrate for nonsense-mediated decay (NMD) will have the overall effect of reducing gene expression. This strategy was utilized for the knock-down of *LRRK2*, the hyperactivation of which is associated with Parkinson's disease [19–22].

Exon skipping can also be induced by ASOs as a way to modify the function of the encoded protein, creating novel protein isoforms that lack or gain functions that would be therapeutically beneficial. For example, ASOs that induce exon 17 skipping of amyloid precursor protein (APP) mRNA encode a protein lacking the region that encodes the amyloid-β (Aβ) producing, γ-secretase cleave site. ASO-based induction of this novel APP isoform reduces the generation of Aβ peptide, which is implicated in Alzheimer's disease, but maintains other potentially important functions of the protein [23].

3.1.2 Blocking Cryptic Splice Sites

Perhaps one of the most straightforward applications of steric blocking ASOs as a therapeutic is for blocking cryptic splicing (Fig. 3). Cryptic or de novo splice sites may be created directly by a mutation or through the creation of splicing enhancer elements that activate a nearby splice site. In these cases, blocking an activated splice site can redirect to the natural splice site. There are many examples of the use of ASOs to correct splicing in this way. One such ASO, milasen, gained world-wide attention for its rapid development and FDA approval for the treatment of a single patient. Milasen was used to treat a patient with CLN7 Batten disease caused by a mutation that created a cryptic splice site. The ASO blocked recognition of the cryptic splice site and reduced the frequency and severity of seizures in the patient [24]. Another ASO splicing corrector designed to block cryptic splicing that has advanced to clinical trials is sepofarsen (QR-110, ProQR). This therapy is designed to block recognition of a cryptic splice site caused by a mutation in the gene *CEP290* to correct visual impairment in patients with Leber's congenital amaurosis 10 [25].

3.1.3 Modulating Alternative Splicing

There has been much research and success in modulating naturally occurring alternative splicing events using ASOs. ASO design for such an approach in many cases must be empirically determined depending on the type of splicing events that are competing as well as the desired outcome. Strategies to promote use of a splice site through an activating approach often involve targeting splicing silencers or weakening the competing splice site by targeting sequences that enhance the competing site (Fig. 2). Alternatively, strategies to block splicing at a particular site are typically achieved by designing ASOs to base-pair at the splice-sites or to splicing enhancer sequences.

The development of a splice-modulating ASO for the treatment of spinal muscular atrophy (SMA) is a now classic example of how ASOs can be used to manipulate alternative splicing for therapeutic benefit. Spinal muscular atrophy (SMA) is a rare, autosomal recessive, progressive neuromuscular disease caused by mutations in the survival of motor neuron 1 (*SMN1*) gene, encoding SMN protein. Humans have a second gene, *SMN2*, that also encodes SMN protein. However, the majority of SMN2 mRNA is alternatively spliced to produce an mRNA lacking exon 7, which encodes a protein isoform that does not compensate for the loss of SMN1 in SMA. Much of SMA therapeutic development has been aimed at increasing exon 7 splicing and restoring SMN protein levels. Currently, there are two SMN2-targeted splicing therapies available in the clinic to treat SMA patients, one of which is an ASO called nusinersen (Spinraza™; Ionis Pharmaceuticals/Biogen). Nusinersen is designed to base-pair with a cis-acting sequence element, ISS-N1, located downstream of SMN2 exon 7 that normally binds to the splicing factor hnRNP A1 [26, 27]. Binding of hnRNP A1 to this intronic RNA sequence inhibits splicing at the upstream 5′ splice site, resulting in skipping of the exon and the production of an mRNA with a termination codon in exon 7. Nusinersen base-pairs to ISS-N1, which blocks the binding and exon 7 splicing repressor activity of hnRNP A1, thereby increasing exon 7 splicing and the production of SMN protein [27–29].

Another example of how ASOs can be used therapeutically to influence alternative splicing is in blocking of naturally occurring, nonproductive splicing that results in mRNAs with premature termination codons that destabilize the mRNA or otherwise limit translation via pathways such as nonsense-mediated decay. Intron retention and splicing in of the so-called poison exons that introduce stop codons of reading frameshifts are two such splicing events that can be potentially targeted by ASOs to reduce target activity. An example of such an approach that is currently being tested in clinical trials (NCT04740476) is an ASO for the treatment of Dravet syndrome, a form of epilepsy in infants most often caused by mutations in *SCN1A*. The steric blocking ASO, STK-001 (Stoke Therapeutics), targets natural AS events that are nonproductive or result in NMD and increases expression of a functional transcript [30].

Tumor-associated changes in alternative splicing are also not uncommon and many are acknowledged as hallmarks of cancer [31]. This deregulation of splicing has been extensively explored as targets for therapy. ASOs have been considered a promising approach to normalizing deregulated alternative splicing in cancers [32].

3.2 Small Molecules

The search for small molecule modulators of splicing has largely struggled with barriers of specificity and potency. Small molecules targeted to proteins involved in splicing would be expected to affect splicing of many, if not all, introns. Indeed, a number of small molecules that block activity of specific splicing factors have general splicing inhibitory activity. However, these molecules may have found utility in downregulation of these proteins in certain cancers where some splicing factors have been found to be commonly mutated in a way that increases their abundance/activity [31, 33]. Small molecules that appear to more specifically affect distinct splicing events have also been discovered. Small molecules designed to recognize RNA structures that influence splicing by either altering access of the sequence to snRNA or protein binding or other mechanisms have the potential for specificity as structures might be more specific to a particular sequence. Indeed, these types of molecules have proven promising as drugs, with one, risdiplam, having been recently approved by the FDA for the treatment of spinal muscular atrophy [34](Table 1).

3.2.1 Activators

Two small molecules, risdiplam (Evrysdi™, RG7916, RO7034067, Roche) and branaplam (LMI070, NVS-SM1, Novartis), which target exon 7 inclusion to increase functional SMN in SMA have recently been discovered [35, 36]. Risdiplam, an orally administered pyrimidine-4-one analog that distributes centrally and peripherally, was recently FDA approved for the treatment of SMA, showing clinically meaningful benefit in SMA patients [37]. Based on experiments with analogs of risdiplam, the molecule is proposed to bind to an AG-rich motif in the middle of SMN2 exon 7 and recruit the splicing factor Far Upstream Element Binding Protein 1 (FUBP1) and its homolog KH-type Splicing Regulatory Protein (KHSRP), displacing the splicing repressor hnRNP G [38, 39]. It has also been proposed that the molecule stabilizes the interaction of U1 snRNP with the 5′ss, thereby promoting exon 7 splicing

[40]. Branaplam, a pyridazine analog, is being tested in Phase 2 clinical trials and acts by binding to the 5′ splice site of intron 7 and improving recognition by U1 snRNP by stabilizing the U1 snRNP and *SMN2* pre-mRNA interaction (NCT02268552) [36, 41]. The success in the discovery of small molecules that can act in a relatively specific manner to improve splicing is promising for the future of designing small molecules that target RNA and RNA–protein interactions.

3.2.2 Inhibitors

Many small molecule inhibitors targeting splicing-related proteins have also been discovered [32]. Such inhibitors have been explored for the treatment of cancer because putative driver mutations have been identified in many proteins involved in splicing and regulators of splicing. Changes in expression of splicing proteins have also been associated with cancer. In addition, transcriptome-wide aberrant splicing in cancer tissue compared to normal tissues has been found [42]. Often, these alternatively spliced isoforms are advantageous to cell proliferation and tumorigenesis. H3B-8800 is a modulator of the core spliceosomal protein SF3b complex that selectively kills tumor cells with mutations in SF3B1 [43]. A phase I clinical trial demonstrated safety with prolonged dosing and decreased red blood cell or platelet transfusion requirements in some patients (NCT02274051) [44]. H3B-8800 is currently in phase I clinical trials to evaluate safety in patients with myelodysplastic syndromes, acute myeloid leukemia and chronic myelomonocytic leukemia (NCT02274051). A number of small molecule therapeutics (GSK3326595, JNJ-64619178, PF-06939999, GSK3368715) that inhibit arginine methyltransferases, PRMT5 or PRMT1, modifiers of snRNP-associated Sm proteins and the splicing regulator RBM15, are also being tested in clinical trials (NCT03614728, NCT02783300, NCT04676516, NCT03573310, NCT03854227, NCT03666988) as treatments for a number of cancers. It is largely unclear how inhibitors of proteins involved in a fundamental step in gene expression can target cancer cells and not cause general cellular toxicity. Nonetheless, the fact that they can be used with apparent target selectivity suggests that this class of compound can be designed for specificity and/or targeted delivery for therapeutic effects. A better understanding of this specificity may lead to the discovery of small molecule inhibitors and modulators of splicing for the treatment of other diseases and pathological conditions.

3.3 Emerging Technologies

Correction of pathogenic mutations by directly changing them in the DNA or RNA is a rapidly growing area of investigation [45]. Base editing of DNA can introduce permanent changes into the genome whereas RNA editors introduce temporary changes that must be continuously employed for a persistent effect. Some of the natural substrates of the endogenous base-editing adenosine deaminase acting on RNA (ADAR) enzymes are splice sites that are edited to regulate splice site use (Fig. 2) [46]. This same activity is being harnessed for therapeutics (Fig. 3). There are a growing number of base editing approaches that have been devised. These different types of editors have different modifications that are typically aimed at optimizing specificity, customization, delivery, and reducing off-target editing, which challenge the translation of this approach to the clinic. There are two types of RNA base editors, RNA adenosine deaminases that carry out adenosine to inosine editing and RNA cytosine deaminases that generate a uridine from cytosine. Base editing has been shown to correct disease-associated mutations [47]. Inosines are biochemically recognized as guanosines, and the functions of the terminal guanosine residues of the 5′ and 3′ss can be replaced by inosines [48, 49]. Thus, mutations of these positions that introduce an A can be edited to an inosine using a base editor to correct splicing at the site. Similarly, U to C mutations in the conserved second position of the intron in the nearly invariant U could be targeted for editing by C-to-U base editors.

An additional approach that has been explored as a means to manipulate alternative splicing involves the use of decoy oligonucleotides the harbor binding sites for a particular RNA-binding protein. These decoys act to sequester the targeted splicing factor to inhibit splicing activity. This strategy was shown to be effective in inducing apoptosis and growth inhibition of cancer cells treated with a decoy containing binding sites for the oncogene *SRSF1* [50]. Given that SRSF1 and any other splicing protein targeted in this manner are core splicing components, careful titration and/or specific delivery to disease-relevant cells would likely be required to achieve therapeutic efficacy without toxicity.

4 Conclusion and Future Perspectives

RNA therapeutics that involve intervening in pre-mRNA splicing have gained attention as a powerful approach in recent years, largely based on the clinical success of the FDA-approved drugs targeting this reaction. There are numerous areas that are currently a focus of research and development or that will be needed to achieve that full potential of technologies targeting splicing. Some of these include: (1) expansion of an ASO tool-kit deployed for splicing mutations including rare mutations and key splicing events or expression pathways disrupted in diseases such as cancer that could, once identified, be targeted for correction; (2) discovery of new approaches for using ASOs to modulate splicing, including identification of new classes of targetable splicing regulatory elements bound by trans-acting factors or regulatory structures and higher order organizational dynamics; (3) design of small molecules targeting features of RNA or splicing proteins that will provide greater specificity for targeted pathogenic splicing events; (4) improvements in chemistry of ASOs and delivery of therapeutics specifically to cells/tissues of interest to improve potency and selectivity of response and reduce toxic effects; (5) strategies to effectively target and correct splicing defects associated with the most common splicing mutations in the catalytic nucleotides of the splicing reaction. Base-editing approaches may have a future for addressing this need. Overall, the clinical success of splicing drugs and the fact that there are still many avenues to pursue to improve these approaches along with the growing pipeline of splicing therapeutics all make for an optimistic view of the future for targeting the splicing reaction to treat disease.

Acknowledgments

M.L.H. is supported by NIH grant AG060195 and NS113233; J.L.C. is supported by NS113233. We thank Matthew Stratton for critical reading of the manuscript.

References

1. Scharner J, Aznarez I (2021) Clinical applications of single-stranded oligonucleotides: current landscape of approved and in-development therapeutics. Mol Ther 29(2): 540–554. https://doi.org/10.1016/j.ymthe.2020.12.022
2. Wang F, Zuroske T, Watts JK (2020) RNA therapeutics on the rise. Nat Rev Drug Discov 19(7):441–442. https://doi.org/10.1038/d41573-020-00078-0
3. Havens MA, Hastings ML (2016) Splice-switching antisense oligonucleotides as therapeutic drugs. Nucleic Acids Res 44(14): 6549–6563. https://doi.org/10.1093/nar/gkw533
4. Lee Y, Rio DC (2015) Mechanisms and regulation of alternative pre-mRNA splicing. Annu Rev Biochem 84:291–323. https://doi.org/10.1146/annurev-biochem-060614-034316
5. Soukarieh O, Gaildrat P, Hamieh M, Drouet A, Baert-Desurmont S, Frebourg T, Tosi M, Martins A (2016) Exonic splicing mutations are more prevalent than currently estimated and can be predicted by using in silico tools. PLoS Genet 12(1):e1005756. https://doi.org/10.1371/journal.pgen.1005756
6. Xiong HY, Alipanahi B, Lee LJ, Bretschneider H, Merico D, Yuen RK, Hua Y, Gueroussov S, Najafabadi HS, Hughes TR, Morris Q, Barash Y, Krainer AR, Jojic N, Scherer SW, Blencowe BJ, Frey BJ (2015) RNA splicing. The human splicing code reveals new insights into the genetic determinants of disease. Science 347(6218):1254806. https://doi.org/10.1126/science.1254806
7. Li YI, van de Geijn B, Raj A, Knowles DA, Petti AA, Golan D, Gilad Y, Pritchard JK (2016) RNA splicing is a primary link between genetic variation and disease. Science 352(6285): 600–604. https://doi.org/10.1126/science.aad9417
8. Kastner B, Will CL, Stark H, Luhrmann R (2019) Structural insights into nuclear pre-mRNA splicing in higher eukaryotes. Cold Spring Harb Perspect Biol 11(11): a032417. https://doi.org/10.1101/cshperspect.a032417
9. Papasaikas P, Valcarcel J (2016) The spliceosome: the ultimate RNA chaperone and sculptor. Trends Biochem Sci 41(1):33–45. https://doi.org/10.1016/j.tibs.2015.11.003
10. Will CL, Luhrmann R (2011) Spliceosome structure and function. Cold Spring Harb Perspect Biol 3(7):a003707. https://doi.org/10.1101/cshperspect.a003707
11. Pan Q, Shai O, Lee LJ, Frey BJ, Blencowe BJ (2008) Deep surveying of alternative splicing complexity in the human transcriptome by high-throughput sequencing. Nat Genet 40(12):1413–1415. https://doi.org/10.1038/ng.259

12. Wang ET, Sandberg R, Luo S, Khrebtukova I, Zhang L, Mayr C, Kingsmore SF, Schroth GP, Burge CB (2008) Alternative isoform regulation in human tissue transcriptomes. Nature 456(7221):470–476. https://doi.org/10.1038/nature07509
13. Ohno K, Takeda JI, Masuda A (2018) Rules and tools to predict the splicing effects of exonic and intronic mutations. Wiley Interdiscip Rev RNA 9(1). https://doi.org/10.1002/wrna.1451
14. Crooke ST, Liang XH, Baker BF, Crooke RM (2021) Antisense technology: a review. J Biol Chem 296:100416. https://doi.org/10.1016/j.jbc.2021.100416
15. Koo T, Wood MJ (2013) Clinical trials using antisense oligonucleotides in Duchenne muscular dystrophy. Hum Gene Ther 24(5): 479–488. https://doi.org/10.1089/hum.2012.234
16. Dhillon S (2020) Viltolarsen: first approval. Drugs 80(10):1027–1031. https://doi.org/10.1007/s40265-020-01339-3
17. Frank DE, Schnell FJ, Akana C, El-Husayni SH, Desjardins CA, Morgan J, Charleston JS, Sardone V, Domingos J, Dickson G, Straub V, Guglieri M, Mercuri E, Servais L, Muntoni F, Group S-NS (2020) Increased dystrophin production with golodirsen in patients with Duchenne muscular dystrophy. Neurology 94(21): e2270–e2282. https://doi.org/10.1212/WNL.0000000000009233
18. Centa JL, Jodelka FM, Hinrich AJ, Johnson TB, Ochaba J, Jackson M, Duelli DM, Weimer JM, Rigo F, Hastings ML (2020) Therapeutic efficacy of antisense oligonucleotides in mouse models of CLN3 batten disease. Nat Med 26(9):1444–1451. https://doi.org/10.1038/s41591-020-0986-1
19. Korecka JA, Thomas R, Hinrich AJ, Moskites AM, Macbain ZK, Hallett PJ, Isacson O, Hastings ML (2020) Splice-switching antisense oligonucleotides reduce LRRK2 kinase activity in human LRRK2 transgenic mice. Mol Ther Nucleic Acids 21:623–635. https://doi.org/10.1016/j.omtn.2020.06.027
20. Korecka JA, Thomas R, Christensen DP, Hinrich AJ, Ferrari EJ, Levy SA, Hastings ML, Hallett PJ, Isacson O (2019) Mitochondrial clearance and maturation of autophagosomes are compromised in LRRK2 G2019S familial Parkinson's disease patient fibroblasts. Hum Mol Genet 28(19):3232–3243. https://doi.org/10.1093/hmg/ddz126
21. Di Maio R, Hoffman EK, Rocha EM, Keeney MT, Sanders LH, De Miranda BR, Zharikov A, Van Laar A, Stepan AF, Lanz TA, Kofler JK, Burton EA, Alessi DR, Hastings TG, Greenamyre JT (2018) LRRK2 activation in idiopathic Parkinson's disease. Sci Transl Med 10(451). https://doi.org/10.1126/scitranslmed.aar5429
22. Korecka JA, Talbot S, Osborn TM, de Leeuw SM, Levy SA, Ferrari EJ, Moskites A, Atkinson E, Jodelka FM, Hinrich AJ, Hastings ML, Woolf CJ, Hallett PJ, Isacson O (2019) Neurite collapse and altered ER Ca(2+) control in human Parkinson disease patient iPSC-derived neurons with LRRK2 G2019S mutation. Stem Cell Rep 12(1):29–41. https://doi.org/10.1016/j.stemcr.2018.11.021
23. Chang JL, Hinrich AJ, Roman B, Norrbom M, Rigo F, Marr RA, Norstrom EM, Hastings ML (2018) Targeting amyloid-beta precursor protein, APP, splicing with antisense oligonucleotides reduces toxic amyloid-beta production. Mol Ther 26(6):1539–1551. https://doi.org/10.1016/j.ymthe.2018.02.029
24. Kim J, Hu C, Moufawad El Achkar C, Black LE, Douville J, Larson A, Pendergast MK, Goldkind SF, Lee EA, Kuniholm A, Soucy A, Vaze J, Belur NR, Fredriksen K, Stojkovska I, Tsytsykova A, Armant M, DiDonato RL, Choi J, Cornelissen L, Pereira LM, Augustine EF, Genetti CA, Dies K, Barton B, Williams L, Goodlett BD, Riley BL, Pasternak A, Berry ER, Pflock KA, Chu S, Reed C, Tyndall K, Agrawal PB, Beggs AH, Grant PE, Urion DK, Snyder RO, Waisbren SE, Poduri A, Park PJ, Patterson A, Biffi A, Mazzulli JR, Bodamer O, Berde CB, Yu TW (2019) Patient-customized oligonucleotide therapy for a rare genetic disease. N Engl J Med 381(17):1644–1652. https://doi.org/10.1056/NEJMoa1813279
25. Cideciyan AV, Jacobson SG, Ho AC, Garafalo AV, Roman AJ, Sumaroka A, Krishnan AK, Swider M, Schwartz MR, Girach A (2021) Durable vision improvement after a single treatment with antisense oligonucleotide sepofarsen: a case report. Nat Med 27(5):785–789. https://doi.org/10.1038/s41591-021-01297-7
26. Singh NK, Singh NN, Androphy EJ, Singh RN (2006) Splicing of a critical exon of human survival motor neuron is regulated by a unique silencer element located in the last intron. Mol Cell Biol 26(4):1333–1346. https://doi.org/10.1128/MCB.26.4.1333-1346.2006
27. Hua Y, Vickers TA, Okunola HL, Bennett CF, Krainer AR (2008) Antisense masking of an hnRNP A1/A2 intronic splicing silencer corrects SMN2 splicing in transgenic mice. Am J Hum Genet 82(4):834–848. https://doi.org/10.1016/j.ajhg.2008.01.014
28. Hua Y, Sahashi K, Rigo F, Hung G, Horev G, Bennett CF, Krainer AR (2011) Peripheral

SMN restoration is essential for long-term rescue of a severe spinal muscular atrophy mouse model. Nature 478(7367):123–126. https://doi.org/10.1038/nature10485

29. Rigo F, Hua Y, Krainer AR, Bennett CF (2012) Antisense-based therapy for the treatment of spinal muscular atrophy. J Cell Biol 199(1): 21–25. https://doi.org/10.1083/jcb.201207087
30. Han Z, Chen C, Christiansen A, Ji S, Lin Q, Anumonwo C, Liu C, Leiser SC, Meena AI, Liau G, Isom LL (2020) Antisense oligonucleotides increase Scn1a expression and reduce seizures and SUDEP incidence in a mouse model of Dravet syndrome. Sci Transl Med 12(558):eaaz6100. https://doi.org/10.1126/scitranslmed.aaz6100
31. Urbanski LM, Leclair N, Anczukow O (2018) Alternative-splicing defects in cancer: splicing regulators and their downstream targets, guiding the way to novel cancer therapeutics. Wiley Interdiscip Rev RNA 9(4):e1476. https://doi.org/10.1002/wrna.1476
32. Bonnal SC, Lopez-Oreja I, Valcarcel J (2020) Roles and mechanisms of alternative splicing in cancer - implications for care. Nat Rev Clin Oncol 17(8):457–474. https://doi.org/10.1038/s41571-020-0350-x
33. Gebauer F, Schwarzl T, Valcarcel J, Hentze MW (2021) RNA-binding proteins in human genetic disease. Nat Rev Genet 22(3): 185–198. https://doi.org/10.1038/s41576-020-00302-y
34. Singh RN, Ottesen EW, Singh NN (2020) The first orally deliverable small molecule for the treatment of spinal muscular atrophy. Neurosci Insights 15:2633105520973985. https://doi.org/10.1177/2633105520973985
35. Ratni H, Ebeling M, Baird J, Bendels S, Bylund J, Chen KS, Denk N, Feng Z, Green L, Guerard M, Jablonski P, Jacobsen B, Khwaja O, Kletzl H, Ko CP, Kustermann S, Marquet A, Metzger F, Mueller B, Naryshkin NA, Paushkin SV, Pinard E, Poirier A, Reutlinger M, Weetall M, Zeller A, Zhao X, Mueller L (2018) Discovery of Risdiplam, a selective survival of motor Neuron-2 (SMN2) gene splicing modifier for the treatment of spinal muscular atrophy (SMA). J Med Chem 61(15):6501–6517. https://doi.org/10.1021/acs.jmedchem.8b00741
36. Palacino J, Swalley SE, Song C, Cheung AK, Shu L, Zhang X, Van Hoosear M, Shin Y, Chin DN, Keller CG, Beibel M, Renaud NA, Smith TM, Salcius M, Shi X, Hild M, Servais R, Jain M, Deng L, Bullock C, McLellan M, Schuierer S, Murphy L, Blommers MJ, Blaustein C, Berenshteyn F, Lacoste A, Thomas JR, Roma G, Michaud GA, Tseng BS, Porter JA, Myer VE, Tallarico JA, Hamann LG, Curtis D, Fishman MC, Dietrich WF, Dales NA, Sivasankaran R (2015) SMN2 splice modulators enhance U1-pre-mRNA association and rescue SMA mice. Nat Chem Biol 11(7):511–517. https://doi.org/10.1038/nchembio.1837
37. Baranello G, Darras BT, Day JW, Deconinck N, Klein A, Masson R, Mercuri E, Rose K, El-Khairi M, Gerber M, Gorni K, Khwaja O, Kletzl H, Scalco RS, Seabrook T, Fontoura P, Servais L, Group FW (2021) Risdiplam in type 1 spinal muscular atrophy. N Engl J Med 384(10):915–923. https://doi.org/10.1056/NEJMoa2009965
38. Wang J, Schultz PG, Johnson KA (2018) Mechanistic studies of a small-molecule modulator of SMN2 splicing. Proc Natl Acad Sci U S A 115(20):E4604–E4612. https://doi.org/10.1073/pnas.1800260115
39. Sivaramakrishnan M, McCarthy KD, Campagne S, Huber S, Meier S, Augustin A, Heckel T, Meistermann H, Hug MN, Birrer P, Moursy A, Khawaja S, Schmucki R, Berntenis N, Giroud N, Golling S, Tzouros M, Banfai B, Duran-Pacheco G, Lamerz J, Hsiu Liu Y, Luebbers T, Ratni H, Ebeling M, Clery A, Paushkin S, Krainer AR, Allain FH, Metzger F (2017) Binding to SMN2 pre-mRNA-protein complex elicits specificity for small molecule splicing modifiers. Nat Commun 8(1):1476. https://doi.org/10.1038/s41467-017-01559-4
40. Campagne S, Boigner S, Rudisser S, Moursy A, Gillioz L, Knorlein A, Hall J, Ratni H, Clery A, Allain FH (2019) Structural basis of a small molecule targeting RNA for a specific splicing correction. Nat Chem Biol 15(12): 1191–1198. https://doi.org/10.1038/s41589-019-0384-5
41. Cheung AK, Hurley B, Kerrigan R, Shu L, Chin DN, Shen Y, O'Brien G, Sung MJ, Hou Y, Axford J, Cody E, Sun R, Fazal A, Fridrich C, Sanchez CC, Tomlinson RC, Jain M, Deng L, Hoffmaster K, Song C, Van Hoosear M, Shin Y, Servais R, Towler C, Hild M, Curtis D, Dietrich WF, Hamann LG, Briner K, Chen KS, Kobayashi D, Sivasankaran R, Dales NA (2018) Discovery of small molecule splicing modulators of survival motor Neuron-2 (SMN2) for the treatment of spinal muscular atrophy (SMA). J Med Chem 61(24):11021–11036. https://doi.org/10.1021/acs.jmedchem.8b01291
42. Kahles A, Lehmann KV, Toussaint NC, Huser M, Stark SG, Sachsenberg T, Stegle O, Kohlbacher O, Sander C, Cancer Genome

Atlas Research N, Ratsch G (2018) Comprehensive analysis of alternative splicing across tumors from 8,705 patients. Cancer Cell 34(2):211–224. e216. https://doi.org/10.1016/j.ccell.2018.07.001

43. Seiler M, Yoshimi A, Darman R, Chan B, Keaney G, Thomas M, Agrawal AA, Caleb B, Csibi A, Sean E, Fekkes P, Karr C, Klimek V, Lai G, Lee L, Kumar P, Lee SC, Liu X, Mackenzie C, Meeske C, Mizui Y, Padron E, Park E, Pazolli E, Peng S, Prajapati S, Taylor J, Teng T, Wang J, Warmuth M, Yao H, Yu L, Zhu P, Abdel-Wahab O, Smith PG, Buonamici S (2018) H3B-8800, an orally available small-molecule splicing modulator, induces lethality in spliceosome-mutant cancers. Nat Med 24(4):497–504. https://doi.org/10.1038/nm.4493
44. Steensma DP, Wermke M, Klimek VM, Greenberg PL, Font P, Komrokji RS, Yang J, Brunner AM, Carraway HE, Ades L, Al-Kali A, Dominguez JMA, Alonso A, Coombs CC, Deeg HJ, Donnellan WB, Foran JM, Garcia-Manero G, Maris MB, McMasters M, Micol J-B, De Oteyza JP, Thol F, Wang ES, Watts JM, Buonamici S, Kim A, Gourineni V, Marino AJ, Rioux N, Schindler J, Smith S, Yao H, Yuan X, Yu K, Platzbecker U (2019) Results of a clinical trial of H3B-8800, a splicing modulator, in patients with myelodysplastic syndromes (MDS), acute myeloid leukemia (AML) or chronic myelomonocytic leukemia (CMML). Blood 134(Supplement_1):673
45. Porto EM, Komor AC, Slaymaker IM, Yeo GW (2020) Base editing: advances and therapeutic opportunities. Nat Rev Drug Discov 19(12):839–859. https://doi.org/10.1038/s41573-020-0084-6
46. Behm M, Ohman M (2016) RNA editing: a contributor to neuronal dynamics in the mammalian brain. Trends Genet 32(3):165–175. https://doi.org/10.1016/j.tig.2015.12.005
47. Katrekar D, Chen G, Meluzzi D, Ganesh A, Worlikar A, Shih YR, Varghese S, Mali P (2019) In vivo RNA editing of point mutations via RNA-guided adenosine deaminases. Nat Methods 16(3):239–242. https://doi.org/10.1038/s41592-019-0323-0
48. Tarn WY (1996) Site-specific substitution of inosine at the terminal positions of a pre-mRNA intron: implications for the configuration of the terminal base interaction. Biochimie 78(11–12):1057–1065. https://doi.org/10.1016/s0300-9084(97)86730-2
49. Parker R, Siliciano PG (1993) Evidence for an essential non-Watson-crick interaction between the first and last nucleotides of a nuclear pre-mRNA intron. Nature 361(6413):660–662. https://doi.org/10.1038/361660a0
50. Denichenko P, Mogilevsky M, Clery A, Welte T, Biran J, Shimshon O, Barnabas GD, Danan-Gotthold M, Kumar S, Yavin E, Levanon EY, Allain FH, Geiger T, Levkowitz G, Karni R (2019) Specific inhibition of splicing factor activity by decoy RNA oligonucleotides. Nat Commun 10(1):1590. https://doi.org/10.1038/s41467-019-09523-0

Chapter 3

An Optimized Protocol for the Mapping of Cell Type–Specific Ribosome-Associated Transcript Isoforms from Small Mouse Brain Regions

Giulia Di Bartolomei and Peter Scheiffele

Abstract

Over the past years, technological advances in transcriptomics provided deep insights into gene expression programs and their role in tissue organization and cellular functions. The isolation of ribosome-associated transcripts is a powerful approach for deep profiling of cell type–specific transcripts, and particularly well-suited for quantitative analysis of transcript isoforms. This method employs conditional ribosome epitope-tagging in genetically defined cell types, followed by affinity-isolation of ribosome-associated mRNAs. Advantages of this approach are twofold: first, the method enables rapid retrieval of mRNAs without tissue dissociation and cell sorting steps. Second, capturing of ribosome-associated mRNAs, enriches for transcripts recruited for active translation, therefore providing an approximation to the cellular translatome. Here, we describe one application of this method for the identification of the transcriptome of excitatory neuronal cells (mitral and tufted cells) of the mouse olfactory bulb, through RiboTag isolation from the vGlut2-IRES-cre mouse line as genetic driver of endogenously tagged ribosome expression.

Key words Ribosome-affinity purification, TRAP, RiboTag, Transcriptomics, Cell-type specificity, Alternative Splicing, mRNA isolation

1 Introduction

The past decade has seen breathtaking advances in transcriptomics which contributed to a detailed molecular dissection and classification of cell types and cell states. Next-generation sequencing approaches eliminated challenges posed by microarray technologies, including high variability especially for lowly expressed genes, dynamic range of detection, and caveats arising from incomplete genome annotations. Single cell sequencing proved particularly powerful for a detailed cell type classification based on gene expression profiles [1, 2]. Advanced transcriptomic studies also uncovered a major role for alternative splicing and alternative transcription start sites in controlling cellular functions in development and

Peter Scheiffele and Oriane Mauger (eds.), *Alternative Splicing: Methods and Protocols*,
Methods in Molecular Biology, vol. 2537, https://doi.org/10.1007/978-1-0716-2521-7_3,

disease states [3–7]. However, accurate mapping of alternative transcripts across cell types remains a significant challenge, as the limited and stochastic transcript coverage from single-cell sequencing results in low sampling of exon spanning junctions. Moreover, there is some debate whether transcript isoforms detected by transcriptomics are indeed recruited for translation to produce protein isoforms that are functionally relevant in vivo [8–10].

A key step for uncovering cell type–specific splicing programs is the access to deep, high quality transcriptomic data from the cell type of interest. The most commonly applied methods include enzymatic tissue dissociation followed by immune-isolation or fluorescence activated cell sorting (FACS), [11, 12]. However, artifacts and variability introduced during the dissociation and sorting process pose a notable challenge [13]. Moreover, in morphologically complex cells—such as neurons or vascular cells—RNA isoforms are lost from cellular processes that are severed during the generation of single cell suspensions. A complementary approach overcoming some of these limitations is the isolation of ribosome-associated mRNAs from genetically defined cell populations. Based on cell type–specific epitope tagging with genetic tools, two closely related methods were developed, named TRAP (Translating Ribosome Affinity Purification) [14] and RiboTag [15]. In both methods, an epitope-tagged core ribosomal protein is expressed in the cell population of interest. This tag then enables the selective affinity purification and analysis of ribosome-associated mRNAs. Whereas TRAP is based on bacterial artificial chromosome (BAC) transgenic mice that overexpress EGFP-tagged ribosomal protein L10 (RPL10A), RiboTag employs cre recombinase–dependent HA epitope tagging of the endogenous ribosomal protein RPL22. Crossing with cell type–specific cre lines or viral delivery of cre recombinase then enables selecting the desired population of cells for tagging and mRNA isolation. Selective capture of cell type–specific mRNAs overcomes many challenges posed by dissociation-based gene expression profiling approaches. Moreover, RNA-sequencing of TRAP/RiboTag samples enables the enrichment of transcripts that are engaged by the ribosomes, a closer approximation to the transcript isoforms recruited for mRNA translation. In the past years, the TRAP/RiboTag approach has proven versatile and effective in characterizing cell type–specific transcriptomes in a broad range of experimental systems, ranging from *Drosophila* and mouse models to cultured human-induced pluripotent stem cells [16–18]. More recent studies have capitalized on the high yield of cell type–specific mRNAs to assess cell type–specific alternative splicing and translational regulation [19–22].

This chapter describes an optimized protocol for RiboTag-isolation of intact mRNA from sparse cell populations in small regions of the mouse brain. This protocol has been used successfully to obtain high quality preparations from a single cortical layer of the neonatal mouse visual cortex (area V1) or a subset of hippocampal interneurons. The example protocol is described for mRNA isolation from vGlut2-positive tufted and mitral cells of the mouse olfactory bulb, but can be readily transferred to other preparations with limiting and highly complex starting materials.

2 Materials

2.1 Preparation of Work Environment

1. RNase AWAY® for molecular biology.
2. Sterile, nuclease-free pipette filter tips.
3. Glass homogenizers (or motor driven Teflon glass homogenizer).
4. Nuclease-free 1.5 mL LowBind "microcentrifuge tubes."
5. Diethylpyrocarbonate (DEPC)-treated water (0.1% DEPC in ultrapure 18.2 MΩ cm water, autoclaved).

2.2 Preparation for Immunoprecipitation of Tagged Ribosome and Associated RNAs

1. Ice-cold sterile, nuclease-free phosphate buffered saline (PBS).
2. Homogenization buffer: 50 mM Tris–HCl, pH 7.4, 100 mM KCl, 12 mM $MgCl_2$ in DEPC-treated water. Good for 2 months when stored at 4 °C.
3. Supplements to homogenization buffer: 1 mg/mL Heparin from porcine intestinal mucosa, 200 units/mL RNasin RNase inhibitor, 100 μg/mL cycloheximide, 1 mM DTT, 1× protease inhibitor (e.g., Roche "complete" EDTA free). The "supplemented homogenization buffer" has to be made fresh on the day of the experiment and always kept on ice.
4. 10% IGEPAL stock (molecular biology grade) in DEPC-treated water (store at 4 °C).
5. High-salt washing buffer: 300 mM KCl, 1% IGEPAL, 50 mM Tris–HCl, pH 7.4, 12 mM $MgCl_2$, 1 mM DTT, 100 μg/mL cycloheximide, in DEPC-treated water. The high-salt washing buffer has to be made fresh on the day of the experiment and always kept on ice.
6. Anti-HA Magnetic Beads, Monoclonal IgG1 antibody, clone 2–2.2.14, (Thermo Scientific). Store at 4 °C.
7. Magnetic rack.
8. Rotating wheel.

2.3 Reagents and Equipment for RNA Purification and Clean-up

1. RNeasy Micro plus kit (Qiagen) containing RLT buffer, wash buffers, gDNA eliminator columns, and RNeasy MinElute Spin Columns.
2. Beta-mercaptoethanol.

3. 70% and 80% ethanol (molecular biology grade, nuclease-free, diluted with RNase free water).
4. DEPC-treated water.

2.4 RNA Quantification and Library Preparation

1. RNA 6000 Pico kit (Agilent).
2. Reverse Transcription System, containing 25 mM $MgCl_2$, Reverse Transcription 10× Buffer, 10 mM dNTP Mixture, Recombinant RNasin Ribonuclease Inhibitor, 15 units AMV Reverse Transcriptase, Oligo(dT)15 Primer, Random Primers, 1.2 kb Kanamycin Positive Control RNA, 1 μg poly (A) + mRNA, nuclease-free water.
3. Agilent BioAnalyzer.
4. NanoDrop spectrophotometer.
5. QuantiFluor RNA System (Promega).
6. TruSeq RNA Library Prep Kit v2 (Illumina) or Smart-Seq library Prep v4 Plus Kit (TakaraBio).

2.5 Transgenic Mouse Lines

The conditional RiboTag allele in C57BL/6J background is available from the Jackson laboratory (B6J.129(Cg)-*Rpl22*$^{tm1.1Psam}$/SjJ; JAX Stock No: 029977). The vGlut2-IRES-cre mouse line in C57BL/6J background used to direct cell class–specific Rpl22-HA expression is B6J.129S6(FVB)-*Slc17a6*$^{tm2(cre)Lowl}$/MwarJ JAX Stock No: 028863). TRAP BAC transgenic mouse lines driving RPL10A-GFP in various cell classes are available from JAX.

2.6 Time Considerations

The preparation of the RNase free environment, set-up of equipment prior the start of the experiment takes about 1 h. One complete round of pull-down takes around 7 h. The subsequent RNA purification and Reverse Transcription and Polymerase Chain Reaction (RT-PCR) require ~3–4 h. One day is required for quality checks of RNA concentration quantification by fluorometry, RNA integrity number (RIN) calculation, and quantitative-Polymerase Chain Reaction (qPCR) enrichment analysis for cell type–specific markers.

3 Methods

3.1 Preparation of Reagents and Sterile Equipment

1. Wipe work surfaces and pipettes with RNase AWAY® for molecular biology or similar.
2. Autoclave glass homogenizers, flush with RNase AWAY just before use, and wash three times with 50 mL of DEPC-treated water (*see* **Note 1**).
3. Prepare the "supplemented homogenization buffer" and the "high-salt buffer" using DEPC-treated water.

4. Use sterile pipette filter tips throughout the entire procedure.
5. Clean dissection tools with water and 70% Ethanol and let them air dry.
6. Prepare a petri dish for tissue dissection with PBS on ice.

3.2 Preparation of Tissue Homogenate

1. Euthanize mice (in this example protocol postnatal day 35, *Rpl22*$^{HA/HA}$*::vGlut2*$^{IRES\text{-}cre}$) according to institutional guidelines, for example, by decapitation under isoflurane anesthesia and rapidly collect the brain in ice-cold PBS. *See* **Note 2** for number of animals and RNA yields from different brain regions.
2. Take a toe biopsy for potential reconfirmation of genotype (*see* **Note 3**).
3. Rapidly dissect the region of interest and transfer the tissue into a new dish with fresh ice cold PBS to remove blood (*see* **Notes 4** and **5**).
4. Immediately transfer tissue into a glass homogenizer containing ice-cold supplemented homogenization buffer. For the olfactory bulbs (~30 mg tissue from 1 animal), we advise to use 0.5 mL of supplemented homogenization buffer (1:20 weight per volume of homogenization buffer). Disrupt the tissue with 40 strokes-fitting homogenizers (*see* **Note 6**).
5. Transfer the homogenate to an ice-cold 1.5 mL microcentrifuge tube (*see* **Note 7** for larger volume samples).
6. Centrifuge at 2000 × *g* for 15 min at 4 °C to pellet nuclei and debris.
7. During the centrifugation step, prepare anti-HA magnetic beads: for RiboTag pulldown from the olfactory bulbs from 1 mouse, use 5 μL of packed beads (equivalent to ~20 μL of beads from the commercial vial). Wash beads three times with 1 mL of supplemented homogenization buffer (vortex at medium speed 5 s—put tube on magnetic rack—remove liquid without disturbing the beads—add buffer). Resuspend beads in 20 μL of ice-cold supplemented homogenized buffer (*see* **Note 8**). Scale up the amount of beads according to the number of samples.
8. Transfer supernatant from **step 5** to a new ice-cold 1.5 mL microcentrifuge tube (*see* **Note 9**).
9. Add 55 μL of 10% IGEPAL (to a final concentration of 1%) to solubilize cellular membranes. Invert 3–4 times until the lysate becomes clear.
10. Incubate on ice for 5 min (*see* **Note 10**).

11. Keep the sample in the same ice-cold 1.5 mL microcentrifuge tube and centrifuge at 13,000 × *g* for 10 min at 4 °C, to separate insoluble material from lysate.

3.3 Immunoprecipitation of Epitope-Tagged Ribosomes

1. After the centrifugation, transfer the supernatant to a new ice-cold 1.5 mL microcentrifuge tube.
2. Take an aliquot (4.5 μL, corresponding to 1% of total) of the lysate supernatant to be analyzed as "input" material and immediately mix this aliquot with 350 μL of RLT plus buffer (Add 10 μL of β-mercaptoethanol per 1 mL Buffer RLT plus). Vortex few seconds and store at −80 °C until RNA purification step.
3. Add 20 μL of washed and resuspended anti-HA-coupled magnetic beads (equivalent of 5 μL of packed beads) to the lysate sample (for bead volumes refer to **Note 8**).
4. Incubate 3–4 h at 4 °C on rotating wheel (*see* **Note 11**), at low speed.
5. Place the microcentrifuge tube on the magnetic rack on ice (*see* **Note 7** for higher volume samples).
6. Wait ~1 min for the beads to be collected at the bottom of the tube.
7. Carefully aspirate the supernatant with the vacuum pump without disturbing the beads, then wash three times with 1 mL of ice-cold high-salt washing buffer: Aspirate supernatant—add high-salt washing buffer—vortex gently until the beads are dispersed in the washing buffer—collect beads again on the magnetic rack (*see* **Note 12**).
8. After the last wash, remove supernatant and resuspend the beads directly in 350 μL of RLT plus buffer supplemented with 3.5 μL of β-mercaptoethanol.

3.4 RNA Extraction and Purification

1. Vortex gently for 5 s at room temperature. Place the tubes back on the magnetic rack and wait for ~1 min.
2. Pipet the supernatant containing the purified nucleic acids in a new ice-cold 1.5 mL microcentrifuge tube. Discard the beads.
3. Proceed with the purification of input (IN) and ribosome-associated RNA (IP) according to the RNeasy Micro plus kit. In brief, the samples are first loaded onto the gDNA columns to ensure elimination of genomic DNA contaminants from the sample (*see* **Note 13**). The flow-through is then loaded on the RNeasy columns for purification (binding capacity of a single column is up to 100 μg RNA). Upon washes of the columns with the buffers provided in the RNeasy Micro plus kit, the RNA is finally eluted in 15 μL of nuclease-free water and stored

at −80 °C until further processing. The RNA purification procedure is to be performed on ice.

3.5 Testing RNA Concentration and Quality

1. Determine RNA concentration in input and IP samples with NanoDrop Spectrophotometer (the typical yield of RNA retrieved from vGlut2-cre positive neurons of the olfactory bulbs from one P35 mouse, two hemispheres, is ~500 ng).
2. Use a few ng of IN and IP RNA for Reverse Transcription with random hexamers according to instructions of the transcriptase manufacturer (*see* **Note 14** for cases with low RNA recovery).
3. Perform qPCR on IN and IP cDNA with primers amplifying transcripts known to be enriched or deenriched in the cell type of interest (*see* **Notes 15** and **16**). Additionally, two or three "housekeeping genes" should be assessed. For example, in pull-downs from the *Vglut2*^{Ribo} mouse line we use *Vglut1 (Slc17a7)*, *Vglut2 (Slc17a6)* and *Tbr2* (*Eomes*) genes to probe for marker enrichment and parvalbumin (*Pvalb*), *Gfap,* and *Vgat (Slc32a1)* to test for deenrichment of contaminants. Enrichment values are calculated based on the ratio of target mRNA over housekeeping genes in the IP relative to IN samples. As housekeeping genes, we use *Gapdh* and *Actb* (*see* **Note 17**).
4. For accurate calculation of RNA concentration perform fluorocytometry analysis, using the QuantiFluor RNA System.
5. Evaluate the quality of isolated RNA by calculating the RNA Integrity Number (RIN) using a pico-chip (Agilent RNA 6000 Pico kit). We used for RNA-sequencing only samples with RIN > 7.5. Moreover, we selected for RNA-sequencing samples displaying RIN values in the same range, thus excluding outlier samples with either too high or too low RIN.
6. Proceed with RNA-sequencing library preparation. At least 50 ng of immune-isolated RNA should be used for PolyA-enrichment TruSeq library preparation for RNA-sequencing (*see* **Note 18**), *see* Fig. 1.

4 Notes

1. It is critical to establish an RNase-free environment and minimize the risk of RNase contamination throughout the procedure.
2. Typical total RNA yields from RiboTRAP purifications from a single mouse (two brain hemispheres), obtained with this protocol are as follows: P35 olfactory bulbs *vGlut2*^{Ribo} ~ 500 ng, P25 neocortex *CamK2*^{Ribo} ~1.5 μg, P25 hippocampi

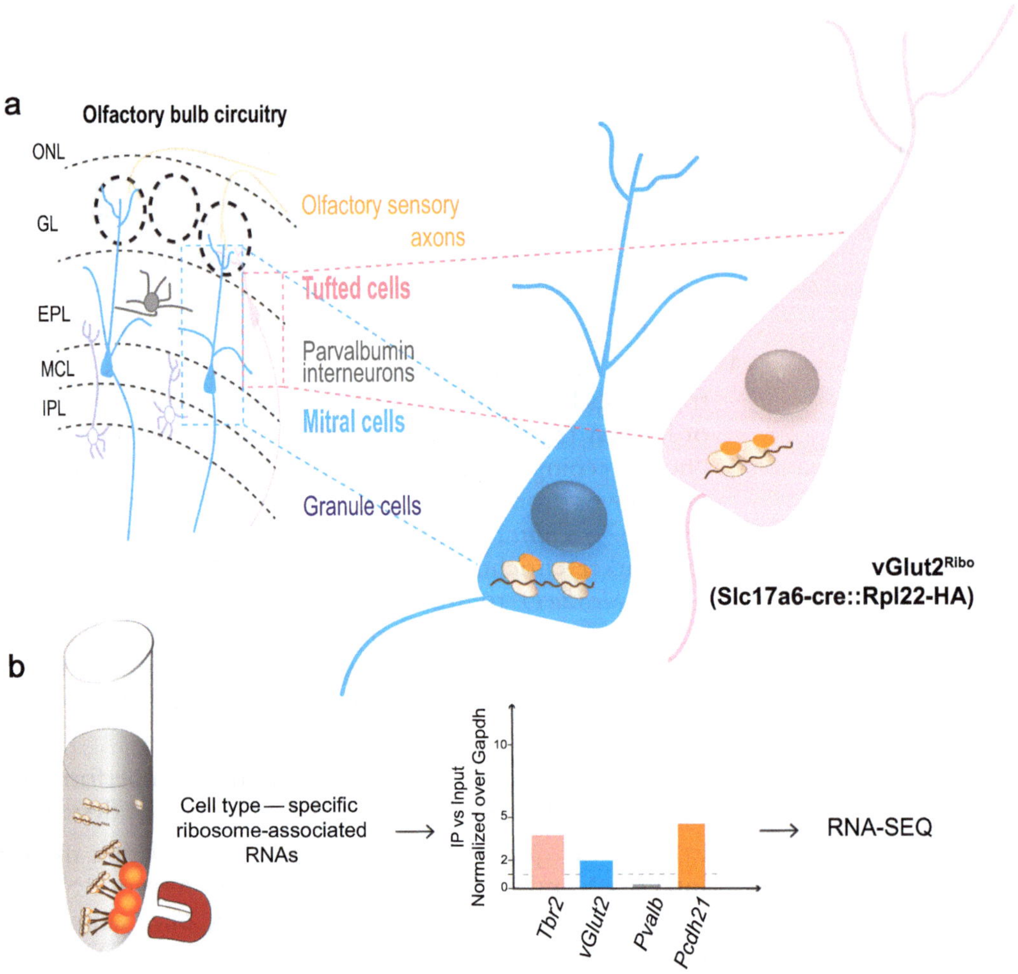

Fig. 1 Illustration of cell type–specific ribosome-associated mRNAs pull-down approach in the olfactory bulb. (**a**) Schematic representation of the olfactory bulb circuitry, where genetically defined vGlut2$^+$ neurons of the olfactory bulb conditionally express HA-tagged ribosomes. Highlight of the different cellular populations defined by the Slc17a6-cre::Rpl22-HA mouse line, namely mitral and tufted neurons of the olfactory bulb. These cells express HA-tagged ribosomes. (**b**) Upon tissue lysis (IN), ribosome-associated mRNAs are immunoprecipitated using magnetic anti-HA beads (IP). (**c**) mRNAs are then purified and used for RNA-sequencing analysis. Expected qPCR fold-enrichment/deenrichment values for cell type–specific markers upon mRNAs pull-down from vGlut2$^+$ neuronal population are displayed on the right panel

CamK2^{Ribo}~450 ng, P35 neocortex *Pvalb*^{Ribo} ~250 ng, P4 visual cortices *Rorb*^{Ribo} 20 ng.

3. Depending on the cre mouse-line used, extra attention needs to be taken to ensure that no aberrant germline recombination of the Rpl22HA-allele has occurred (*see* [23] for discussion of this issue). Thus, it is strongly encouraged to use genotyping primers that detect germline recombination (e.g., the primer combination used to detect the floxed RiboTag allele shows a

band of ~320 bp size when germline recombination has occurred (RiboTag Forward primer: 5-′-GGGAGGCTTGCTGGATATG-3′; RiboTag Reverse primer: 5′-TTTCCAGACACAGGCTAAGTACAC-3′). In developmental studies, the onset of cre-recombinase expression needs to be carefully evaluated to assess the efficiency of Rpl22 HA-tagging at early developmental timepoints.

4. During tissue dissection, it is critical to minimize carrying over blood as much as possible, given its content of RNase activity. Perform the dissection step on ice in prechilled PBS as quickly as possible. Upon dissection, tissue needs to be placed immediately into the ice-cold homogenization buffer containing RNase inhibitors.
5. Alternatively, tissue can be immediately flash-frozen in liquid nitrogen upon dissection and stored at −80 °C until use. Upon thawing, homogenize the tissue immediately, when it is still frozen. Note that quality and total mRNA yield retrieved from frozen tissue is reduced compared to tissue processed immediately after dissection.
6. Wherever possible, we recommend using tissue from single animals as replicates. Typical tissue weights for P30-P40 mouse brain regions from a single animal (2 hemispheres) are: visual cortices (~10 mg), olfactory bulbs (~30 mg), hippocampi (~70 mg), whole neocortex (~300 mg). Adjust the volume of homogenization buffer to 1:20 weight per volume (e.g., 20 mL of buffer per 1 g of tissue) for optimal tissue homogenization. However, when using very small amounts of tissue, work with a minimal volume of 500 μL of homogenization buffer to improve the handling of the sample.

   For replicates, always use the same number of strokes to homogenize the tissue. Upon homogenization, the sample should appear dense and viscous, without any remaining tissue chunks visible. Efficiency of homogenization can be evaluated for an aliquot of the homogenate under a bright field microscope.
7. When working with larger amounts of starting material and the corresponding higher volumes of homogenization buffer, transfer the homogenate to a 15 mL plastic tube for low-speed centrifugation (2000 × *g* for 10 min at 4 °C to pellet nuclei). Then transfer the supernatant to a new 15 mL plastic tube for tissue lysis with IGEPAL (adapt the amount of IGEPAL to add, depending on the resulting volume of supernatant). For high-speed centrifugation, distribute the lysate into prechilled 2 mL microcentrifuge tubes and spin for 10 min at 4 °C at 13,000 × *g*. Transfer supernatant to new prechilled 2 mL microcentrifuge tubes and incubate with the

HA-magnetic beads (5 μL packed beads per tube) for 3–4 h. Use 1.5 mL high-salt buffer per tube for washes and collect beads sequentially in a total of 350 μL of RLT buffer supplemented with 3.5 μL of β-mercaptoethanol.

8. Carefully choose the amount of magnetic beads used for immunoisolation when applying the protocol to a new tissue and/or cell type. Bead concentration should be optimized for maximum ribosome recovery with low unspecific mRNA recovery. When trying to perform pulldown from a new type of sample, we recommend to test the correct amount of beads to use in a pilot experiment (background increases when an excess of beads is employed).
9. Ensure that no nuclear pellet is carried over during the transfer of the clarified supernatant to a clean microcentrifuge tube. It is preferable to leave and discard few microliters of homogenized tissue, instead of risking carry over of nuclear contaminants.
10. Incubation time with IGEPAL is necessary in this step to achieve optimal membrane solubilization (assessed by lysate clearing). Instead of IGEPAL, 1% NP-40 can also be used.
11. We observed that 3 h incubation of beads with the lysate are sufficient for ribosome binding and yield a good amount of retrieved RNA. We observed no differences between 3 h and 5 h incubation. We do not recommend overnight incubation as prolonged incubation time frequently results in increased RNA degradation and unspecific binding.
12. Three washes with 1 mL of high-salt buffer and gentle vortexing for 5 s is sufficiently stringent to minimize unspecific binding. Longer washing times (15–25 min on a rotating wheel) significantly decrease the quality and integrity of the isolated RNA. The washing step needs to be performed on ice.
13. DNA removal on a gDNA column (RNeasy Micro plus Qiagen kit) is usually sufficient to eliminate DNA contamination. However, in some cases, additional DNA digestion by DNase improves RNA recovery and purity. This is recommended in particular when performing pull-down from small neuronal structures/small brain regions, where low yield of total RNA recovery is expected.
14. When working with very small brain regions or sparse cell populations, perform fluorometric RNA quantification, which is more reliable at low concentrations. When RNA yield is too low to quantify, take a fixed percentage (e.g., 10%) of the eluted RNA to perform the reverse transcription. For experimental questions where higher RNA recovery is expected (e.g., an abundant cell population of neocortex), elute RNA in 25 μL of DEPC-treated water and use ~25 ng RNA for reverse transcription.

15. Input aliquots are necessary in order to assess the quality of the pull-down through enrichment of cell type–specific markers by qPCR. For our experiments, input RNA aliquots were not analysed for RIN and were not used for RNA sequencing. We recommend diluting resulting cDNA samples at least 1:5 before proceeding for qPCR. Use 3 μL of diluted cDNA in a total volume of 20 μL qPCR reaction. Markers of interest used to assess by qPCR the enrichment of mitral and tufted neurons in the olfactory bulb using the vGlut2-cre mouse line are: *vGlut2*, *Pcdh21*, *Tbr2*. (vGlut2 Forward primer: GCATGGTCTGGTACATGTTCTG; vGlut2 Reverse primer: GACGGGCATGGATGTGAAAAAC; Tbr2 Forward primer: ATAAACGGACTCAACCCCACC; Tbr2 Reverse primer: CCCTGCATGTTATTGTCCGC; Pcdh21 Forward primer: ATCACTGTCAACGACTCAGACC; Pcdh21 Reverse primer: GTCAATGGCAGCTGAGTTTTCC).
16. Note, that RiboTag isolation of mRNAs from abundant cell types will result in only modest marker enrichment (e.g., for broad classes of abundant glutamatergic cells such as CamK2-cre defined neurons in neocortex). By contrast, pulldowns for sparse cell classes (such as VIP interneurons) results in a very high enrichment of markers.
17. Many genes that are widely considered "housekeeping genes" are differentially expressed across neuronal cell types and cell states. Thus, multiple such "housekeeping genes" should be assessed to allow for accurate enrichment analysis of target mRNAs.
18. In experiments aimed at quantifying transcript isoforms, we recommend to use paired-end sequencing with read length of 100–150 bases and a sequencing depth of 50 to 100 million reads per sample (depending on what type and complexity of splice variants will be examined). For low input RNA samples <15 ng RNA, we recommend using the SMART-seq v4 library preparation kit.

Acknowledgments

We are thankful to O. Mauger and E. Furlanis for constructive comments on the chapter manuscript and to T.M. Nguyen, L. Traunmüller, E. Furlanis, and S. Falkner for discussions and help in setting up the protocol for RiboTRAP purifications. Work in the laboratory was supported by funds to P.S. from the Swiss National Science Foundation, a European Research Council Advanced Grant (SPLICECODE), EU-AIMS and AIMS-2-TRIALS which are supported by the Innovative Medicines Initiatives from the European Commission. The results leading to this publication has received funding from the Innovative Medicines Initiative 2 Joint Undertaking under grant agreement no. 777394. This Joint Undertaking receives support from the European Union's Horizon 2020 research and innovation programme and EFPIA and AUTISM SPEAKS, Autistica, SFARI. The Scheiffele Laboratory is an associate member of the Swiss National Science Foundation's National Competence Centre for Research (NCCR) RNA and Disease.

References

1. Zeisel A, Hochgerner H, Lonnerberg P, Johnsson A, Memic F, van der Zwan J, Haring M, Braun E, Borm LE, La Manno G, Codeluppi S, Furlan A, Lee K, Skene N, Harris KD, Hjerling-Leffler J, Arenas E, Ernfors P, Marklund U, Linnarsson S (2018) Molecular architecture of the mouse nervous system. Cell 174(4):999–1014 e1022
2. Zeng H, Sanes JR (2017) Neuronal cell-type classification: challenges, opportunities and the path forward. Nat Rev Neurosci 18(9): 530–546
3. Nguyen TM, Schreiner D, Xiao L, Traunmuller L, Bornmann C, Scheiffele P (2016) An alternative splicing switch shapes neurexin repertoires in principal neurons versus interneurons in the mouse hippocampus. eLife 5:e22757
4. Wamsley B, Jaglin XH, Favuzzi E, Quattrocolo G, Nigro MJ, Yusuf N, Khodadadi-Jamayran A, Rudy B, Fishell G (2018) Rbfox1 mediates cell-type-specific splicing in cortical interneurons. Neuron 100(4):846–859.e7
5. Gonatopoulos-Pournatzis T, Niibori R, Salter EW, Weatheritt RJ, Tsang B, Farhangmehr S, Liang X, Braunschweig U, Roth J, Zhang S, Henderson T, Sharma E, Quesnel-Vallieres M, Permanyer J, Maier S, Georgiou J, Irimia M, Sonenberg N, Forman-Kay JD, Gingras AC, Collingridge GL, Woodin MA, Cordes SP, Blencowe BJ (2020) Autism-Misregulated eIF4G microexons control synaptic translation and higher order cognitive functions. Mol Cell 77(6):1176–1192 e1116
6. Parikshak NN, Swarup V, Belgard TG, Irimia M, Ramaswami G, Gandal MJ, Hartl C, Leppa V, Ubieta LT, Huang J, Lowe JK, Blencowe BJ, Horvath S, Geschwind DH (2016) Genome-wide changes in lncRNA, splicing, and regional gene expression patterns in autism. Nature 540(7633):423–427
7. Furlanis E, Scheiffele P (2018) Regulation of neuronal differentiation, function, and plasticity by alternative splicing. Annu Rev Cell Dev Biol 34:451–469
8. Vogel C, Marcotte EM (2012) Insights into the regulation of protein abundance from proteomic and transcriptomic analyses. Nat Rev Genet 13(4):227–232
9. Liu Y, Beyer A, Aebersold R (2016) On the dependency of cellular protein levels on mRNA abundance. Cell 165(3):535–550
10. Weatheritt RJ, Sterne-Weiler T, Blencowe BJ (2016) The ribosome-engaged landscape of alternative splicing. Nat Struct Mol Biol 23(12):1117–1123
11. Crouch EE, Doetsch F (2018) FACS isolation of endothelial cells and pericytes from mouse brain microregions. Nat Protoc 13(4): 738–751
12. Tischfield DJ, Anderson SA (2017) Differentiation of mouse embryonic stem cells into cortical interneuron precursors. J Vis Exp 130: 56358
13. van den Brink SC, Sage F, Vertesy A, Spanjaard B, Peterson-Maduro J, Baron CS, Robin C, van Oudenaarden A (2017) Single-cell sequencing reveals dissociation-induced gene expression in tissue subpopulations. Nat Methods 14(10):935–936
14. Heiman M, Schaefer A, Gong S, Peterson JD, Day M, Ramsey KE, Suarez-Farinas M, Schwarz C, Stephan DA, Surmeier DJ, Greengard P, Heintz N (2008) A translational profiling approach for the molecular characterization of CNS cell types. Cell 135(4):738–748
15. Sanz E, Yang L, Su T, Morris DR, McKnight GS, Amieux PS (2009) Cell-type-specific isolation of ribosome-associated mRNA from complex tissues. Proc Natl Acad Sci U S A 106(33): 13939–13944
16. Thomas A, Lee PJ, Dalton JE, Nomie KJ, Stoica L, Costa-Mattioli M, Chang P, Nuzhdin S, Arbeitman MN, Dierick HA (2012) A versatile method for cell-specific profiling of translated mRNAs in drosophila. PLoS One 7(7):e40276

17. Gregory JA, Hoelzli E, Abdelaal R, Braine C, Cuevas M, Halpern M, Barretto N, Schrode N, Akbalik G, Kang K, Cheng E, Bowles K, Lotz S, Goderie S, Karch CM, Temple S, Goate A, Brennand KJ, Phatnani H (2020) Cell type-specific in vitro gene expression profiling of stem cell-derived neural models. Cell 9(6):1406

18. Mardinly AR, Spiegel I, Patrizi A, Centofante E, Bazinet JE, Tzeng CP, Mandel-Brehm C, Harmin DA, Adesnik H, Fagiolini M, Greenberg ME (2016) Sensory experience regulates cortical inhibition by inducing IGF1 in VIP neurons. Nature 531(7594):371–375

19. Thomson SR, Seo SS, Barnes SA, Louros SR, Muscas M, Dando O, Kirby C, Wyllie DJA, Hardingham GE, Kind PC, Osterweil EK (2017) Cell-type-specific translation profiling reveals a novel strategy for treating fragile X syndrome. Neuron 95(3):550–563 e555

20. Gonzalez C, Sims JS, Hornstein N, Mela A, Garcia F, Lei L, Gass DA, Amendolara B, Bruce JN, Canoll P, Sims PA (2014) Ribosome profiling reveals a cell-type-specific translational landscape in brain tumors. J Neurosci 34(33):10924–10936

21. Furlanis E, Traunmuller L, Fucile G, Scheiffele P (2019) Landscape of ribosome-engaged transcript isoforms reveals extensive neuronal-cell-class-specific alternative splicing programs. Nat Neurosci 22(10):1709–1717

22. Lee H, Fenster RJ, Pineda SS, Gibbs WS, Mohammadi S, Davila-Velderrain J, Garcia FJ, Therrien M, Novis HS, Gao F, Wilkinson H, Vogt T, Kellis M, LaVoie MJ, Heiman M (2020) Cell type-specific transcriptomics reveals that mutant huntingtin leads to mitochondrial RNA release and neuronal innate immune activation. Neuron 107(5): 891–908 e898

23. Luo L, Ambrozkiewicz MC, Benseler F, Chen C, Dumontier E, Falkner S, Furlanis E, Gomez AM, Hoshina N, Huang WH, Hutchison MA, Itoh-Maruoka Y, Lavery LA, Li W, Maruo T, Motohashi J, Pai EL, Pelkey KA, Pereira A, Philips T, Sinclair JL, Stogsdill JA, Traunmuller L, Wang J, Wortel J, You W, Abumaria N, Beier KT, Brose N, Burgess HA, Cepko CL, Cloutier JF, Eroglu C, Goebbels S, Kaeser PS, Kay JN, Lu W, Luo L, Mandai K, McBain CJ, Nave KA, Prado MAM, Prado VF, Rothstein J, Rubenstein JLR, Saher G, Sakimura K, Sanes JR, Scheiffele P, Takai Y, Umemori H, Verhage M, Yuzaki M, Zoghbi HY, Kawabe H, Craig AM (2020) Optimizing nervous system-specific gene targeting with Cre driver lines: prevalence of germline recombination and influencing factors. Neuron 106(1):37–65 e35

Chapter 4

FACS-Based Neuronal Cell Type–Specific RNA Isolation and Alternative Splicing Analysis

Zicheng Wei, Yuxin Qin, Gordon Fishell, and Boxing Li

Abstract

Alternative splicing of pre-mRNAs expands the coding abilities of genomes by generating distinct transcription variants from individual genes. It contributes to the marvelous complexity of the transcriptome in neurons. Given the differential expression of alternative splicing regulators and diversity in alternative splicing programs in neuronal subpopulations, it is urgent and necessary to develop methods to efficiently isolate diverse subgroups of neurons and analyze their transcriptomic diversity. Here, we describe a protocol to isolate RNA from specific neuronal types using a fluorescence-activated cell sorting (FACS)-based method to analyze alternative splicing events in a cell type–specific manner. The method is universally applicable to analyze alternative splicing in fluorescent protein-labeled neuronal types. It was optimized to preserve the transcription state and improve efficiency in cell suspension purification. With our protocol, fluorescent protein-labeled neurons could be efficiently purified. The transcription states suitable for gene expression and alternative splicing analysis could be well-preserved.

Key words Pre-mRNA splicing, Cell type–specific, FACS

1 Introduction

Alternative splicing is one of the fundamental post-transcription modification mechanisms in eukaryotes. Pre-mRNA transcribed from a single gene undergoing alternative splicing results in distinct transcript variants with different stability, localization distribution, and molecular function [1, 2]. This mechanism plays an essential role in the nervous system and nerve-derived tissue sources [3]. In neurons, alternative splicing occurs at every step of neuronal development, thus contributing to many biological processes, including axon guidance and synaptogenesis [3]. Furthermore, alternative splicing alters the properties of crucial proteins related to cell excitation, such as ion channel and neurotransmitter receptors, affecting neuronal functions, including neuronal plasticity and neuronal excitability [4, 5].

Peter Scheiffele and Oriane Mauger (eds.), *Alternative Splicing: Methods and Protocols*,
Methods in Molecular Biology, vol. 2537, https://doi.org/10.1007/978-1-0716-2521-7_4,

Elegant works have shown that many splice regulators are differentially regulated in excitatory and inhibitory neurons. For example, Slm1 expression level is high in dentate granule cells and VIP$^+$ interneurons but low in CA pyramidal and SST$^+$ neurons [6]. Similarly, Rbfox1 expression levels are higher in PV$^+$ and SST$^+$ interneurons than in VIP$^+$ interneurons during development [7]. Meanwhile, genome-wide analyses revealed that many alternative splicing programs were non-overlapping within different types of neurons [8]. In particular, they are markedly distinct within neural progenitor cells (NPCs) versus neurons and promot NPC-neuron transition [9]. Therefore, analyzing alternative splicing events in a cell type–specific manner provides considerable insight into how alternative splicing is regulated and, in turn contributes to neuronal functions. Although single-cell sequencing and analysis methods are rapidly developing, they are limited by the low-throughput, reduced transcript coverage, and the associated high financial cost. Thus, reliable analysis and quantification of alternative splicing from single-cell transcriptome remain a significant challenge [10].

To study alternative splicing programs in different neuron subpopulations, we developed a method for efficient isolation of neuronal subpopulations from murine brain tissue. This method is mainly used to isolate neuron subpopulations labeled with fluorescent proteins driven by neuronal cell type–specific promoters. The protocol was modified from Saxena, A et al. 2012 [11] and optimized to preserve the transcription state during digestion and improve cell suspension purification efficiency. During the digestion and dissociation process, trehalose is added to a final concentration of 5% (wt/v) to the buffer to stabilize the cell membrane. To preserve the transcription state and reduce excitotoxicity, small-molecule inhibitors are added during digestion and dissociation. For cell suspension purification, 37% Percoll solution is used to separate myelin and cell debris from intact viable cells [12]. If TRIzol reagent is used for RNA extraction after cell sorting, additional glycogen is recommended to act as a carrier to facilitate RNA precipitation for RNA extracted from less than 20,000 cells [13].

In this chapter, we use Thy1-YFP mice (the yellow fluorescent protein expression was driven by a mouse Thy1 promoter preferentially labeling motor and sensory neurons) [14] as an example. However, the method can be adapted to all other fluorescent protein-labeled neuron types by changing the gating strategy of cell sorting. With our protocol, fluorescent protein-labeled neuron subpopulations could be significantly enriched (Fig. 1). High-quality RNA with a well-preserved transcription state could be obtained for subsequent analysis of gene expression and alternative splicing.

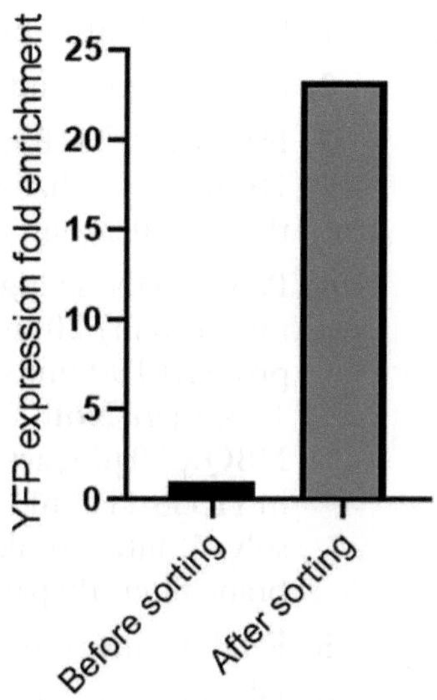

Fig. 1 Fold enrichment of YFP mRNA expression after cell sorting. After sorting, the YFP mRNA expression level is much higher than before

2 Materials

Purchase RNase-free and sterile plasticware (e.g., tubes, transfer pipette, pipette tips). Prepare all solutions with ultrapure water. Filter sterilize indicated solution through 0.22 μm high protein-binding nitrocellulose filter units. Autoclave sterilize surgery tools for brain dissection.

2.1 Brain Tissue Harvest

1. Mouse strains for isolation of different neuron subtypes: Thy1-YFP mice for cortex layer V excitatory neurons. Other commonly used mouse strains containing cell type–specific Cre-dependent fluorescent proteins (e.g., tdTomato Cre reporter mice crossed to PV-Cre, SST-Cre, VIP-Cre mice).
2. Ketamine–xylazine solution: ketamine 10 mg/mL, xylazine 1.6 mg/mL, in sterile water. Store at 4 °C for up to 2 weeks.
3. 1 mL syringe with 27G needle.
4. 18G blunt perfusion needle.
5. Surgery tools: scissors, iris scissors, curved-tip forceps, razor blade.
6. Blot paper.
7. Petri dish.
8. DPBS: calcium- and magnesium-free, store at 4 °C. Oxygenated with Sterile 95% O_2 + 5% CO_2 before use.
9. 75% ethanol.
10. Sterile 95% O_2 + 5% CO_2.

2.2 Tissue Digestion and Dissociation

1. 0.22 μm high protein-binding nitrocellulose filter units.
2. 0.22 μm low binding PES filter units.
3. HBSS-HT (HBSS–HEPES–Trehalose): 1× HBSS, HEPES 10 mM, Trehalose 5%(wt/v), pH 7.2–7.4. Filter-sterilize through 0.22 μm nitrocellulose membrane filter. Store at 4 °C.
4. Papain–DNase digestion solution: papain (Sigma, ≥10 units/mg protein) 20 units/mL, DNase I (Sigma, ≥2000 unit/mg protein) 100 units/mL (*see* **Note 1**), L-cysteine hydrochloride 1 mM (*see* **Note 2**), EDTA 0.5 mM, TTX 1 μM, AP5 50 μM, NBQX 10 μM, actinomycin D 5 μg/mL (*see* **Note 3**), dissolved in HBSS-HT, pH 7.2–7.4. After all components are fully dissolved, filter-sterilize the solution with a 0.22 μm PES membrane filter. Prepare the solution prior to use and store at 4 °C.
5. Fetal bovine serum (FBS).
6. Dissociation solution: DNase I 100 units/mL, TTX 1 μM, AP5 50 μM, NBQX 10 μM, actinomycin D 5 μg/mL, dissolved in HBSS-HT, pH 7.2–7.4. Filter with 0.22 μm PES membrane filter. Prepare the solution before use and store at 4 °C.
7. 50 mL conical tube, 15 mL conical tube.
8. 5 mL plastic transfer pipette.
9. 1000 μL, 200 μL, 10 μL manual pipettes and pipette tips.
10. Rocker platform (for 37 °C incubation).
11. Eppendorf 5810R centrifuge and rotor packages for 15 and 50 mL tube (or equivalent).

2.3 Cell Suspension Purification

1. Cell strainer, 70 μm.
2. HBSS-HT with 4% BSA: 4% BSA (wt/v) dissolved in HBSS-HT. Filter with 0.22 μm PES membrane filter. Store at 4 °C.
3. 37% Percoll solution: 100% Percoll (GE healthcare) 3.7 mL (*see* **Note 4**), 10× HBSS 1 mL, add sterile H_2O to 10 mL, adjust pH to 7.2–7.4.
4. 90% Percoll solution: 100% Percoll 0.9 mL, 10× HBSS 0.1 mL, pH 7.2–7.4.
5. Sorting solution: HBSS-HT, 1% FBS(v/v).
6. 5 mL round-bottom polystyrene tube.
7. Mini cell strainer for 5 mL round-bottom polystyrene tube, 70 μm.

2.4 Cell Sorting

1. DAPI solution for flow cytometry (1.0 mg/mL).
2. MoFlo Astrios EQs cell sorter (Beckman Coulter) (or equivalent).
3. 1.5 mL Eppendorf tube.
4. TRIzol LS (Thermo Fisher) (or equivalent).

2.5 RNA Extraction

1. Chloroform.
2. Glycogen (20 mg/mL) (optional).
3. Isopropanol.
4. 100% and 75% ethanol.
5. DEPC-treated ddH_2O.
6. Eppendorf centrifuge 5424R with fixed angle rotor (or equivalent).
7. Vortex.
8. Qubit 4 Fluorometer (Invitrogen), Qubit RNA HS Assay Kit.
9. 2100 BioAnalyzer Instrument (Agilent), Agilent RNA 6000 Nano Kit.

3 Methods

Always wear disposable gloves to prevent the sample from being contaminated by RNase on the skin.

3.1 Brain Tissue Harvest

1. Anesthetize the mouse with ketamine–xylazine solution. Intraperitoneally inject 0.1 mL per 10 gram of body weight. Spray the body of mice with 75% ethanol.
2. Transcardially perfuse the mice with ice-cold oxygenated HBSS-HT through 18G blunt perfusion needle for about 5 min until liver appeals pale.
3. Dissect the whole brain. Place the brain on a cold, wet blot paper rinsed with ice-cold oxygenated HBSS-HT. Extract tissue of interest.
4. Fill the petri dish with ice-cold oxygenated HBSS-HT. Transfer the tissue into the petri dish. Cut the tissue with iris scissors (*see* **Note 5**).
5. Transfer minced tissue to a 50 mL conical tube (*see* **Note 6**) with a plastic transfer pipette. Discard excess HBSS-HT.

3.2 Tissue Digestion and Dissociation

1. Add 5 mL papain–DNase digestion solution to the tube (*see* **Note 7**). Gently mix tissue suspension with a plastic transfer pipette.

2. Equilibrate the solution by dispensing sterile 95% O_2 + 5% CO_2 above the solution through a 0.22 μm PES filter(*see* **Note 8**). Immediately cap the tube.
3. Incubate the tube for 45 min (*see* **Note 9**) at 37 °C with constant gentle shaking on a rocker platform (about 100 rpm).
4. After digestion incubation, put the tube on ice. Add 500 μL of FBS and gently mix to stop digestion. Avoid bubbles during mixing. Centrifuge the tissue suspension at 300 × *g* for 1 min, 4 °C.
5. Discard the supernatant and add 3 mL dissociation solution to the tube. Pipette tissue suspension with 1000 μL pipette tips 10 times, avoid bubbles.
6. Briefly centrifuge tissue suspension at 300 × *g* and transfer cells suspension supernatant to another 15 mL conical tube.
7. Add 2 mL dissociation solution to the remaining chunks of tissue. Pipette tissue suspension with 1000 μL pipette tips for 15 times, avoid bubbles. If any chunks of tissue remain, attach a 200 μL tip to a 1000 μL pipette tip. Pipette up and down for another 15 times. Transfer cell suspension to the 15 mL conical tube. Keep the tube on ice.

3.3 Cell Suspension Purification

1. Rinse a 70 μm cell strainer with cold HBSS-HT. Put cell strainer on a 50 mL conical tube.
2. Filter cell suspension with the cell strainer. Add 5 mL cold HBSS-HT to the cell strainer to wash off residue cells.
3. To the tube containing filtered cell suspension, load 1 mL cold HBSS-HT with 4% BSA at the bottom to act as a cushion.
4. Centrifuge cell suspension at 300 × *g* for 5 min, 4 °C.
5. Discard the supernatant and resuspend the cell pellet with 10 mL 37% Percoll solution. Mix by inverting the tube several times until cell suspension appears homogenous.
6. Transfer 2 mL cell suspension in 37% Percoll solution to a clean 15 mL conical tube. Load 25 μL 90% Percoll solution at the bottom of the tube to act as a cushion underneath the cell suspension in 37% Percoll solution (*see* **Note 10**). Carefully transfer the remaining 8 mL cell suspension to the tube.
7. Centrifuge the cell suspension at 500 × *g* for 10 min. After centrifugation, the top layer will be myelin and debris. The cell pellet layer is between the 90% Percoll solution and the 37% Percoll solution.
8. Discard the top layer and most of the 37% Percoll solution. Leave about 50 μL solution. Avoid disturbing the cell pellet layer.

9. Resuspend cell pellet layer with 5 mL cold HBSS-HT. Mix by inverting the tube several times.
10. Centrifuge cell suspension at 300 × *g* for 5 min, 4 °C.
11. Discard the supernatant and resuspend the cell pellet with 1 mL sorting solution.
12. Put a mini cell strainer on a 5 mL round-bottom polystyrene tube. Filter cell suspension with mini cell strainer (*see* **Note 11**).

3.4 Cell Sorting

1. Add DAPI solution to the cell suspension to final concentration at 0.05–0.2 μg/mL DAPI for live/dead staining right before running the sample.
2. Run sample by FACS.
3. Gate cells on forward scatter (FSC-A), and sideward scatter (SSC-A) to exclude cell debris.
4. Plot FSC-A against FSC-W to discriminate single cells from aggregated cells and doublets.
5. Exclude DAPI-positive (dead) cells.
6. Gate on specific YFP fluorescence positive cells to collect the desired cell population (here: a cortex layer V excitatory neuron sub-population) (*see* **Note 12**).
7. Sort the first few cells on a slide and examine the cells under a fluorescent microscope to confirm the isolation of the cell population of interest.
8. Prepare a 1.5 mL Eppendorf tube containing 750 μL TRIzol LS. Cells collected for bulk RNA sequencing (RNA seq) should be sorted directly into the ice-cold TRIzol LS (*see* **Note 13**). Sample volumes of up to 250 μL are acceptable for 750 μL TRIzol LS.
9. After sorting, immediately cap the Eppendorf tube and invert the tube 10 times. Incubate the lysate on ice for 5 min.

3.5 RNA Extraction

1. Add 0.2 mL chloroform to the Eppendorf tube. Cap the tube and mix vigorously by vortex for 15 s. The liquid appeals milky and pink after vortex.
2. Incubate on ice for 2 min.
3. Centrifuge the sample at 12,000 × *g* for 15 min, 4 °C.
4. Carefully transfer the aqueous upper phase to a 1.5 mL Eppendorf tube (*see* **Note 14**).
5. (optional) Estimate the approximate concentration of RNA in the sample. If RNA concentration is lower than 10 ng/μL, add glycogen (*see* **Note 15**) to a final concentration of 20 μg/mL and invert the tube to mix.

6. Add 0.5 mL of isopropanol to the tube, per 750 μL TRIzol LS used for cell lysis. Mix by inverting the tube several times.
7. Incubate for 10 min, RT.
8. Centrifuge at 12,000 × *g* for 10 min. Discard supernatant.
9. Wash the RNA pellet by resuspending pellet in 1 mL 75% ethanol followed by brief vortex and 7500 × *g* centrifugation for 5 min, 4 °C.
10. Carefully discard the supernatant. Do not touch the pellet.
11. Air-dry RNA pellet for 5–10 min (*see* **Note 16**).
12. Dissolve RNA pellet in 15-30 μL DEPC-treated ddH$_2$O (*see* **Note 17**). Vortex and centrifuge briefly to collect RNA solution at the bottom of the tube.
13. Measure the final RNA concentration with Qubit 4 Fluorometer.
14. Measure total RNA integrity by 2100 Bioanalyzer. RNA with RIN ≥ 7 is considered good enough for library preparation and next-generation sequencing (NGS).
15. Purified RNA should be either stored at −80 °C after snap-freezing in liquid nitrogen or reverse transcribed into cDNA (*see* **Note 18**).

3.6 Analysis of Alternative Splicing

1. For medium-throughput alternative splicing analysis, fluorescent-labeled RT-PCR is recommended. This method is previously described by Ryan et al. [15]. In brief, RNA is reverse transcribed into cDNA with RNase-H(−) reverse transcriptase. PCR is performed with Taq polymerase master mix with fluorescent dye-labeled (e.g., Dy-682) primers.
2. For alternative splicing events such as skipped exon, the primer pair should be located on upstream and downstream constitutive exon. PCR amplicons can be separated and visualized by acrylamide gel electrophoresis and fluorescent imager. Inclusion and exclusion amplicon intensity can be quantified by freely available ImageJ software.
3. For high throughput alternative splicing analysis, RNA should be further purified with an mRNA isolation kit or rRNA depletion kit for library preparation and sequencing. Library preparation methods should follow the guide from manufacturers such as Illumina and Roche.
4. Conducting RNA seq on the cell type–specific RNA samples facilitates identifying potential differential alternative splicing events on a genome-wide scale. It performs well in accuracy and data amount. The raw data of RNA seq records only cDNA fragments and requires further computational implementation such as TopHat, Hisat2, and STAR to assign those cDNA fragments back to genomics. Recently, with the development

of RNA seq techniques, long-read sequencing methods, such as Pacific Biosciences (PacBio) and Oxford Nanopore Technologies (ONT), allow users to sequence full-length cDNAs, providing a more accurate estimation for isoform abundance during alternative splicing analysis.

5. Many computational methods have been developed to detect differential alternative splicing from RNA seq data. According to the detection principle, these methods can be classified into three major types: count-based methods, isoform resolution methods, and intron excision-based methods [16, 17]. Each method can excel others in some particular aspects. The requirement of experiments determines the appropriate computational method to detect differential alternative splicing. Based on the Bayesian model, rMATS performs well in real data to detect all basic kinds of alternative splicing events [16]. It can detect five types of alternative splicing events, including skipped exon, alternative 5′ spliced site, alternative 3′ splice site, mutually exclusive exons, and retained intron [18]. However, alternative splicing events can be far more complicated than these five basic events. LeafCutter is a better choice for complex alternative splicing because it detects alternative splicing events from intron splicing [17]. If the experiment intends to discover novel alternative splicing sites, other computational methods, such as Cufflink, are recommended [19].

4 Notes

1. DNase I is sensitive to shear denaturation. Mix gently while preparing the solution.
2. Cysteine is needed for papain activation. It is enhanced when EDTA is also presented. Thus, 1 mM cysteine and 0.5 mM EDTA are added to activate papain.
3. To prevent excitotoxicity and preserve cell transcription state during digestion and dissociation, a cocktail of small molecule inhibitors (TTX, NBQX, AP5, and actinomycin D) is added. TTX (sodium channel blocker), NBQX (AMPA and kainate receptor antagonist), AP5 (NMDA receptor antagonist) silence the neurons. Actinomycin D intercalates DNA, stabilizes topoisomerases I and II, and inhibits transcription.
4. As Percoll is a liquid with high viscosity, use the reverse pipetting technique to transfer the precise volume of Percoll. Press the pipette operating button to the second stop and dip the tip to a depth of 1 cm. Slowly release the button to let liquid fill the tip. Dispense the liquid into the receiving tube by depressing the operating button steadily to the first stop.

5. Tissue should be slightly minced so a plastic transfer pipette can aspirate the cut tissue suspension. Overminced tissue may result in cell loss when excess HBSS-HT is discarded.
6. A 50 mL conical tube is better than a 15 mL conical tube for tissue derived from one or more cortices. Minced tissue in a larger tube is more comfortable to be stirred during constant shaking incubation.
7. 5 mL papain–DNase digestion solution is sufficient for a whole cortex of the adult mouse brain. If more brains are used, the volume of digestion solution could be modified accordingly.
8. Do not bubble gas through the solution to avoid contamination, protein denaturation, and enzyme activity loss.
9. Incubation time should be determined empirically. Incubation time varies from 30 min to 90 min. In general, for brain tissue of 8- to 10-week-old mice, 45 min incubation time is sufficient.
10. Always load 90% Percoll solution at the bottom of the tube after transferring 2 mL cell suspension. If loading 90% Percoll solution first, the subsequent transfer of cell suspension may disturb the 90% Percoll solution cushion.
11. The filtration may result in cell loss. However, it is necessary as cells after centrifugation tend to be a bit clumpy.
12. Prepare cell suspension from wild-type mice as a negative fluorescent control for gating. Before loading the fluorescent cell sample on the cell sorter, prerun the wild-type cell sample and gate the wild-type cell cluster in the YFP channel (or other fluorescence channels). Thus, during neuron subpopulation sorting, fluorescence-positive cells can be easily discriminated from others with gating parameters collected from the prerun. Typically, capture the top 1–10% of fluorescent events if there is no negative control sample.
13. If TRIzol LS is not available, ordinary TRIzol is also acceptable. However, 1 mL TRIzol can only lyse liquid samples within 0.1 mL.
14. Be careful not to aspirate the interphase and organic layer to prevent DNA and protein contamination.
15. During isopropanol precipitation, glycogen acts as a nucleic acid carrier and co-precipitate with RNA. It can increase the efficiency of low concentration RNA precipitation. Glycogen does not interfere with most downstream applications such as cDNA synthesis.
16. Do not overdry RNA pellet as it may become difficult to dissolve.
17. DNA contamination happens from time to time during RNA extraction. For highly sensitive downstream applications such as sequencing. DNase I digestion is recommended to eliminate

genome DNA contamination after RNA extraction. Do not heat inactivate DNase I after digestion because chemicals in the digestion buffer could affect RNA integrity under heat treatment. Use phenol/chloroform extraction to eliminate DNase I and other chemicals in RNA solution. Do use recombinant DNase I (RNase-free grade) for genome DNA elimination.

18. Use high-efficiency thermal stable reverse transcriptases such as SuperScript IV (Invitrogen) for cDNA synthesis. Higher reaction temperature facilitates RNA secondary structure to unfold, therefore increasing reverse transcription efficiency and cDNA yield.

References

1. Mauger O, Scheiffele P (2017) Beyond proteome diversity: alternative splicing as a regulator of neuronal transcript dynamics. Curr Opin Neurobiol 45:162–168. https://doi.org/10.1016/j.conb.2017.05.012
2. Keren H, Lev-Maor G, Ast G (2010) Alternative splicing and evolution: diversification, exon definition and function. Nat Rev Genet 11(5):345–355. https://doi.org/10.1038/nrg2776
3. Lee CJ, Irizarry K (2003) Alternative splicing in the nervous system: an emerging source of diversity and regulation. Biol Psychiatry 54(8): 771–776. https://doi.org/10.1016/S0006-3223(03)00375-5
4. Li Q, Lee J-A, Black DL (2007) Neuronal regulation of alternative pre-mRNA splicing. Nat Rev Neurosci 8(11):819–831. https://doi.org/10.1038/nrn2237
5. Li B, Suutari BS, Sun SD, Luo Z, Wei C, Chenouard N, Mandelberg NJ, Zhang G, Wamsley B, Tian G, Sanchez S, You S, Huang L, Neubert TA, Fishell G, Tsien RW (2020) Neuronal inactivity co-opts LTP machinery to drive Potassium Channel splicing and homeostatic spike widening. Cell 181(7): 1547–1565 e1515. https://doi.org/10.1016/j.cell.2020.05.013
6. Traunmüller L, Bornmann C, Scheiffele P (2014) Alternative splicing coupled nonsense-mediated decay generates neuronal cell type-specific expression of SLM proteins. J Neurosci 34(50):16755–16761. https://doi.org/10.1523/JNEUROSCI.3395-14.2014
7. Wamsley B, Jaglin XH, Favuzzi E, Quattrocolo G, Nigro MJ, Yusuf N, Khodadadi-Jamayran A, Rudy B, Fishell G (2018) Rbfox1 mediates cell-type-specific splicing in cortical interneurons. Neuron 100(4):846–859.e847. https://doi.org/10.1016/j.neuron.2018.09.026
8. Huntley MA, Srinivasan K, Friedman BA, Wang T-M, Yee AX, Wang Y, Kaminker JS, Sheng M, Hansen DV, Hanson JE (2020) Genome-wide analysis of differential gene expression and splicing in excitatory neurons and interneuron subtypes. J Neurosci 40(5): 958. https://doi.org/10.1523/JNEUROSCI.1615-19.2019
9. Zhang X, Chen MH, Wu X, Kodani A, Fan J, Doan R, Ozawa M, Ma J, Yoshida N, Reiter JF, Black DL, Kharchenko PV, Sharp PA, Walsh CA (2016) Cell-type-specific alternative splicing governs cell fate in the developing cerebral cortex. Cell 166(5):1147–1162.e1115. https://doi.org/10.1016/j.cell.2016.07.025
10. Furlanis E, Scheiffele P (2018) Regulation of neuronal differentiation, function, and plasticity by alternative splicing. Annu Rev Cell Dev Biol 34(1):451–469. https://doi.org/10.1146/annurev-cellbio-100617-062826
11. Saxena A, Wagatsuma A, Noro Y, Kuji T, Asaka-Oba A, Watahiki A, Gurnot C, Fagiolini M, Hensch TK, Carninci P (2012) Trehalose-enhanced isolation of neuronal sub-types from adult mouse brain. BioTechniques 52(6):381–385. https://doi.org/10.2144/0000113878
12. Legroux L, Pittet CL, Beauseigle D, Deblois G, Prat A, Arbour N (2015) An optimized method to process mouse CNS to simultaneously analyze neural cells and leukocytes by flow cytometry. J Neurosci Methods 247:23–31. https://doi.org/10.1016/j.jneumeth.2015.03.021
13. Rio DC, Ares M Jr, Hannon GJ, Nilsen TW (2010) Ethanol precipitation of RNA and the use of carriers. Cold Spring Harb Protoc 5(6):

pdb.prot5440. https://doi.org/10.1101/pdb.prot5440

14. Feng G, Mellor RH, Bernstein M, Keller-Peck C, Nguyen QT, Wallace M, Nerbonne JM, Lichtman JW, Sanes JR (2000) Imaging neuronal subsets in transgenic mice expressing multiple spectral variants of GFP. Neuron 28(1):41–51. https://doi.org/10.1016/s0896-6273(00)00084-2
15. Percifield R, Murphy D, Stoilov P (2014) Medium throughput analysis of alternative splicing by fluorescently labeled RT-PCR. In: Hertel KJ (ed) Spliceosomal pre-mRNA splicing: methods and protocols. Humana Press, Totowa, NJ, pp 299–313. https://doi.org/10.1007/978-1-62703-980-2_22
16. Liu R, Loraine AE, Dickerson JA (2014) Comparisons of computational methods for differential alternative splicing detection using RNA-seq in plant systems. BMC Bioinformatics 15(1):364. https://doi.org/10.1186/s12859-014-0364-4
17. Li YI, Knowles DA, Humphrey J, Barbeira AN, Dickinson SP, Im HK, Pritchard JK (2018) Annotation-free quantification of RNA splicing using LeafCutter. Nat Genet 50(1):151–158. https://doi.org/10.1038/s41588-017-0004-9
18. Shen S, Park JW, Lu ZX, Lin L, Henry MD, Wu YN, Zhou Q, Xing Y (2014) rMATS: robust and flexible detection of differential alternative splicing from replicate RNA-seq data. Proc Natl Acad Sci U S A 111(51): E5593–E5601. https://doi.org/10.1073/pnas.1419161111
19. Trapnell C, Hendrickson DG, Sauvageau M, Goff L, Rinn JL, Pachter L (2013) Differential analysis of gene regulation at transcript resolution with RNA-seq. Nat Biotechnol 31(1): 46–53. https://doi.org/10.1038/nbt.2450

Chapter 5

Quantitative Measurement of Alternatively Spliced RNA Isoform Levels

Zhelin Jeff Li and Sika Zheng

Abstract

Conventional approaches to quantify alternative splicing are exon-centric and derive a ratio based on relative levels of the isoforms (or isoform groups) that include versus exclude a particular alternative RNA segment. The ratio measurement to study alternative splicing regulation can be confounded when alternative isoforms undergo differential RNA decay, for example, nonsense-mediated mRNA decay (NMD). Isoform-centric quantification is more informative for functional studies of alternative splicing, but challenges remain in distinguishing specific isoforms. Here, we provide a practical guide on addressing the specificity of isoform quantification and describe a simple sensitive method. Quantitative measurement of alternatively spliced RNA isoforms can be used to differentiate splicing regulation from transcriptional control and isoform-specific RNA decay regulation.

Key words AS-NMD, Alternative splicing, RT-qPCR, Isoform specific, Quantitation, Junction primers, Exon, Intron, Reporter

1 Introduction

Alternative splicing (AS) determines the final composition of mature RNA transcripts and produces isoform variants of different structure and function from a single pre-mRNA transcript [1, 2]. AS is an essential gene regulatory mechanism to development, metabolism, growth, and apoptosis [3–5]. Around 90% of human multiexon genes are subject to alternative splicing regulation [6].

In addition to increasing proteome diversity, alternative splicing can modulate gene expression outputs by coupling with nonsense mediated RNA decay (AS-NMD) and producing NMD sensitive isoforms [7–9]. Many splicing regulators and RNA binding proteins (RBP) control their own splicing and use AS-NMD to maintain homeostatic expression [8–11]. AS-NMD has also been documented to control non-RBP genes and even enforce tissue

Peter Scheiffele and Oriane Mauger (eds.), *Alternative Splicing: Methods and Protocols*,
Methods in Molecular Biology, vol. 2537, https://doi.org/10.1007/978-1-0716-2521-7_5,

specific expression, for example, of postsynaptic density protein 95 (Psd-95) and Bakl [11–14]. Skipping Psd-95 exon 18 and including Bakl exon 5 trigger NMD of respective transcripts. Psd-95 exon 18 is included only in neurons leading to neural specific expression of PSD-95 proteins. By contrast, Bakl exon 5 is excluded only in neurons leading to neural specific loss of BAKl proteins [12].

To measure the splicing ratio of an alternative exon, conventional alternative splicing assay simultaneously amplifies both the inclusion and exclusion isoforms in one PCR with primers annealing to the flanking constitutive exons. The two PCR amplicons (representing the two isoforms) are resolved by electrophoresis and their relative abundances are quantified [15]. Typically, a percent spliced in (PSI) value is computed as the ratio of the molar concentration of inclusion isoform divided by the sum of molar concentrations of the inclusion and exclusion isoform, which marks the splicing efficiency of a target exon.

The accuracy of this simple calculation depends on simultaneous measurement of both isoforms in the linear dynamic range of the detection method and mitigating technical biases of their uneven detection. These should not be taken for granted because PCR has amplification bias toward shorter amplicons [16].

Assessing the expression level of an isoform is essential to understanding the context of its biological functions and the relevance of alternative splicing regulation. Here, we describe a broadly applicable RT-qPCR based assay using isoform-specific primers to monitor isoform expression. Because of the degenerative sequences around splice sites, isoform-specific primers can be particularly difficult to design for exclusion isoforms, for which the exon–exon junction primer spanning across two constitutive exons would be the only choice. In some scenarios, an exon-exon junction primer presumably specific to the exclusion isoform can still amplify the inclusion isoform (nonspecifically) with a couple mismatched nucleotides during the first several PCR cycles. We lay out a step-by-step protocol to design isoform-specific primers and a strategy to test for their specificity.

Theoretically, isoform-centric quantification can be used to distinguish alternative splicing regulation from transcriptional and NMD regulation. For example, in Fig. 1a, under alternative splicing regulation, the expression level of one isoform increases while the other isoform decreases. If only one isoform alters expression level, it is likely due to isoform-specific regulation (e.g., NMD) (Fig. 1b, c). If both isoforms increase or decrease in the same direction, transcriptional regulation occurs (Fig. 1d, e). In this method, we use AS-NMD targets, *Psd-95*, *Tra2b*, *Hnrnpl*, and *Ptbp2* to demonstrate the utility of isoform quantification in distinguishing alternative splicing from NMD regulation [17]. The method can be further used to monitor cellular NMD activity and

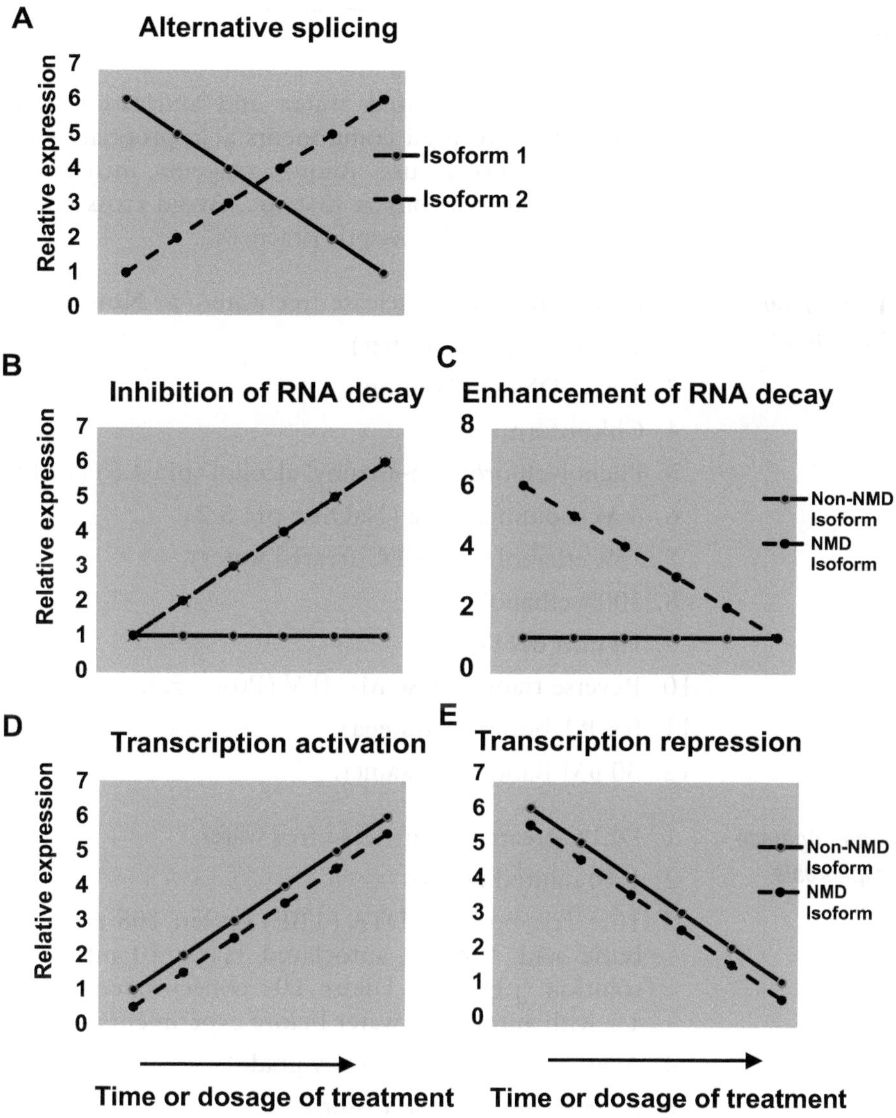

Fig. 1 Illustrative RT-qPCR readouts demonstrate isoform-specific quantification can be used to distinguish splicing, isoform-specific RNA decay (NMD), and transcriptional regulation. (**a**) Under alternative splicing regulation, the two isoforms change expression in the opposite direction. (**b**) NMD sensitive RNA isoform level increases while the non-NMD isoform remains relatively constant, indicating inhibition of NMD. (**c**) On the other hand, NMD activity is enhanced when only the NMD isoform decreases. (**d**, **e**) In case where multiple isoforms increase or decrease in the same direction, suggesting transcriptional activation or repression, respectively

to study mechanisms modulating NMD [17]. To allow future improvement and standardization of the method by the community, we followed the guidelines of the international Real-time PCR Data Markup Language (RDML) consortium and provided detailed information about the junction primers and RT-qPCR assays.

2 Materials

Autoclaved DEPC-treated water and nuclease-free tubes are recommended. Keep all components at appropriate storage temperature when not in use. Aliquot reagents, including enzymes, into appropriate portions at first use. Avoid cross contamination by strictly following lab aseptic practices.

2.1 RNA Extraction and cDNA Synthesis

1. DEPC-treated or nuclease-free water (*see* **Note 1**).
2. TRIzol (Thermo Fisher).
3. Turbo DNase (Ambion).
4. Chloroform.
5. Phenol–chloroform–isoamyl alcohol (pH 4.5) (*see* **Note 2**).
6. 3 M sodium acetate (NaOAc; pH 5.2).
7. 75% ethanol (in DEPC-treated water).
8. 100% ethanol.
9. 10 mM dNTP.
10. Reverse transcriptase M-MLV (Promega).
11. 5× RT buffer (Promega).
12. 30 μM Random hexamer.

2.2 PCR and Agarose Gel Electrophoresis

1. DEPC-treated or nuclease-free water.
2. 1:10 diluted cDNA.
3. 10× Tris–borate–EDTA (TBE) buffer: 108 g Tris base, 55 g boric acid, 900 ml autoclaved H_2O, 40 ml 0.5 M EDTA solution (pH 8.0). Dilute 10x concentrated TBE buffer to 1× with autoclaved water before experiments (*see* **Note 3**).
4. Agarose (molecular biology grade).
5. Agarose gel cast and cassette.
6. Ethidium bromide dissolved in H_2O (EtBr 1 mg/1 ml).
7. 10 mM dNTP.
8. 10 μM junction primers.
9. 5 units/μl Taq polymerase.

2.3 RT-qPCR

1. DEPC-treated or nuclease-free water.
2. 2x SYBR Green Master Mix (Thermo Fisher).
3. 10 μM junction primers.
4. RT-qPCR QuantStudio 6 microplate (Thermo Fisher).
5. Microplate seal (Thermo Fisher).
6. QuantStudio 6 Real-Time PCR System (Thermo Fisher).

2.4 Computational and Bioinformatic Resources

1. UCSC genome browser: https://genome.ucsc.edu/
2. Primer 3: https://bioinfo.ut.ee/primer3/

3 Methods

3.1 Design of RT-qPCR Junction Primers for an Exclusion Isoform

1. In an artificial transcript structure containing exon 1, intronl, exon 2, intron 2, and exon 3, for which exon 2 is the alternative exon (Fig. 2a), a forward junction primer spanning exon 1 and exon 3 can be separated into the 5′ part (exon 1) and the 3′ part (exon 3) (*see* **Note 4**). The reverse primer can be ~80–150 nts downstream from the forward primer.
2. The nucleotide number of the 3′ part should be between one to six nucleotides. Longer than six increases the possibility where the 3′ part alone is sufficient for seeding a round of nonspecific PCR amplification from the 5′ end of exon 3 (i.e., detecting both the inclusion and exclusion isoforms).
3. Double check the 3′ part does not anneal to intron 1 or exon 2 (Fig. 2a). The 3′ part should not match the 5′ end of intron 1 to avoid amplifying unspliced transcripts (*see* **Note 6**), although depending on the size of intron 1 the unspliced transcripts may be too long to be amplified. The 3′ part should not match exon 2, otherwise the inclusion isoform will be amplified.
4. Retrieve sequences of exon 1 and exon 3 from UCSC genome browser or equivalent databases (*see* **Note 7**).
5. Input sequences into the Primer 3 software following the instruction manual (*see* **Note 8**). We use *NPsd-95* and *Hnrnpl* exclusion isoforms as examples of designing forward junction primers (Fig. 2b). Use "{}" to include 20 nts sequences upstream of the exon junction (Psd-95 exon 17 and Hnrnpl exon 5, respectively) and 6 nts on the following exon after the exon junction (Psd-95 exon 19 and Hnrnpl exon 7, respectively) (*see* **Note 9**).
6. Set the amplicon size to be around 80 nts to 150 nts.
7. Keep the primer melting temperature at 60 °C (*see* **Note 10**).
8. Use default settings for other parameters in Primer 3 to generate recommended primers.
9. If no optimal forward junction primers are identified, repeat **steps 1–8** to design a reverse junction primer. In this case, the 5′ part of the reverse junction primer is reverse complementary to the 5′ end of exon 3, and the 3′ part of the primer is reverse complementary to the 3′ end of exon 1 (Fig. 2c). We use *Tra2b* and *NPtbp2* as examples of designing reverse junction primers (Fig. 2d).

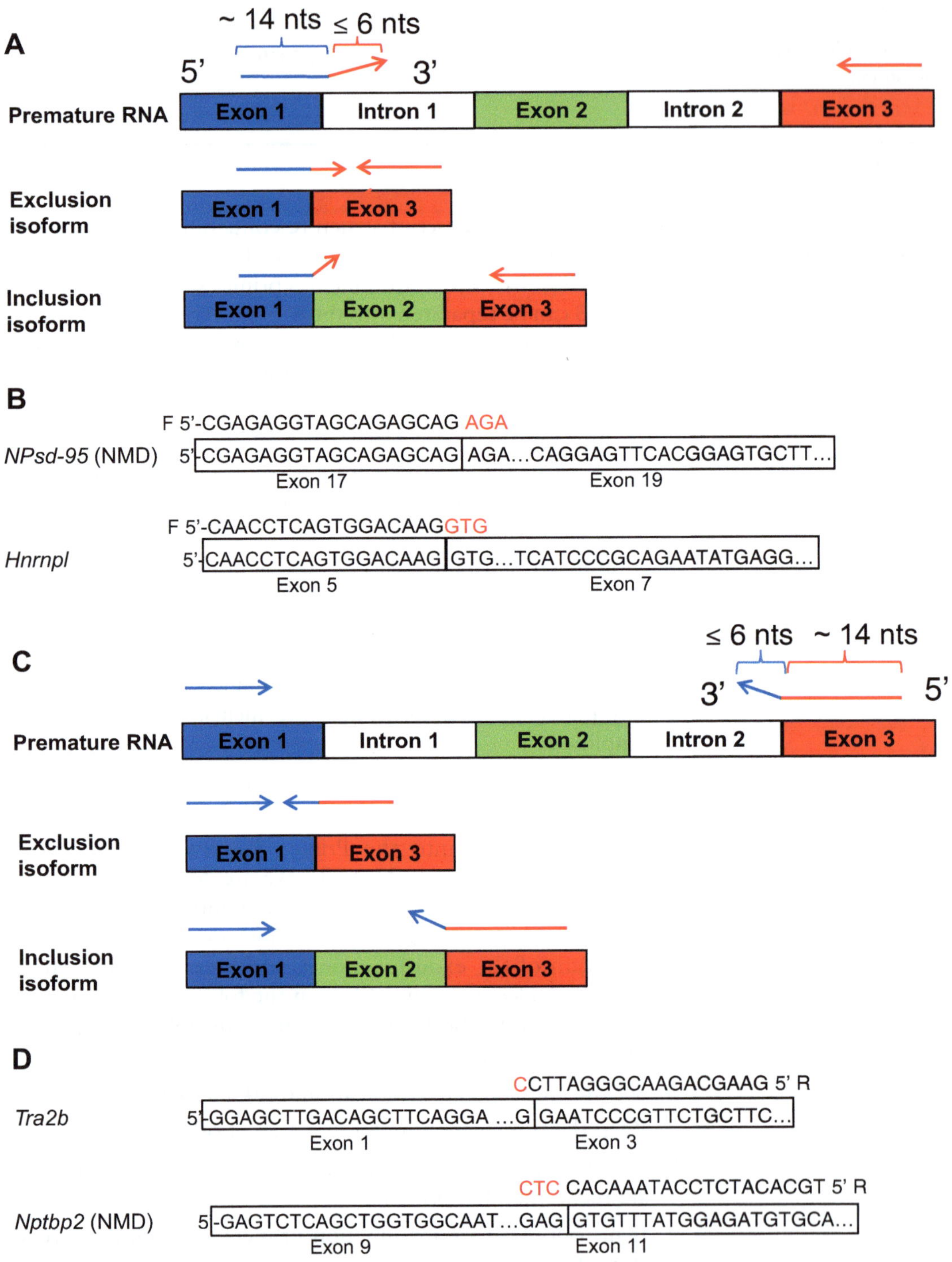

Fig. 2 Schematics of designing a junction primer for an exclusion isoform. (**a**) Design a PCR forward primer specific to an exclusion isoform. The 3′ part of a forward junction primer should be no longer than six nucleotides. It is specific to the exclusion isoforms because the 3′ part matches only to exon 3. (**b**) Examples of forward junction primer designs using *NPsd-95* (NMD sensitive exclusion isoform) and *Hnrnpl*. For *NPsd-95*, its 3′ part contains three nucleotides in exon 19 and the 5′ part matching exon 17. For *Hnrnpl*, three nucleotide sequences at the 3′ part of the forward primer match exon 7. (**c**) Design a PCR reverse primer specific to an

3.2 Design of RT-qPCR Primers for an Inclusion Isoform

1. To detect an inclusion isoform, ideally design a primer completely residing within the alternative exon (exon 2 in Fig. 3a, c).
2. Retrieve sequences of exon 2 and exon 3 from UCSC genome browser or equivalent databases.
3. Input sequences into the Primer 3 software following the instruction manual (*see* **Note 8**). We use *NHnrnpl* exclusion isoform as an example of designing a forward junction primer. Use "[ ]" across the exon junction sequences exon 6–7 to define a "must amplify" region and set forward primers to match the alternatively exon (*see* **Note 11**).
4. If no optimal forward primers are identified, repeat steps 1–3 to design a reverse primer within the alternative exon (Fig. 3c). We use *Psd-95, NTra2b,* and *Ptbp2* inclusion isoforms as examples (Fig. 3d).
5. If the alternative exon is too short, or Primer 3 is unable to find optimal primers, follow steps of junction primer design in Subheading 3.1 (*see* **Note 12** for modifications applied to inclusion isoforms).

3.3 RNA Extraction and cDNA Synthesis

1. Add TRIzol directly to cells or tissues to extract total RNA following the manufacturing standard protocol.
2. Isolated RNA is treated with 4 units of Turbo DNase at 37 °C for 35 min to degrade all remaining genomic DNA (*see* **Note 13**).
3. After DNase treatment, purify RNA using phenol–chloroform–isoamyl alcohol following the standard extraction procedure [18]. Measure the concentration of DNase-treated RNA using Nanodrop 2000c.
4. Convert 1 μg freshly isolated DNA-free RNA to cDNA following the Promega M-MLV protocol in a 20 μl reaction.
5. To set up reverse transcription, prepare an RNA mix for each sample. Include a no-template control as a negative control.

RNA	1 μg
10 mM dNTP	1 μl
30 μM Random Hexamer	1 μl
DEPC-treated H_2O	fill up to total volume 15 μl

Fig. 2 (continued) exclusion isoform. In this scenario, the 3′ part of the reverse junction primer should be reverse complementary to exon 1. (**d**) Reverse junction primers for *Tra2b* and *NPtbp2.* One nucleotide of *Tra2b* and three nucleotides of *NPtbp2* reverse primers are across the exon junctions. F indicates forward primer and R indicates reverse primer (*see* **Note 5**)

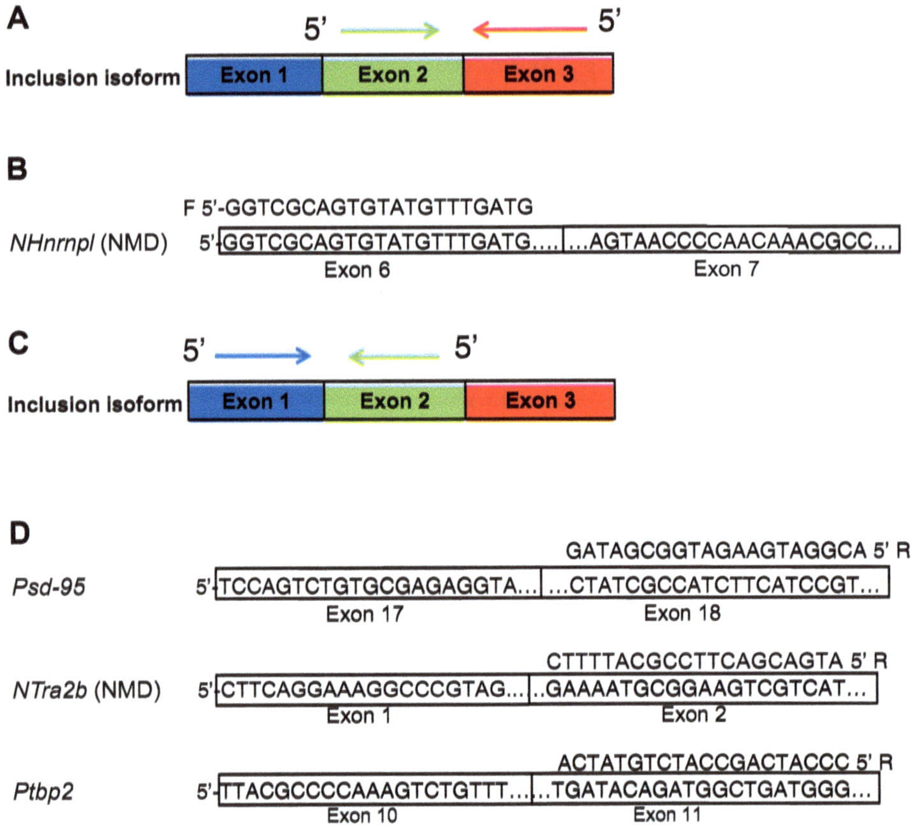

Fig. 3 Schematics of primer design strategy for inclusion isoforms. (**a**) Forward primer resides within the alternative exon. (**b**) An example of (**a**) where the *NHnrnpl* forward primer is designed using *Hnrnpl* alternative exon 6. (**c**) Reverse Primer specific to the inclusion isoform annealing to the alternative exon. (**d**) Examples of (**c**) using *Psd-95*, *NTra2b*, and *Ptbp2*. For the *Psd-95* inclusion isoform, the reverse primer is reverse complementary to the sequence of exon 18. For *NTra2b* and *Ptbp2*, the reverse primers are in their exon 2 and exon 11, respectively

6. Denature at 70 °C for 5 min in a thermal cycler.
7. Place immediately on ice for 2 min.
8. Prepare cDNA synthesis mix as the following. Vortex the completely thawed 5x RT buffer briefly. Take out reverse transcriptase (RT) right before use and put it back to −20 °C right after use.

5× RT buffer	4 μl
RT (200 units/μl)	1 μl

9. Add 5 μl cDNA synthesis mix to each RNA mix. Pipet up and down five times to mix (*see* **Note 14**).
10. Gently spin down the tubes with a microfuge.

11. Put in a thermal cycler.

25 °C	10 min (*see* **Note 15**)
42 °C	60 min
85 °C	5 min
4 °C	Forever

12. Dilute the 20 μl cDNA reactions with 180 μl (1:10) nuclease-free or DEPC-treated H_2O before use. For detailed RT protocol using different reverse transcriptase, please refer to your RT supplier's manual.

3.4 Validation and Quality Control of Isoform-Specific Primers

It is important to perform gel electrophoresis, a RT-qPCR standard curve test, and a melting curve test to validate the specificity and sensitivity of in silico designed primers. Use control cDNA to examine these primers before using them to study experimental samples. Make sure all regent tubes are completely thawed and reagents spun down before use. Autoclave PCR tubes and strips are recommended.

1. For each 20 μl PCR reaction, prepare a PCR master mix as the following (*see* **Note 16**).

10× Taq buffer	2 μl
10 mM dNTP	0.4 μl
10 μM forward primer	1 μl
10 μM reverse primer	1 μl
5 units/μl Taq (μl)	0.1 μl
Nuclease-free H_2O	8.5 μl

2. Pipet up and down gently to mix the master mix. Spin down the master mix using a microcentrifuge.
3. Allocate 13 μl master mix to each PCR tube.
4. Add 7 μl of 1:10 diluted cDNA to each PCR tube.
5. Set the following PCR cycles on the thermal cycler.

94 °C	3 min
94 °C, 30 s; 60 °C, 30 s; 72 °C, 30 s. Repeat for 30 cycles (*see* **Note 17**)	
72 °C	5 min
4 °C	Forever

6. Make a 2% agarose gel using 1× 150 ml TBE and 12 μl EtBr.
7. Perform gel electrophoresis until bands separate clearly (Fig. 4a).
8. If agarose gel electrophoresis shows only the expected PCR bands, perform a RT-qPCR standard curve test to examine the sensitivity of primers and a melting curve test to further validate the specificity.
9. Follow the RT-qPCR procedure in Subheading 3.5 but select the QuantStudio 6 standard curve program instead of gene expression program. Here, a standard curve of 1:4, 1:40, 1:400, 1:4000 of cDNA dilution was constructed. This range encompasses the 1:10 dilution, which will be used to calculate PCR amplification efficiency (Fig. 4b) (*see* **Note 18**).
10. Before starting the standard curve run, make sure the melting curve option is selected by checking the melting curve box on the QuantStudio 6 program. A melting curve test starts from 60 °C to 95 °C at a 0.05 °C/s measuring rate. Examples of melting curve test results are shown in Fig. 4c (*see* **Note 19**).

3.5 Quantitative Analysis of Alternative Isoforms Using RT-qPCR

Perform RT-qPCR assays using Power SYBR Green on a QuantStudio 6 Real-Time PCR platform. Please adapt the procedure for other RT-qPCR instrumentations accordingly.

1. Prepare 2× Power SYBR Green RT-qPCR master mix for a 10 μl reaction:

2× SYBR mix	5 μl
0.3 nM Forward primer	0.3 μl
0.3 nM Reverse primer	0.3 μl
Nuclease-free H_2O	1.4 μl

2. Mix the master mix well and pipet 7 μl to each well directly on the bottom of the microplate. Do not touch the walls with pipette.
3. Next, pipet 3 μl diluted cDNA to each well with the master mix. Tilt pipette tips against the wall of the microplate wells to avoid touching master mix.
4. Seal microplate with microplate film and spin down for 2 min.
5. Set the QuantStudio 6 RT-qPCR run program as follows.

50 °C	2 min
95 °C for 15 s; 60 °C for 1 min, with the 95 °C and 60 °C steps repeated for 40 cycles	

A

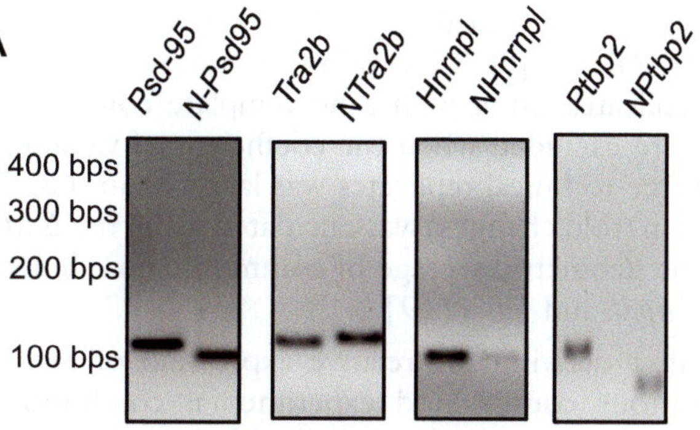

B

Targets	Size	Slope	R^2	Eff%
Psd-95	116	-3.275	0.990	100%
NPsd-95	105	-3.296	0.990	100%
Tra2b	105	-3.340	0.998	100%
NTra2b	116	-3.458	0.998	95%
Hnrnpl	92	-3.211	0.999	104%
NHnrnpl	100	-3.394	0.996	97%
Ptbp2	107	-3.417	0.988	96%
NPtbp2	72	-3.346	0.990	99%

C

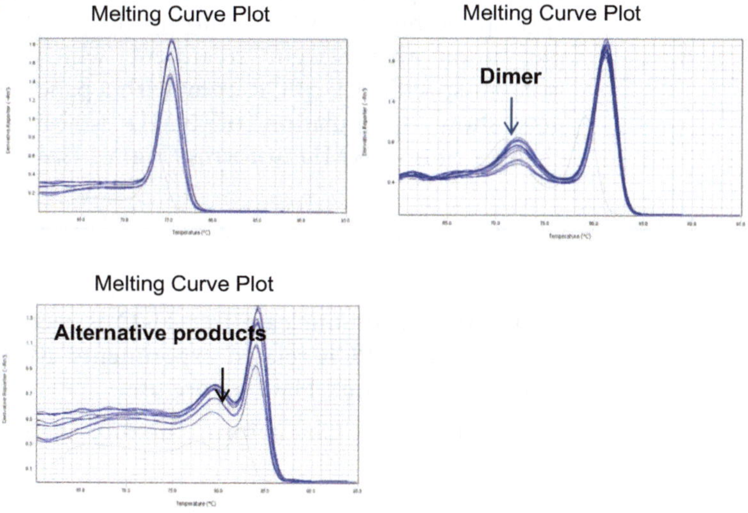

Fig. 4 Validation of isoform-specific primers using PCR gel electrophoresis, melting curve test, and standard curve test for RT-qPCR efficiency. (**a**) Gel images of *Psd-95*, *Tra2b*, *Hnrnpl*, and *Ptbp2* inclusion and exclusion primers amplifying one clear single band of the expected sizes using control cDNA from neuro-2A (N2a) cells. (**b**) RT-qPCR standard curve results calculated from a series dilution of 1:4, 1:40, 1:400, and 1:4000 cDNA samples. A slope of −3.3 indicates 100% replication efficiency which will be used to calculate $\Delta\Delta C_t$. (**c**) Generic interpretations of melting curves. When all RT-qPCR products are denatured at the same temperature (one peak), it means there is only one homogenous amplicon product. When multiple peaks are detected in a melting curve, primer dimers (usually a peak below 75 °C) or alternative amplicons are present

6. QuantStudio Real-Time PCR software was used for the analysis. All RT-qPCR reactions are conducted with three technical replicates along with a no template control (NTC). Outliers were excluded when the coefficient of variation of Ct for the three technical replicates was larger than 0.3. Relative expression (fold changes) was calculated using the $\Delta\Delta C_t$ method with the geometric average of commonly used housekeeping genes *Gapdh* and *Sdha* [19].
7. After deriving the relative expression of each isoform under various control and experimental conditions, interpret the data using principles described in introduction and Fig. 1.
8. Figure 5 shows the versatility of this RT-qPCR based isoform quantification vs the traditional splicing assay that measures the ratio of alternative isoforms. Traditional splicing assay cannot distinguish alternative splicing regulation of Psd-95 exon 18 from NMD inhibition of exon 18 exclusion isoform (*NPsd-95)* (Fig. 5a). In contrast, by quantifying expression of each isoform (i.e., *Psd-95* and *NPsd-95),* UPF2 mediated NMD can be distinguished from PTBP2 mediated alternative splicing of exon 18 (Fig. 5b, c) (*see* **Note 20**) ([20]).
9. One application of RT-qPCR quantification of AS-NMD isoforms (such as *Tra2b, Hnrnpl,* and *Ptbp2*) is to monitor cellular NMD activity (*see* **Note 21**). Changes in cellular NMD efficiency is expected when NMD isoforms change while the corresponding non-NMD isoforms do not change. For example, upon thapsigargin treatment which inhibits NMD, isoform-specific RT-qPCR assay with a panel of AS-NMD targets show upregulation of NMD isoforms and marginal changes of non-NMD isoforms (Fig. 6) (*see* **Note 22**).

4 Notes

1. DEPC-treated or nuclease-free H_2O is recommended for all RNA and RT-qPCR related experiments. Make sure to autoclave DEPC-treated H_2O before use.
2. Low pH Phenol-Chloroform mixture is recommended for RNA extraction.
3. The final solution should contain: 0.13 M Tris (pH 7.6), 45 mM Boric acid, and 2.5 mM EDTA. If precipitation is present, warm to 37 °C and mix with a stirring bar until completely dissolved prior to dilution. Autoclave the diluted 1× TBE is recommended but not required.
4. The junction primer should be around 18 to 22 nts in length with the 3′ part no longer than 6 nts.

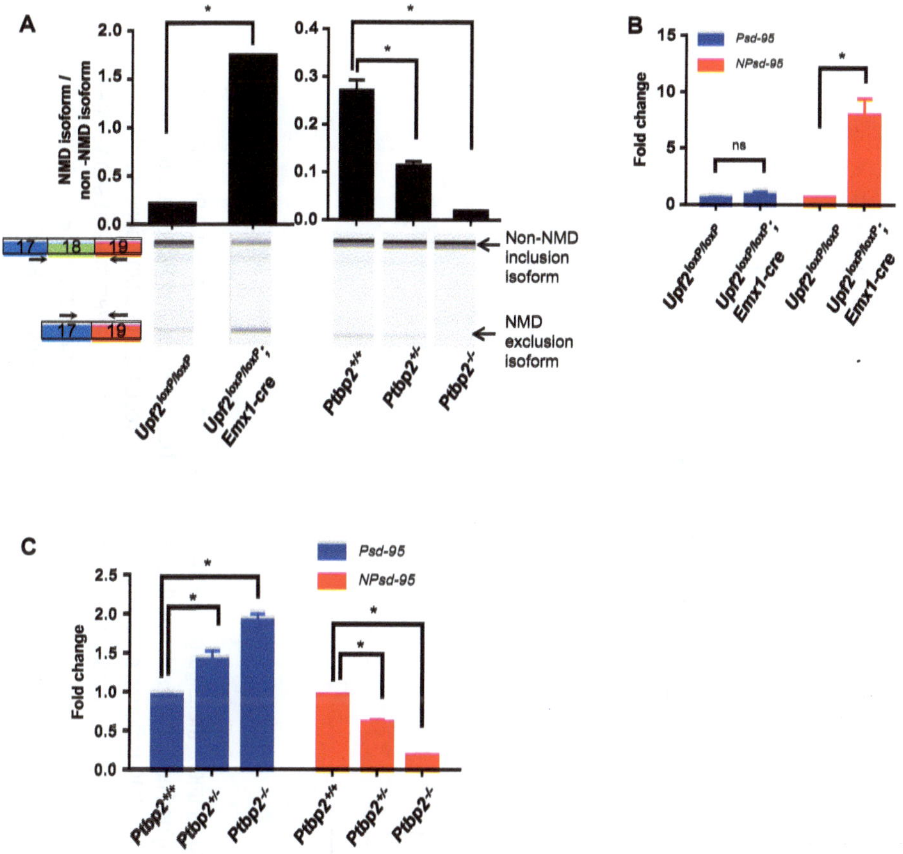

Fig. 5 Difference between the traditional alternative splicing assay and isoform-specific quantification assay (reproduced from Ref. 17 with permission from Cold Spring Harbor Press). (**a**) Conventional splicing assay to derive expression ratios of the *Psd-95* NMD exclusion isoform relative to the non-NMD inclusion isoform in *Upf2* conditional knockout cortices (*left* panel) and *Ptbp2* null cortices (*right* panel). The ratios by themselves do not distinguish NMD regulation from alternative splicing regulation. Representative digital gel images are shown in the *lower* panel. Statistics were calculated using a two-tailed unpaired Student's *t*-test for the *Upf2* knockout samples ($N = 2$) and a one-way ANOVA test followed by Tukey's multiple comparison test for the *Ptbp2* mutant samples (*Ptbp2*$^{+/+}$ and *Ptbp2*$^{+/-}$, $N = 2$; *Ptbp2*$^{-/-}$, $N = 4$). (**b**) The same *Upf2* knockout samples in (**a**) were analyzed using isoform-specific RT-qPCR assay. *Psd-95* non-NMD isoform and *Psd-95* NMD (*NPsd-95*) isoform as in Figs. 2 and 3 are quantified separately. (**c**) The same *Ptbp2* null samples as in (**a**) were analyzed using the isoform-specific RT-qPCR assay. A two-way ANOVA test followed by Dunnett's multiple comparison tests was used to determine significant expression changes between samples. (*) $P < 0.05$; (ns) not significant. Error bars represent mean ± SEM

5. These designs assure that PCR amplifications only occur with perfect binding of the 3′ part of the forward or reverse junction primer to the template.
6. For example, if the starting sequence of intron 1 is GTGA, then a forward junction primer with a 3′ end GGGA is sufficient for polymerase extension with the unspliced transcript.

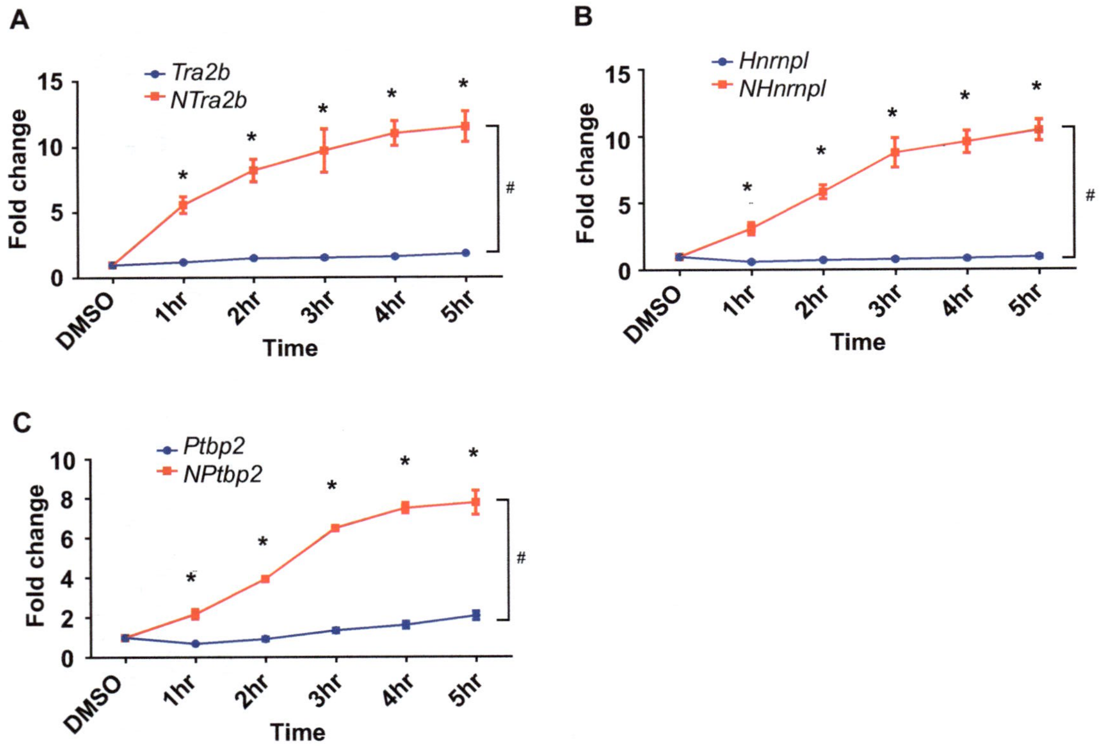

Fig. 6 Quantitative isoform-specific RT-qPCR assay analyzing AS-NMD targets under thapsigargin treatment (reproduced from Ref. 17 with permission from Cold Spring Harbor Press). (**a–c**) Temporal expression of NMD isoforms (red lines) and non-NMD isoforms (blue lines) of *Tra2b*, *Hnrnpl*, and *Ptbp2* treated with 0.2 μM of thapsigargin. Expression levels are normalized to DMSO-treated samples. The NMD isoforms of all three genes are upregulated in a time dependent manner while the non-NMD isoforms showed no significant changes in expression. This indicates that thapsigargin can inhibit NMD activity overtime and shows one practical use of isoform-specific quantification method in monitoring NMD. A two-way ANOVA test was used to determine a significant difference of expression between the two isoforms. (#) $P < 0.05$. Dunnett's multiple comparison tests were used to determine significant expression changes of individual isoforms over time relative to DMSO treatment. (*) $P < 0.05$. $N = 3$. Error bars represent mean ± SEM

7. Visit UCSC genome browser to identify exons of interest and obtain the sequences of the correct strand.
8. It is highly recommended for users to read the Primer 3 manual thoroughly to understand command lines for accurate primer designs.
9. "{}" in Primer 3 indicates a sequence_included_region where the program selects primers. In this step, "{}" instructs Primer 3 to must use those sequences spanning across the exon junctions as part of the forward junction primer design.
10. 60 °C is the melting temperature we prefer during primer design. Adjust melting temperatures accordingly for your PCR or RT-qPCR protocols.
11. "[]" is code for sequence_target in Primer 3. It refers to the must amplify region. In this case, the must amplify region is the junction of Exon 17 and 18. The program then picks the best primers flanking that region.
12. In these cases, the forward junction primer should span over exon 1 and exon 2 (3′ part). If using reverse junction primer, it should span over exon 3 and exon 2 (3′ part). All junction primer designing rules apply (Fig. 2).
13. This step increases data quality and confidence by eliminating residual genomic DNA.
14. Use a multichannel (set to 15 μl) to handle multiple samples at the same time.
15. The 25 °C step is needed for primer annealing using random hexamer. For oligodT primers, skip this step.
16. Always include 1 or 2 more reactions in the master mix to ensure there are enough master mix, for example, make an 18-sample master mix for a 16-sample experiment.
17. Polymerase typically extends 1 kb/min. When testing primers, use 30 cycles for sensitivity and a parallel PCR of 35 cycles for specificity. Taq polymerase extension temp depends on polymerase ranging from 68 °C to 72 °C. For detailed PCR troubleshooting or procedures, please refer to published PCR guides.
18. Standard curves increase accuracy of gene expression calculation. An ideal standard curve should be created by data points ranging at least 1000 folds. The dilution used in actual experiments should be a data point (concentration) within the standard curve range. Calculate the qPCR efficiency using $10^{(-1/\text{slope})}$.

19. Melting curve is more sensitive than PCR gel electrophoresis in detecting nonspecific amplicons. It is used to double check primer specificity. Multipeak or shoulder-peak melting curves indicate the primer is not suited for qPCR analysis (Fig. 4c).
20. Exon 18 is repressed by polypyrimidine tract binding protein 2 (PTBP2) [16]. Knocking out NMD core factor UPF2 increases the expression level of *NPsd-95* by suppressing NMD.
21. The inclusion isoforms of genes did not use junction primers.
22. By analyzing the reporter gene expression changes over time, one can determine the effects of thapsigargin inhibiting NMD. This is one example of using isoform-specific quantification being for mechanistic studies.

Acknowledgments

This work was supported by the National Institutes of Health (grant number R01NS104041 and R01MH116220 to S.Z.).

References

1. Lee Y, Rio DC (2015) Mechanisms and regulation of alternative pre-mRNA splicing. Annu Rev Biochem 84:291–323
2. Scotti MM, Swanson MS (2016) RNA mis-splicing in disease. Nat Rev Genet 17: 19–32
3. Furlanis E, Scheiffele P (2018) Regulation of neuronal differentiation, function, and plasticity by alternative splicing. Annu Rev Cell Dev Biol 34:451–469
4. Manning KS, Cooper TA (2017) The roles of RNA processing in translating genotype to phenotype. Nat Rev Mol Cell Biol 18:102–114
5. Vuong CK, Black DL, Zheng S (2016) The neurogenetics of alternative splicing. Nat Rev Neurosci 17:265–281
6. Wang ET, Sandberg R, Luo S et al (2008) Alternative isoform regulation in human tissue transcriptomes. Nature 456:470–476
7. Baralle FE, Giudice J (2017) Alternative splicing as a regulator of development and tissue identity. Nat Rev Mol Cell Biol 18:437–451
8. Lewis BP, Green RE, Brenner SE (2003) Evidence for the widespread coupling of alternative splicing and nonsense-mediated mRNA decay in humans. Proc Natl Acad Sci U S A 100:189–192
9. Zheng S, Black DL (2013) Alternative pre-mRNA splicing in neurons: growing up and extending its reach. Trends Genet 29: 442–448
10. Spellman R, Llorian M, Smith CWJ (2007) Crossregulation and functional redundancy between the splicing regulator PTB and its paralogs nPTB and ROD1. Mol Cell 27: 420–434
11. Zheng S (2016) Alternative splicing and nonsense-mediated mRNA decay enforce neural specific gene expression. Int J Dev Neurosci 55:102–108. https://doi.org/10.1016/j.ijdevneu.2016.03.003
12. Lin L, Zhang M, Stoilov P et al (2020) Developmental attenuation of neuronal apoptosis by neural-specific splicing of Bak1 microexon. Neuron 107:1180–1196.e8
13. Zheng S, Damoiseaux R, Chen L, Black DL (2013) A broadly applicable high-throughput screening strategy identifies new regulators of Dlg4 (Psd-95) alternative splicing. Genome Res 23:998–1007
14. Zheng S, Gray EE, Chawla G et al (2012) PSD-95 is post-transcriptionally repressed during early neural development by PTBP1 and PTBP2. Nat Neurosci 15:381–388

15. Harvey SE, Cheng C (2016) Methods for characterization of alternative RNA splicing. Methods Mol Biol 1402:229–241
16. Shagin DA, Lukyanov KA, Vagner LL, Matz MV (1999) Regulation of average length of complex PCR product. Nucleic Acids Res 27:e23
17. Li Z, Vuong JK, Zhang M et al (2017) Inhibition of nonsense-mediated RNA decay by ER stress. RNA 23:378–394
18. Richter DJ, Fozouni P, Eisen MB, King N (2018) Gene family innovation, conservation and loss on the animal stem lineage. Elife 7:e34226. https://doi.org/10.7554/eLife.34226
19. Schmittgen TD, Livak KJ (2008) Analyzing real-time PCR data by the comparative C(T) method. Nat Protoc 3:1101–1108
20. Zhang M, Ergin V, Lin L et al (2019) Axonogenesis is coordinated by neuron-specific alternative splicing programming and splicing regulator PTBP2. Neuron 101:690–706.e10

Chapter 6

Analysis of Splicing Regulation by Third-Generation Sequencing

Eric Allemand and Fabrice Ango

Abstract

In Metazoa, the diversity of transcripts produced by the RNA Polymerase II is generated essentially through post-transcriptional processing of the nascent transcripts. The regulation of exon inclusion by alternative splicing is one of the main sources of this diversity, which leads to the expansion of the proteome. The portfolio of alternative transcripts remains largely underestimated. Improvement of the sequencing technologies has enhanced the characterization of RNA isoforms and led to the perpetual incrementation of gene expression diversity. Here, we describe a high throughput approach to assess in-depth the splicing regulation of target gene(s) using the third-generation sequencing (TGS) technologies.

Key words Alternative splicing, RNA isoforms, Long read sequencing, ONT, PacBio

1 Introduction

The transfer of genetic information from DNA to mRNA is not contiguous but splits in multiple exons, and multiple post-transcriptional molecular mechanisms affect the transcript diversity [1, 2]. Among them, alternative splicing stands as a key mechanism for extending coding capacity. The development of sequence technologies has been an essential contributor to our assessment of the genetic complexity. In the 1970s, the Sanger sequencing method was able to detect only a few alternative transcripts for a given gene, and in the 1980s and 1990s the use of polymerase chain reaction (PCR) has significantly expanded the portfolio of genes known to be subjected to alternative splicing regulation [3]. The introduction at the beginning of the twenty-first century of the Next Generation Sequencing (NGS) and the genome annotation have established that alternative splicing affects almost every gene, demonstrating that gene expression is ubiquitously regulated by this mechanism [4]. Over the last decade, the use of Third-Generation Sequencing (TGS) developed by Pacific Bioscience

Peter Scheiffele and Oriane Mauger (eds.), *Alternative Splicing: Methods and Protocols*,
Methods in Molecular Biology, vol. 2537, https://doi.org/10.1007/978-1-0716-2521-7_6,

(PacBio) and Oxford Nanopore (ONT) uncovered a greater diversity of transcripts than previously thought. Although the number of isoforms per gene listed in RefSeq databases is on average about 10, recent TGS approaches to analyze the target gene(s) expression highlighted that it may be closer to hundreds or thousands [5–10]. Currently, the *DSCAM* (Down's syndrome cell adhesion molecule) gene from *Drosophila* holds the record with the identification of 18,496 splice variants, out of 38,016 theoretically possible [5, 10].

The whole transcriptome sequencing by NGS approaches is a powerful strategy to determine the global level of gene expression. However, the short length of reads (<600 bp) does not decipher the comprehensive arrangement of all exons included in a transcript. In contrast, the TGS technologies provide sequences of up to 100 kb, thereby allowing the identification of the complete exon architecture included in individual transcripts. However, they are less accurate at the single-base level than NGS. Moreover, analysis of TGS data to identify multiple splice variants remains a difficult task. In particular, because bioinformatic tools that allow an accurate clustering of reads corresponding to distinct splice variants are still poorly efficient, many laboratories are forced to develop their own bioinformatics pipelines [8, 9, 11–13]. In this chapter, we include our bioinformatics approach to tackle these difficulties. Moreover, the increasing number of studies using TGS to analyze splicing indicates that this technology will become a new standard releasing in the near future several automated bioinformatics pipelines. The sequencing depth is a parameter that has a high impact on the characterization of transcript isoforms. Indeed, the whole transcriptome is tremendously complex and the limits of sequencing technologies associated with different levels of expression between genes or splice variants lead essentially to the detection of most abundant transcripts. Therefore, focusing the analysis on a specific genetic target is the optimal strategy to reduce the complexity of samples and improve the detection of splice-regulated transcripts [5, 8–10]. Currently, two approaches have been developed to analyze the pool of transcripts expressed by a specific target gene: capture of its transcripts by affinity with exon probes, and RT-PCR amplification performed with primers pairing the first and last exons. These two techniques do not allow an exhaustive characterization of all potential splice isoforms; but they have both led to the identification of many new transcripts. Besides, the multiplexing of libraries allows efficient analysis of several genes together. To date, the expression up to 30 genes has been successfully analyzed in one sequencing run [9].

Here, we will describe the procedure that we have recently used to assess the mRNA transcript diversity of gephyrin (*GPHN*) in several samples using the PacBio and ONT technologies. First, total RNAs are isolated with a procedure recovering all transcripts.

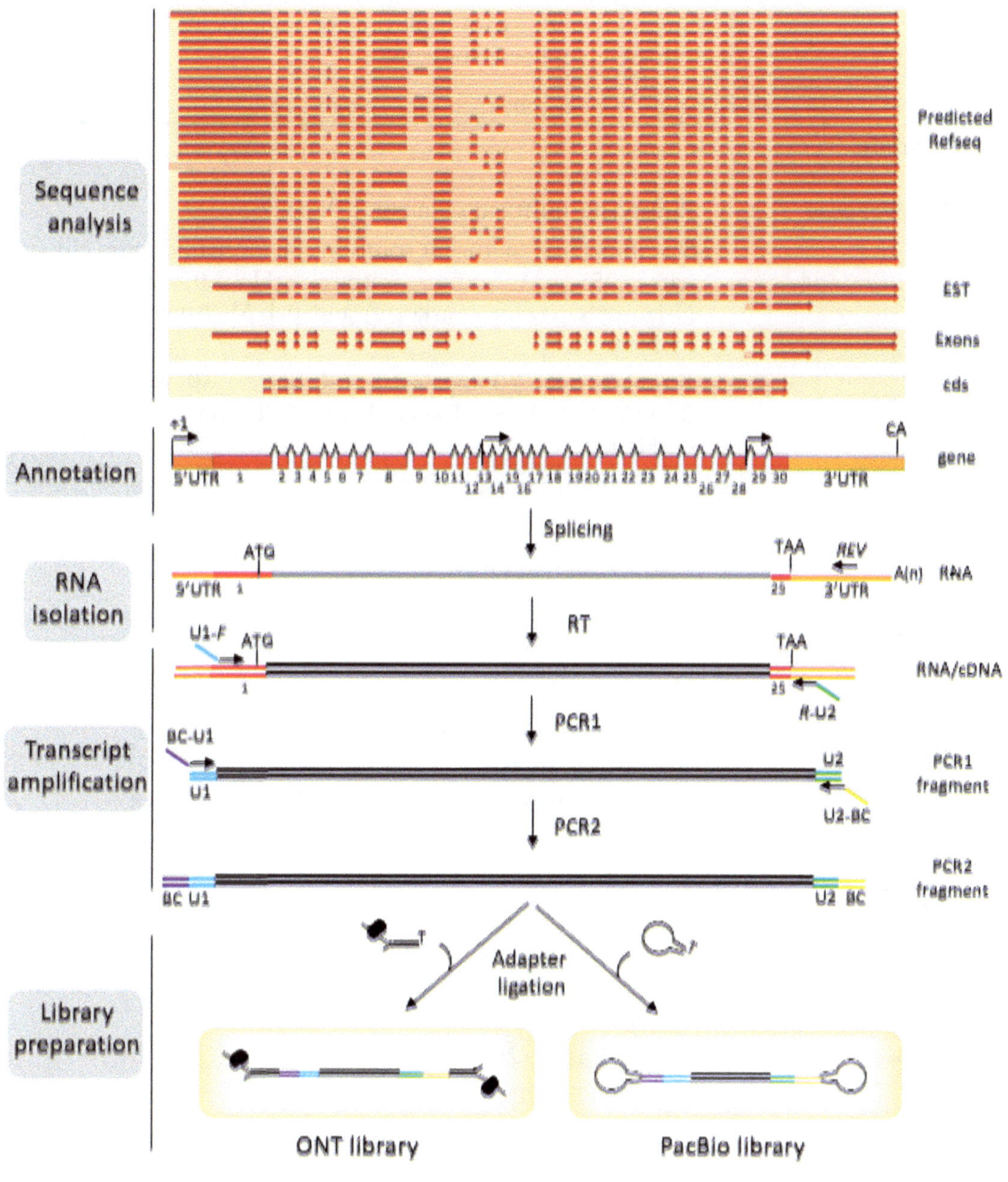

Fig. 1 Library preparation guideline

Second, *GPHN* mRNAs are amplified by RT-PCR with primers targeting the first and last exons. Third, each sample is barcoded with a unique identifier. Fourth, PCR samples are stoichiometrically mixed to build a unique multiplexed sequencing library. A general schema is outlined in Fig. 1 to report the library preparation guideline. Fifth, the library is sequenced using TGS, and we describe the principle of each technology. Sixth, bioinformatic pipeline allows the demultiplexing of samples and the clustering of alternatively spliced isoforms.

2 Materials

2.1 RNA Extraction

1. Phosphate-buffered saline (1× PBS).
2. Solution D (Denaturing solution): 4 M guanidinium thiocyanate, 25 mM sodium citrate, pH 7.0, 0.5% (w/vol) N-lauroylsarcosine (Sarkosyl) and 0.1 M 2-mercaptoethanol; *see* **Note 1**.
3. Proteinase K (20 mg/mL), PK buffer: 30 mM Tris–HCl; 30 mM EDTA; 5% Tween-20; 0.5% Triton X-100; 800 mM GuHCl.
4. 3 M sodium acetate, pH 4.0.
5. Chloroform/isoamyl alcohol (49:1).
6. Water-saturated phenol.
7. Isopropanol.
8. Ethanol.
9. Dounce homogenizer.
10. DNase I recombinant, RNase-free (10 U/μL).
11. RNase free water.
12. TE buffer: 1 mM EDTA, 10 mM Tris–HCl, pH 8.0.
13. Nanodrop or Qubit fluorometer.

2.2 Reverse Transcription and PCR

1. SuperScript III RT (200 U/μL; ThermoFisher Scientific).
2. dNTPs (10 mM and 25 mM),
3. RNaseOUT Recombinant RNase Inhibitor (40 U/μL; ThermoFisher scientific).
4. Exonuclease I (20 U/μL).
5. High-fidelity DNA Polymerase.
6. Forward primer coupled to universal U1 sequence (U1-*F* primer; 10 μM): 5′AmMC6- GCAGTCGAACATGTAGCT GACTCAGGTCAC-*F*-3′.
7. Reverse primer coupled to universal U2 sequence (*R*-U2 primer; 10 μM): 5′AmMC6-*R*- TGGATCACTTGTGCAAG CATCACATCGTAG3′.
8. NGS cleanup and size selection (Macherey-Nagel).
9. Magnetic Separation Rack.
10. Thermal cycler.

2.3 Barcode Tagging and Library Preparation

1. Barcoded Universal F/R Primers (Pacific Bioscience).
2. ONT: Ligation Sequencing Kit (Oxford Nanopore Technologies).
3. PacBio: SMRTbell Express Template Prep Kit 2.0 (PacBio).

2.4 Sequencing

1. ONT: MinION Mk1C or MinION connected to a dedicated computer.
2. PacBio: Sequel II System.

2.5 Bioinformatic Program Repositories

1. Nanopipe 2: https://github.com/IOB-Muenster/nanopipe2
2. NCBI Blast+: https://ftp.ncbi.nlm.nih.gov/blast/executables/blast+/LATEST/
3. Snapgene: https://www.snapgene.com/
4. STARlong: https://github.com/alexdobin/STAR
5. Bedtools: https://github.com/arq5x/bedtools2

3 Methods

3.1 Total RNA Isolation

Our standard protocol to isolate total RNA is a modified version of the single-step method from Chomczynski and Sacchi [14]. This procedure is suitable for a large panel of samples: tissues, cultured cells, plants, yeasts, and bacteria. This method does not use ion exchange columns and allows quantitative isolation of RNAs independently of their length. In addition, proteins corresponding to RNA samples can be analyzed by Western blots (*see* **Note 2**). After homogenization in solution D, samples can be stored at −80 °C for years giving an opportunity to perform RNA isolation later. We describe here the procedure for approximately 1.2 × 10^6 cells which are cultivated in a dish of 35 mm. To isolate RNAs from tissue, the volume of solution D should be adjusted to 1 mL per 100 mg and other solutions should be adjusted proportionally (*see* **Note 1**).

1. Discard the culture medium and add directly in the dish 0.4 mL of freshly prepared solution D, resuspend the cells by pipetting (tissues and rigid specimens can be homogenized with a Dounce or a glass-Teflon homogenizer).
2. To extract RNA add to lysate 3 M sodium acetate to 10% (v/v), mix and add 1 volume of phenol and 0.2 volume of chloroform/isoamyl alcohol. Shake the tubes, incubate 15 min on ice then centrifuge at 10,000 × *g* for 20 min at 4 °C.
3. Transfer the aqueous phase in a new tube without disturbing the interphase, add 1 volume of isopropanol, shake vigorously and incubate for 1 h at −20 °C. Centrifuge at 10,000 × *g* for 20 min at 4 °C, put tubes on ice, and discard the supernatant,

repeat this centrifugation for 1 min to remove the traces of supernatant.

4. To digest the residual proteins, resuspend the pellet in 50 μL of PK buffer supplemented with 2 μg of proteinase K, incubate during 15 min at 50 °C.
5. To extract RNA add 150 μL of RNase free water to the 50 μL solution from (4), 10% (v/v) of 3 M sodium acetate, 1 volume of phenol, 0.2 volume of chloroform/isoamyl alcohol. Shake the tubes and centrifuge at 10,000 × *g* during 20 min at 4 °C.
6. Transfer the aqueous phase in a new tube and precipitate the RNA by adding 3 volumes of ethanol, shake vigorously and incubate for 1 h at −20 °C. Centrifuge at 10,000 × *g* for 20 min at 4 °C, put tubes on ice, discard the supernatant, wash the pellet with 800 μL of 80% ethanol, repeat the centrifugation for 1 min, discard the ethanol and centrifuge again for 1 min to remove the traces of ethanol.
7. To digest the residual DNA resuspend the pellet in 50 μL of DNAse I digestion buffer supplemented with 10 U of DNase I and incubate 15 min at 37 °C.
8. Repeat **steps 5** and **6**.
9. Dissolve the RNA pellet in 30 μL of TE, measure the concentration by Nanodrop or Qubit fluorometer. Analyze the quality of RNAs by running agarose gel or in the Bioanalyzer system (Agilent). High quality total RNAs give 28S rRNA, 18S rRNA (with a ratio 2:1) and 5S rRNA.

3.2 Amplification of Specific Transcripts

We describe here the RT-PCR strategy applied to enrich *GPHN* transcripts from total RNAs extracted at different developmental stages (see schema in Fig. 1; RT and PCR1 steps). Note that each sample is processed in parallel until the preparation of the sequencing library (*see* Subheading 3.3, **step 2**). The primers are designed after an analysis of available RNA sequences, which are aligned to the gene locus to determine the first and the last exon (*see* **Note 4**). The reverse transcription (RT) reaction is performed with a specific reverse primer (*REV*) to reduce the complexity of the cDNAs mixture and improve the quality of PCR1 amplification. Note that oligo(dT) primers can also be used for RT in contrast to random primers that are not adapted (*see* **Notes 5** and **6**). Second, PCR1 reactions are performed with primers fused to universal sequences (*F*-U1 and *R*-U2) to amplify sequences interposed between the first and last exons (*see* **Note** 7). Note that each PCR1 amplicon is finally flanked with U1/U2 sequences. Moreover, a low number of cycles is used during the amplification to avoid template switching that generates artificial sequences [10]. A description of alternative procedure to capture and enrich specific transcripts using exon probes affinity is described in Ray et al. ([9]; *see* **Note 3**).

1. RT reactions are assembled in a final volume of 20 μL and processed in parallel for each sample. Mix 1 μg of total RNA, 2 pmol of specific reverse primer (*REV*), 1 μL of 10 mM dNTP mix, supplement up to 13 μL with RNAse free water and heat the mixture to 65 °C for 5 min. Spin and keep the tube on ice. Add 4 μL of 5× First-Strand Buffer, 1 μL of DTT, 1 μL RNaseOUT, and 1 μL of SuperScriptTM III RT. Incubate the tube at 55 °C for 60 min. Inactivate the reaction by heating at 70 °C for 15 min (*see* **Note 7**).
2. Perform the PCR1 amplification to enrich the sequences between the first and last exons. Dilute the RT reaction 5 folds with RNAse free water (for easy pipetting). The PCR1 mix is prepared at room temperature in a final volume of 50 μL. Add 2 to 5 μL of diluted RT reaction, 5 μL of 10× reaction buffer, 0.5 μL of 25 mM dNTP mix, 1 μL of each primer (U1-*F* and *R*-U2; *see* **Note 4**), 1 μL of high fidelity DNA polymerase (2.5 U) and set the thermocycler for 15 to 18 cycles of amplification. To assess the efficiency of PCR1, process an identical reaction during 38 cycles of amplification and analyze the PCR product (s) by separation in agarose gel (*see* Fig. 2).

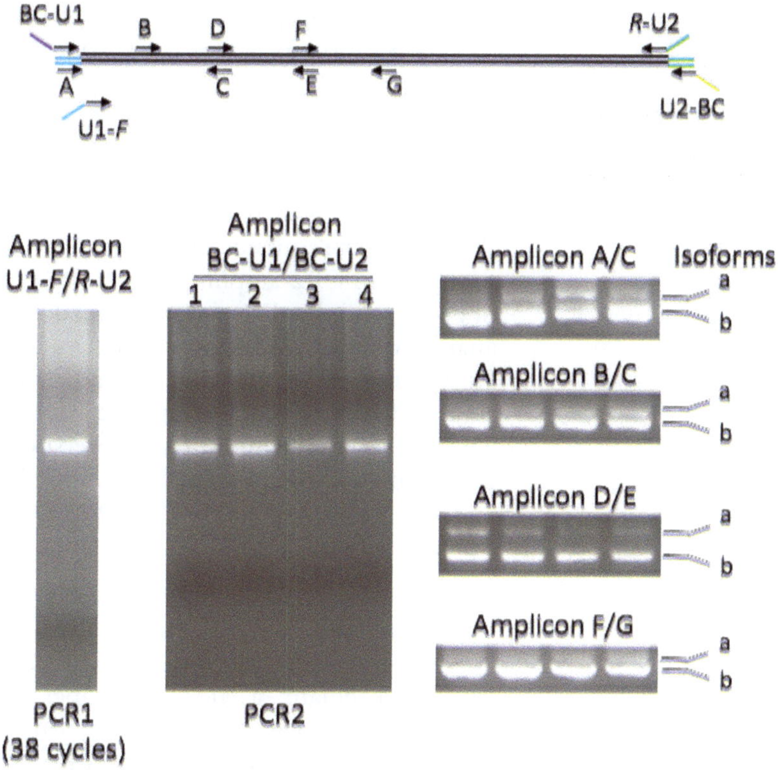

Fig. 2 PCR1 and PCR2 amplification and checking of library diversity

3. Digest the excess of primers in PCR1 by adding 1 μL of Exonuclease I (20 U) directly to the PCR tube. Mix by pipetting and incubate at 37 °C for 15 min, then inactivate the enzyme by incubation at 80 °C for 15 min.
4. Purify PCR1 reactions at room temperature with the NGS cleanup and size selection kit. The PCR1 DNA fragments are captured and cleaned using magnetic beads. Homogenize the bead solution in a shaker for approximately 30 min at room temperature before the beginning of the procedure. Mix by pipetting each PCR1 reaction with an equal volume of bead solution (50 μL). Then, leave the mix for 5 min at room temperature to allow binding of the PCR1 fragments. Concentrate the beads on the magnetic separation rack until the solution is clear, and discard the supernatant by pipetting to avoid disturbing the beads (tubes will stay on magnet until the elution step). Add 200 μL of 80% ethanol, wash by slow pipetting with the tip at the bottom of the tube to avoid disturbance of beads, discard ethanol and repeat the wash one more time with 80% ethanol. Discard the second wash, dry the beads at room temperature for 5 min, remove tubes from magnet, resuspend the beads in 20 μL of the elution buffer and leave for 5 min at room temperature. Put the tubes on the magnet rack for 5 min and transfer 19 μL of clear PCR1 eluate to a new tube. At the end of this step, the PCR1 samples can be proceed to Subheading 3.3 or stored at −20 °C for several months.

3.3 Barcoding and Preparation of a Multiplexed TGS Library

In this procedure, specific barcodes are symmetrically incorporated to distinct PCR1s, all samples are stoichiometrically mixed before the ligation of sequencing adapters. First, individual PCR1 amplicons are processed in a PCR2 reaction to tag their extremities with the identifier sequence (BC-U1 and U2-BC; *see* **Note 8**). As for PCR1, the PCR2 amplifications are performed with a low number of cycles, and the diversity of amplicons can be checked later by the analysis of known splicing events (*see* **Note 9**; Fig. 2). Second, the PCR2 samples are mixed at 1:1 molecular weight ratio to ensure a similar depth of sequencing (we have tested up to 22 samples). Third, the sequencing library is prepared by ligation of TGS adapters to the mixture of amplicons. Note that the principles of PacBio and ONT sequencing are different and therefore library preparation also varies.

1. Add a specific barcode to each PCR1 amplicon. Assemble individual PCR2 reactions like in Subheading 3.2, **step 2** using barcoded primers (BC-U1 and U2-BC), 2 μL of PCR1 eluate, a melting temperature of 61 °C and 20 to 24 PCR cycles. After amplification, treat with the Exonuclease I and purify the amplicons with the NGS cleanup and size selection kit like in Subheading 3.2, **steps 3** and **4**. Measure

the concentration of PCR2 amplicons by Nanodrop or Qubit fluorometer and analyze 50–200 ng of PCR products on agarose gel (*see* Fig. 2). Repeat the PCR2 amplification in case the amount of DNA is too low to proceed to **step 2** in order to reach enough PCR2 to prepare the library as recommended by TGS library kits.

2. Mix together the PCR2 amplicons in a ratio 1:1. Prepare the sequencing library by ligation of adapters following the instructions of the PacBio or ONT ligation sequencing kits. The multiplexed PCR2 amplicon is treated to repair DNA damage, end-repair and adding of A-tailing, then the sequencing adapters are ligated (*see* **Note 10**)

3.4 PacBio and ONT Sequencing Principles

Third-generation sequencing is all about DNA read length. Unlike NGS, where DNA is broken down into short pieces, amplified and then sequenced, TGS does not break down the DNA: it directly sequences a single DNA molecule. The two main TGS devices, Sequel system and MinION, were released respectively by Pacific Biosciences and Oxford Nanopore in 2015.

3.4.1 PacBio (Left Panel Fig. 3)

The Sequel system uses flow cells made of thousands of tiny wells in which each one has a polymerase molecule attached inside that will sequence a unique DNA molecule. Polymerase replicates a single-stranded piece of DNA and sequencing relies on the single-molecule real-time detection of fluorescence issued by the incorporation of nucleotides during the DNA replication. DNA polymerase immobilized at the bottom of optical nanostructures (ZMWs) binds the hairpin adapter present at both extremities of circular DNA and replicates one strand of the target dsDNA. If the lifetime of the polymerase is long enough, both strands can be sequenced multiple times making the concatenation of several copies separated by adapters (CLR). PacBio software recognizes the adapters, and cuts the CLR, and aligns together the multicopies of sequences to generate a unique circular consensus sequence (CCS) that has a higher accuracy. The characteristics of CLR that you can expect to obtain with the PacBio sequel sequencing system are the following: number of reads between $5x10^5$ and $4x10^6$, length of CLR up to 30 kb, and CCS accuracy of >99.999% (https://www.pacb.com/wp-content/uploads/2015/09/Pacific-Biosciences-Glossary-of-Terms.pdf).

3.4.2 ONT (Right Panel Fig. 3)

The MinION system is slightly bigger than an USB stick and uses a flow cell made with an electrical resistant membrane. This membrane has thousands of pores, each with a diameter of one nanometer (nanopore). Steady current is applied to the nanopore and sequencing relies on the changes in electrical conductivity generated when nucleic acid passes through nanopore. The sequence is

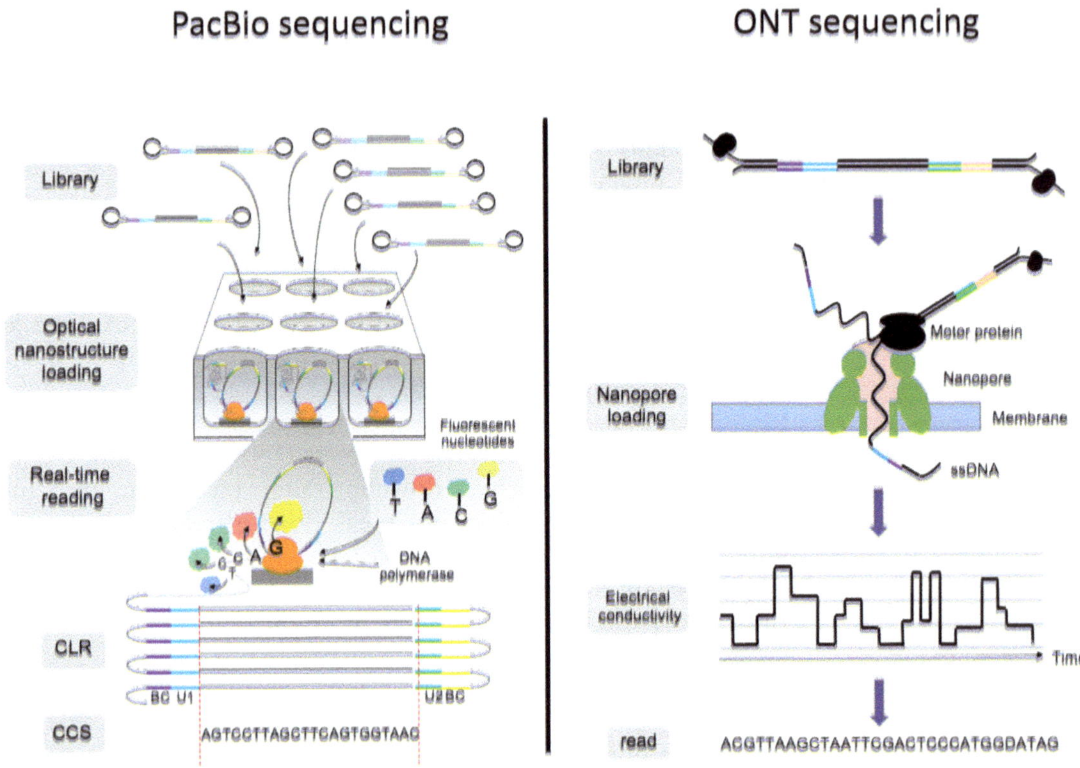

Fig. 3 Principle of PacBio and ONT sequencing

determined by matching the voltage variations to the nucleotides passing through the nanopores. A motor protein attached to the sequencing adapter recruits nucleic acid to the nanopore and unzips the dsDNA. A measure of voltage at individual nanopores is streamed online. Thus, the length of sequencing is correlated to time with the possibility to acquire data during 72 h. Distinct nanopore flow cells include various numbers of pores that influence the depth of sequencing. Currently, the MinION flow cell is the most popular with 512 channels, each one connected to 4 nanopores. The current record for longest fragment sequenced is 4.2 Mb with a read accuracy of 99.1% (https://nanoporetech.com/about-us/news/ncm-announcements-include-single-read-accuracy-991-new-chemistry-and-sequencing).

An important advantage of PacBio is the accuracy of CCS which is currently better than ONT reads, however the sequel system is an expensive investment. In contrast, the MinION uses a minimal space main advantage of ONT is the inexpensive starting package and the minimal space used by the MinION, in addition it provides a higher and longer number of reads on average than PacBio.

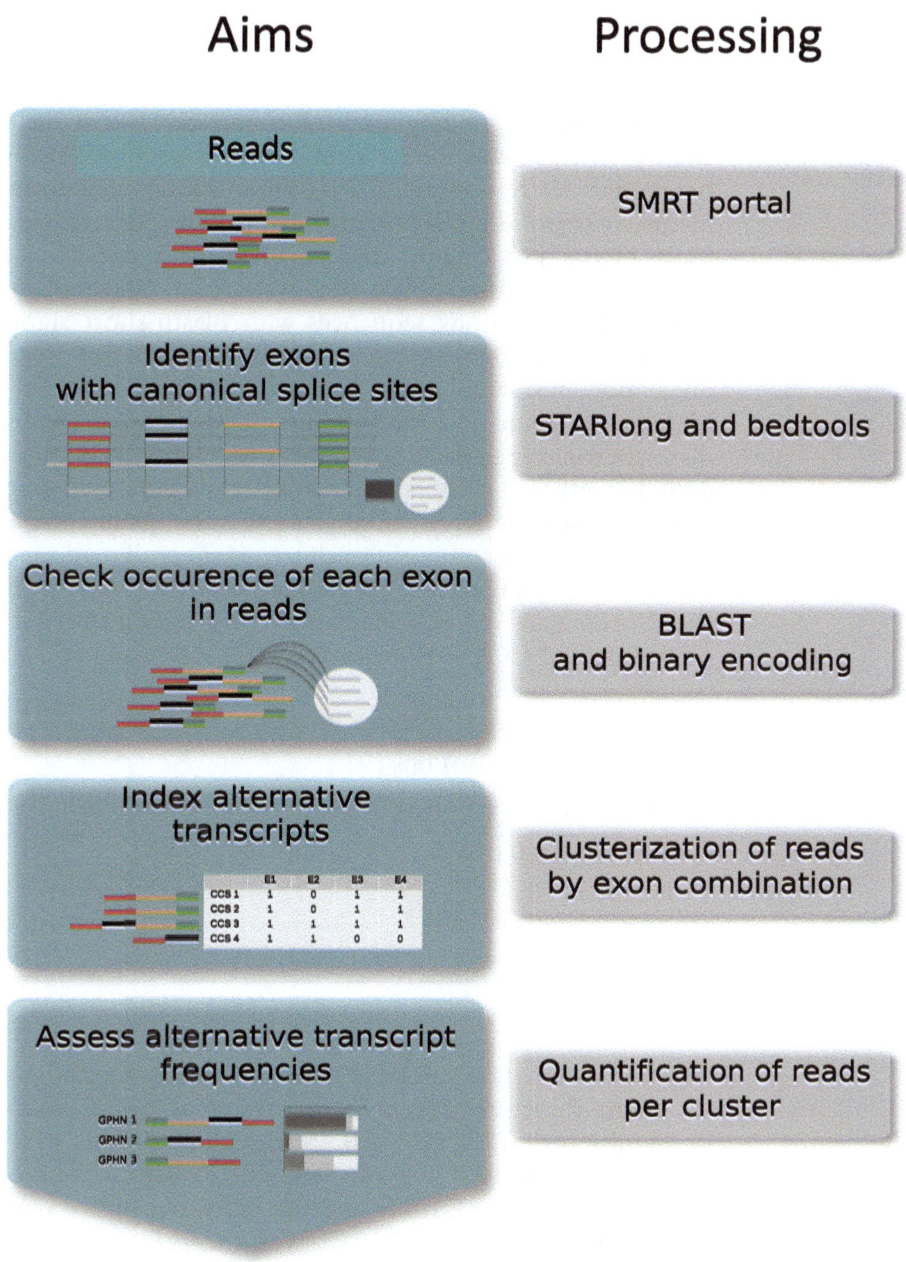

Fig. 4 Bioinformatic pipeline to clusterize splice isoforms

3.5 Demultiplexing and Clustering of Alternative Isoforms

In our experience, software provided by Pacific Bioscience or Oxford Nanopore community did not give an homogeneous clustering of identical sequences. Thus, we describe here our home-made pipeline (Fig. 4), and a list of several other pipelines recently published.

3.5.1 Demultiplexing

We have used Porechop to isolate sequences corresponding to each sample [15].

3.5.2 Clustering of Identical Sequences

Fastq files corresponding to PacBio and ONT sequencing were aligned to the *Gphn* mouse genome locus using STARlong [16] or alternatively Nanopipe 2 [17]. Genomic intervals showing alignments corresponding to potential exons were extracted using bedtools merge [18]. Only those framed with canonical GT-AG dinucleotides at their splice sites were kept. Retained exons are further aligned to raw sequences with Blastn-short algorithm of Blastn [19](using an e-value threshold of 0.001 and a penalty for nucleotide mismatch of −2) to annotate each sequence with a binary code 1 and 0 corresponding to the presence or absence of distinct exon respectively. Sequences with identical annotation were pulled together in clusters and theoretical sequences corresponding to each cluster were generated. To assess the quality of clustering, the raw sequences composing a cluster are next aligned against each theoretical sequence (using a gapped global alignment, similar to the Needleman–Wunsch algorithm with Exonerate 2.4.0) and poorly aligned CCS were removed [20].

3.5.3 Other Pipelines Which Can Be Used for the Analysis

- isONclust: https://github.com/ksahlin/isONclust [11]
- ToFU, Quiver, IsoPops [9]
- Mandalorion version II, FLAIR, minimap2, Pinfish [12]
- Homemade pipeline to study CACNA1C [8]
- UNAGI [13]

4 Notes

1. A stock solution can be prepared by dissolving 250 g guanidinium thiocyanate in 293 mL water at 65 °C followed by addition of 17.6 mL of 0.75 M sodium citrate, pH 7.0 and 26.4 mL of 10% (w/vol) Sarkosyl. The stock solution is aliquoted in 50 mL tubes and can be stored up to 3 months at room temperature and several years at −20 °C. To prepare the working solution D, add 0.36 mL of 98% 2-mercaptoethanol to 50 mL of stock solution.
2. Proteins present in lysate can be prepared for Western blot analysis as follows: dilute 80 μL of solution D lysate with 120 μL H_2O, supplement with 30 μg of glycogen, add 800 μL of acetone, shake vigorously and store at −80 °C for at least 20 min. Centrifuge at 10,000 × *g* for 20 min at 4 °C, place tubes on ice, remove supernatant, repeat the centrifugation for 1 min to remove traces of acetone and resuspend the pellet in 60 μL of buffer containing 100 mM Tris–HCl,

pH 6.8, 2.5% SDS, 100 mM NaCl, 1% Triton X-100. For the resuspension of big pellets additional vortexing, heating at 85 °C, and sonication of samples might be required.

3. The RNA capture approach allows an enrichment of RNA independently of the transcript annotation, and therefore, can lead to the identification of new untranslated regions (UTRs). However, this procedure includes a SMART (Switching Mechanism at 5′ End of RNA Template) step during the reverse transcription that can generate artificial 5′ UTRs. In addition, the lengths of splice variants and the exon architecture are two parameters that can influence the capture and the efficiency of reverse transcription.

4. To determine the first and last exon of transcripts, we align the RNA sequences available in the public database with the gene locus. Annotation of these exons can change depending on the activation of alternative promoters and polyadenylation sites in specific tissues or cells. The sequences *F* and *R* used in the U1-*F* and *R*-U2 primers target the first and last annotated exon (*see* Fig. 1). The primer design is established using Primer-BLAST (https://www.ncbi.nlm.nih.gov/tools/primer-blast/) by retaining those with a score of 0 for the "thermodynamic oligo alignment" and "thermodynamic template alignment" options in the advanced parameters.

5. The 3′UTR of transcripts often includes multiple polyadenylation sites (CAs) that can remove the sequence targeted by the primer *REV*, therefore this primer has to be chosen upstream to the first CA present in the last exon. To identify the CAs in a sequence use the database: https://polyasite.unibas.ch/search.

6. Reverse transcriptions performed at elevated temperatures (50–55 °C) can improve the cDNA synthesis by release of RNA structures that block the reverse transcriptase. The primer *REV* is therefore designed with a melting temperature around 60 °C.

7. To improve the amplification of PCR products above 1 kb, the reactions of reverse transcription can be treated for 20 min with 2 U of RNase H at 37 °C. The RNase H cleaves RNA in RNA–DNA duplexes. Note that the 5′-end of the *F*-U1 and *R*-U2 primer is modified (e.g., 5′AmMC6) to prevent the undesirable ligation of PCR1 amplicons during library construction. Failure to block the 5′-end reduces yield of barcoded reads.

8. The kit of barcoded universal F/R primers proposed by PacBio is suitable for preparation of samples for ONT. Each barcode sequence contains 16 bases that allows sufficient number of combinations for discrimination during sequencing analysis despite the error rate of TGS technologies. Note that the primers (BC-U1 and U2-BC) are aliquoted by pairs allowing

only symmetric tagging of PCR2 with the same identifier. The synthesis of BC-U1 and U2-BC is a cost-effective alternative for their asymmetric incorporation in order to increase the multiplexing possibilities. Barcode sequences are available at the PacBio website.

9. The two steps of amplification might affect the initial diversity present in the expression of the target gene. Analysis of PCR2 amplicons can be performed by using internal primers targeting exons upstream and downstream of the alternative splicing regulation (*see* Fig. 2).
10. The ONT ligation kit generates a DNA fragment with open ends while the PacBio ligation kit creates loops at the DNA ends (*see* Fig. 1). These structures are associated with the sequencing principle of each technology (*see* Figs. 1 and 3).

Acknowledgments

We thank Svetlana Dokudovskaya for comments on the manuscript.

References

1. Berget SM, Moore C, Sharp PA (1977) Spliced segments at the 5' terminus of adenovirus 2 late MRNA. Proc Natl Acad Sci U S A 74: 3171–3175
2. Chow LT, Gelinas RE, Broker TR, Roberts RJ (1977) An amazing sequence arrangement at the 5' ends of adenovirus 2 messenger RNA. Cell 12:1–8
3. Sanger F, Nicklen S, Coulson AR (1977) DNA sequencing with chain-terminating inhibitors. Proc Natl Acad Sci U S A 74:5463–5467
4. Wang ET, Sandberg R, Luo S, Khrebtukova I, Zhang L, Mayr C, Kingsmore SF, Schroth GP, Burge CB (2008) Alternative isoform regulation in human tissue transcriptomes. Nature 456:470–476
5. Sun W, You X, Gogol-Döring A, He H, Kise Y, Sohn M, Chen T, Klebes A, Schmucker D, Chen W (2013) Ultra-deep profiling of alternatively spliced Drosophila Dscam isoforms by circularization-assisted multi-segment sequencing. EMBO J 32:2029–2038
6. Treutlein B, Gokce O, Quake SR, Südhof TC (2014) Cartography of neurexin alternative splicing mapped by single-molecule long-read MRNA sequencing. Proc Natl Acad Sci U S A 111:E1291–E1299
7. Schreiner D, Nguyen T-M, Russo G, Heber S, Patrignani A, Ahrné E, Scheiffele P (2014) Targeted combinatorial alternative splicing generates brain region-specific repertoires of neurexins. Neuron 84:386–398
8. Clark MB, Wrzesinski T, Garcia AB, Hall NAL, Kleinman JE, Hyde T, Weinberger DR, Harrison PJ, Haerty W, Tunbridge EM (2020) Long-read sequencing reveals the complex splicing profile of the psychiatric risk gene CACNA1C in human brain. Mol Psychiatry 25:37–47
9. Ray TA, Cochran K, Kozlowski C, Wang J, Alexander G, Cady MA, Spencer WJ, Ruzycki PA, Clark BS, Laeremans A et al (2020) Comprehensive identification of MRNA isoforms reveals the diversity of neural cell-surface molecules with roles in retinal development and disease. Nat Commun 11:3328
10. Bolisetty MT, Rajadinakaran G, Graveley BR (2015) Determining exon connectivity in complex MRNAs by nanopore sequencing. Genome Biol 16:204
11. Sahlin K, Medvedev P (2020) De novo clustering of long-read transcriptome data using a greedy, quality value-based algorithm. J Comput Biol 27:472–484
12. Tang AD, Soulette CM, van Baren MJ, Hart K, Hrabeta-Robinson E, Wu CJ, Brooks AN (2020) Full-length transcript characterization of SF3B1 mutation in chronic lymphocytic leukemia reveals downregulation of retained introns. Nat Commun 11:1438

13. Al Kadi M, Jung N, Ito S, Kameoka S, Hishida T, Motooka D, Nakamura S, Iida T, Okuzaki D (2020) UNAGI: an automated pipeline for nanopore full-length CDNA sequencing uncovers novel transcripts and isoforms in yeast. Funct Integr Genomics 20: 523–536
14. Chomczynski P, Sacchi N (2006) The single-step method of RNA isolation by acid guanidinium thiocyanate-phenol-chloroform extraction: twenty-something years on. Nat Protoc 1:581–585
15. Wick RR, Judd LM, Holt KE (2018) Deepbinner: demultiplexing barcoded Oxford nanopore reads with deep convolutional neural networks. PLoS Comput Biol 14:e1006583
16. Dobin A, Davis CA, Schlesinger F, Drenkow J, Zaleski C, Jha S, Batut P, Chaisson M, Gingeras TR (2013) STAR: ultrafast universal RNA-Seq aligner. Bioinforma Oxf Engl 29:15–21
17. Shabardina V, Kischka T, Manske F, Grundmann N, Frith MC, Suzuki Y, Makałowski W (2019) NanoPipe-a web server for nanopore MinION sequencing data analysis. GigaScience 8:giy169
18. Quinlan AR, Hall IM (2010) BEDTools: a flexible suite of utilities for comparing genomic features. Bioinforma Oxf Engl 26:841–842
19. Camacho C, Coulouris G, Avagyan V, Ma N, Papadopoulos J, Bealer K, Madden TL (2009) BLAST+: architecture and applications. BMC Bioinformatics 10:421
20. Slater GSC, Birney E (2005) Automated generation of heuristics for biological sequence comparison. BMC Bioinformatics 6:31

Chapter 7

Computational Analysis of Alternative Splicing Using *VAST-TOOLS* and the *VastDB* Framework

André Gohr, Federica Mantica, Antonio Hermoso-Pulido, Javier Tapial, Yamile Márquez, and Manuel Irimia

Abstract

Alternative splicing (AS) can vastly expand animal transcriptomes and proteomes. Two main open questions in the field are how AS is regulated across cell/tissue types and disease, and what roles different AS events play. To facilitate AS research, we have created the computational *VastDB* framework, which comprises a series of complementary software and resources that we describe in this chapter. The *VastDB* framework is especially designed to aid biomedical researchers without a strong computational background. It offers tools and resources to: (a) quantify AS and identify differentially spliced AS events using RNA-seq data (*vast-tools*), (b) perform multiple genomic and sequence analyses for investigating AS events (*Matt*), (c) identify AS events with genomic and regulatory conservation among species (*ExOrthist*), and (d) help with the biological interpretation of the results, and, ultimately, with the identification of interesting AS events to design wet-lab experiments (*VastDB* and *PastDB*).

Key words Alternative splicing, Bioinformatics, Cassette exons, Intron retention, Orthology, Database

1 Introduction to Alternative Splicing and the *VastDB* Framework

Alternative splicing (AS) is the differential processing of introns and exons in eukaryotic genes to generate multiple transcript isoforms per gene. Potentially, these isoforms can exert two major functions at the molecular level. Firstly, they may produce alternative protein isoforms, when both the inclusion and exclusion of the alternative sequence maintains the open reading frame. Secondly, one of the isoforms may disrupt the open reading frame, leading to mRNA degradation, nonsense mediated decay and/or protein truncation, thus effectively causing downregulation of gene activity. On the other hand, transcript diversity by splicing may not have any

André Gohr and Federica Mantica contributed equally to this work.

Peter Scheiffele and Oriane Mauger (eds.), *Alternative Splicing: Methods and Protocols*,
Methods in Molecular Biology, vol. 2537, https://doi.org/10.1007/978-1-0716-2521-7_7,

function, being the result of suboptimal splicing or missplicing (i.e., errors during the process of intron removal). The latter is particularly common when spliceosomal components are mutated or experimentally depleted.

Two of the major challenges of AS research are elucidating how it is regulated and what roles it plays across different physiological and disease conditions. The advent and generalization of high throughput RNA sequencing (RNA-seq) technologies revolutionized the study of AS over the past decade. Transcriptome-wide AS analyses have revealed that up to 95% of multiexonic genes in humans show splicing variation [1, 2], and that nearly half of them generate multiple splice variants with distinct expression patterns across cell and tissue types [3]. Furthermore, many splice isoforms are highly cell and tissue-specific, particularly to nervous system and muscle [3–5]. In combination with gene depletion technologies, it has become possible to identify important AS regulators as well as their networks of exon targets. These AS networks are usually coexpressed in specific cell and tissue types and impart their function in a coordinated and synergistic manner [6–10]. Although it is still unclear how many AS variants are functional [11–13], hundreds of single-case studies have proven the functionality of many AS events in different cellular contexts [14–18]. These functions range from changes in molecular properties of mRNAs and proteins (subcellular location, stability, DNA/RNA/protein binding specificity and affinity, etc.) to the modulation of nearly every possible cellular process (proliferation, apoptosis, migration, adhesion, polarity, membrane depolarization, etc.).

To facilitate these lines of AS research, we have developed different publicly available tools and resources that together constitute the *VastDB* framework (Fig. 1 and Table 1). The explicit goal of this framework, as detailed throughout this chapter, is to provide resources to biomedical researchers without a strong computational background to: (a) quantify AS and identify differentially spliced AS events using RNA-seq data (*vast-tools*), (b) perform multiple genomic and sequence analyses for sets of AS events (*Matt*), (c) identify AS events with genomic and regulatory conservation among species (*ExOrthist*), and (d) help with the biological interpretation of the results, and, ultimately, with the identification of interesting AS events to design wet-lab experiments (*VastDB* and *PastDB*).

2 *vast-tools*: Quantification and Differential Alternative Splicing Analyses

2.1 Introduction

vast-tools is a set of computational tools to identify and quantify AS events of four major types: (a) exon skipping (including microexons) (EX), (b) intron retention (INT), (c) alternative donor choices (ALTD), and (d) alternative acceptor choices (ALTA). *vast-tools* has

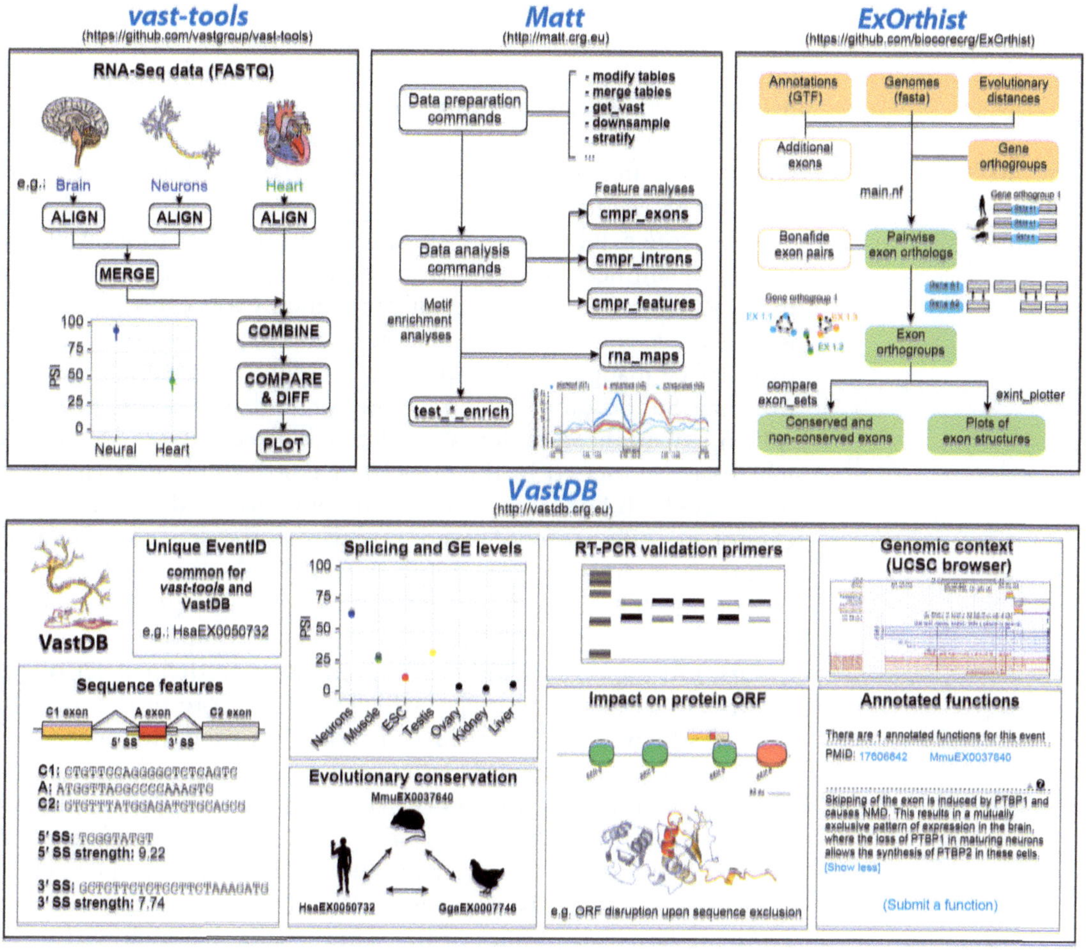

Fig. 1 *The VastDB framework.* Schematic representation of the different modules and commands of vast-tools, Matt and ExOrthist, as well as the different layers of information covered in VastDB and PastDB

Table 1
Summary of software and resources discussed in this chapter

Name	Brief description	Public URL
vast-tools	AS quantification and differential AS analysis	github.com/vastgroup/vast-tools
Matt	Sequence analysis of AS events	gitlab.com/aghr/matt
ExOrthist	Exon homology inference	github.com/biocorecrg/ExOrthist
VastDB	Information about genes and AS events in animal species	vastdb.crg.eu
PastDB	Information about genes and AS events in *A. thaliana*	pastdb.crg.eu

already been applied to study AS in multiple vertebrate and non-vertebrate animals, as well as other eukaryotes with high validation rates [3, 19–27]. *vast-tools* comprises several modules, meant to be used consecutively, to quantify AS inclusion levels (*align*), combine these levels into individual tables (*combine* and *tidy*), perform differential splicing analyses (*compare* and *diff*), and plot the inclusion levels of AS events of interest (*plot*).

vast-tools exhibits two main differences to other tools available for the quantification of AS from RNA-seq data. First, it directly maps RNA-seq reads to sets of predefined exon-exon and exon-intron junctions (EEJs); this allows *vast-tools* to be independent from splice mappers and to perform a unique mappability correction (*see* Subheading 2.2). Second, for each species, it provides splicing quantification for a predefined, fixed set of AS events, which are annotated in the associated *vast-tools* libraries (referred to as VASTDB libraries). This has the obvious disadvantage that it cannot identify novel AS events; however, the sets of annotated events are built using large numbers of RNA-seq and EST datasets from many cell/tissue types and developmental stages per species, ensuring that *vast-tools* is capable of quantifying most relevant AS events. On the other hand, a comprehensive predefined set of AS events has a major advantage, which is at the core of the *VastDB* framework: every AS event in *vast-tools* has a fixed identifier (VastID) and is quantified in the exact same way (i.e., probing the same set of junctions) for any processed RNA-seq sample. Similar to Ensembl IDs, VastIDs (e.g., HsaEX0070258) contain information about the species (Hsa = human) and the type of event (EX, exon skipping). Importantly, these fixed VastIDs, which are stable and constant across assemblies of the same species, have allowed us to build the database *VastDB* (*see* Subheading 5).

In the following subsections, we discuss each *vast-tools* module. A detailed description of each of them, as well as of input/output formats, is provided in the README file on GitHub (https://github.com/vastgroup/vast-tools) and in the original publications [3, 19].

2.2 vast-tools align

This module processes RNA-seq data in FASTQ or FASTA format to generate intermediate output files with read count information, which will be then used by *vast-tools combine* to generate the final output tables (*see* Subheading 2.4). As mentioned above, *vast-tools align* maps RNA-seq reads to a predefined set of EEJs through different submodules (splice site-based, transcript-based, microexon, and intron retention) [3]. To do so, *vast-tools align* first trims reads into 50 nucleotide (nt) fragments in a 25-nt sliding window, and these fragments are then aligned to the EEJ libraries using *bowtie* [28]. Only uniquely mapping fragments are then considered for the splicing quantification. Consequently, *vast-tools* does not rely on splice mappers (*STAR*, *hisat2*, etc.), nor can it accept BAM files as inputs, which is a limitation. Another limitation of this approach is that, given the latest developments of splice mappers and, even more so, with the advent of k-mer-based quantification tools, *vast-tools* is slower than most currently used software (e.g., *rMATS* [29], *MAJIQ* [30], *SUPPA2* [31], *Whippet* [32]). However, this issue can be minimized by parallelizing *vast-tools align* in multiple threads, which means that any RNA-seq sample can be analyzed in a few hours. Moreover, as mentioned above, the *VastDB* framework is to a large extent tailored to wet-lab researchers; if it takes several months to obtain the right biological material for RNA-seq, a few hours to process the data is not a limiting step. Nevertheless, the use of *vast-tools* for computational analyses involving large datasets, such as those from the GTEx or TCGA projects, is less recommended.

On the other hand, the unique mapping approach of *vast-tools* has some clear advantages. Since it does not rely on splice mappers, AS events are always quantified based on the same set of EEJs and using only 50-nt fragments, which makes inclusion level estimates more consistent across samples. Furthermore, since the EEJs are known and predefined, *vast-tools* minimizes quantification biases by applying a mappability correction: positions in each EEJ to which 50-nt read-fragments cannot be mapped uniquely (given the genome-wide set of all EEJs) are discarded and the splicing quantification for each event is corrected to the number of mappable positions (described in detail in [33, 34]). In addition, *vast-tools align* has a unique submodule to specifically quantify microexons of length $\leq$ 15 nt, which are captured only partially by most splice mappers. These unique features together render *vast-tools* splicing quantification comparable to (or even better than) those of other tools when benchmarked against real RT-PCR data [3].

vast-tools align generates several files in the output subfolder *to_combine*, which will then be used by the next module, *vast-tools combine*, to generate the final output tables. Two additional features are worth mentioning about the outputs of *vast-tools align*. First, it prints a series of statistics about the mapping that can be useful for quality control of the input RNA-seq data. In particular, it will report: (a) the number of valid reads that could be trimmed to at least one 50-nt fragment, (b) the percentage of fragments that map to a library of one representative mRNA per gene (*see* Subheading 2.6), (c) the percentage of fragments that map to the genome (excluding spliced fragments), and (d) the percentage of fragments that map to each of the EEJ libraries. Second, one intermediate file in the folder *to_combine* is particularly useful for custom analyses: *Sample_X.eej2*. This file contains the read counts for each possible splice site, in the following format:

ENSG00000000003	7-8	15	NA	8:4,13:1,17:2,18:1,27:1,32:1, 34:1,36:4
ENSG00000000003	9-9	29	NA	2:1,5:3,7:2,10:3,13:1,15:4, 16:1,17:1,19:3,21:3,28:1, 30:4,37:2
ENSG00000000419	0-1	38	NA	6:4,7:1,8:1,9:5,10:2,18:1, 24:5,28:1,29:2,33:1,35:1, 36:14
ENSG00000000419	11-9	78	NA	8:18,10:1,13:32,14:5,18:3, 20:2,26:10,29:1,30:2,31:4
ENSG00000000419	13-12	31	NA	7:7,8:3,10:2,11:1,12:1,3:10, 15:4,20:2,22:1

where column 1: gene ID; column 2: EEJ ID, number of the donor and acceptor for each gene (sorted from 5′ to 3′; 0-based); column 3: NA; column 4: position within the EEJ (0-based) and number of reads starting at that position.

2.3 vast-tools merge

This module merges the outputs of two or more individual RNA-seq *vast-tools align* runs on related subsamples into a table for the supersample. This might be desired, for example, in cases where one RNA-seq sample was split across several sequencing runs and should be merged into the original sample for downstream analyses, or to pool individual replicates to increase read depth (see below). The resulting merged files are equivalent to concatenating the FASTQ files before running *vast-tools align*. *vast-tools merge* requires a config file with two columns: (a) original sample name and (b) merged super-sample. Since the data is stored in memory, when many samples need to be merged into multiple super-samples, we recommend splitting the config file (always keeping the samples from the same supersample together) and running *vast-tools merge* independently for each resulting config file.

A major reason for this module is to be able to merge samples of similar nature to increase read coverage. Quantification of AS relies solely on reads mapping locally to specific EEJs. Therefore, deeper RNA-seq compared to gene expression analyses is usually needed to obtain reliable splicing quantification for AS events in non–highly expressed genes. Although the fraction of AS events whose inclusion levels can be quantified reliably (VLOW/LOW or higher, *see* Subheading 2.4) increases with read length and depth up to ~200 million reads [3], we recommend using at least 70–90 million ≥100-nt paired-end reads for a differential AS analysis of sufficient quality without strong biases in the identified AS events toward highly expressed genes. For comparison, a good differential gene expression analysis with standard Illumina RNA-seq data can be obtained with 20–30 million 50-nt single-end reads.

2.4 vast-tools combine *and* vast-tools tidy

vast-tools combine takes any set of intermediate output files generated by *vast-tools align* and *vast-tools merge* and builds the main output table of *vast-tools*: INCLUSION_LEVELS_FULL-species-N.tab. This table consists of one row per AS event, each with six columns with basic information about the AS event and two columns for each of the N samples combined: (a) the inclusion level, using the Percent Spliced-In (PSI) metric, and (b) a comma-separated array of five scores indicating the reliability of the PSI measure, as well as the normalized count of reads supporting inclusion and exclusion. Information about each score and the exact definitions for each AS event type can be found in https://github.com/vastgroup/vast-tools#combine-output-format. For simplicity, we will only discuss here the most relevant ones. The main score used by all *vast-tools* modules is the first one, which is a qualitative score based on the read coverage supporting the PSI calculation. From less to more reliable: N < VLOW < LOW < OK < SOK. *vast-tools* normally works with scores VLOW or higher, unless the option *--noVLOW* is specified, in which case it will only consider PSIs associated with LOW, OK or SOK scores. In addition, for intron retention events, the option *--p_IR* can be used to filter them by the p-value of a binomial test assessing the read imbalance between both exon-intron junctions (fifth score). For alternative donor and acceptor sites, the inclusion level of the host exon (third score) is also assessed using the option *--min_ALT_use*.

To further simplify INCLUSION_LEVELS_FULL-species-N.tab, *vast-tools tidy* can produce a table containing only the VastID and the PSIs for those samples that have a minimum coverage support (e.g., ≥VLOW or ≥ LOW). Else, the PSI value will be converted into NA. *vast-tools tidy* also serves to select the events that are more likely relevant for downstream analyses (e.g., unsupervised clustering of samples based on inclusion levels), for instance by requiring that at least a fraction of samples have sufficient read coverage and the standard deviation of the PSIs is above a certain value.

2.5 vast-tools compare, vast-tools diff, *and* vast-tools plot

vast-tools contains two modules to perform differential AS analyses between two groups of samples, *compare* and *diff*. *vast-tools diff* was developed by Tim Sterne-Weiler, who also developed *Whippet* [32]. It performs Bayesian inference followed by differential analysis of posterior distributions with respect to inclusion levels. A PSI posterior distribution per sample group is estimated with all replicates per group using maximum-likelihood fitting and used for inference of change in PSIs (ΔPSI) between groups. It outputs a table with a list of events with the average PSI for each of the two compared groups, the ΔPSI and the minimum value (MV) for |ΔPSI| at 0.95, which means there is a 0.95 probability that |ΔPSI| is greater than that minimum value.

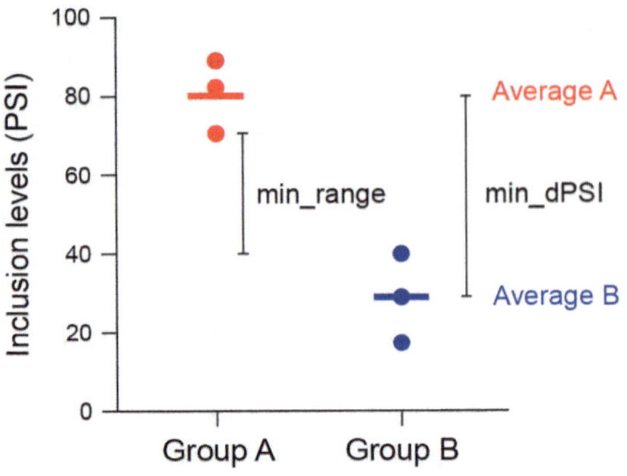

Fig. 2 Logic of the differential AS analysis by vast-tools compare. For AS events with sufficient read coverage for all compared samples within each group, the separating margin (minimum PSI of the group with the highest average PSI - the maximum PSI of the other group) and average ΔPSI (average PSI of group B - average PSI of group A) are calculated. Then, both parameters are required to match the cutoffs specified: *min_range* (default = 5) and *min_dPSI* (default = 15), respectively

In the case of *vast-tools compare*, differentially spliced AS events are identified based on a nonstatistical approach. Since most statistical tests for differential AS analyses rely on the number of inclusion vs exclusion reads, AS events in highly expressed genes have more statistical power. The strategy of *vast-tools compare* is to first select the AS events with a minimum read coverage as to be reliable (e.g., VLOW or higher) across all compared samples, and then compare the PSI distributions using cutoffs for the average ΔPSI and the size of the margin that separates the two PSI value ranges (difference between the largest and lowest PSI values in each distribution) (Fig. 2). *vast-tools diff* and *vast-tools compare* do not need replicates, although at least two replicates per compared group are strongly advised. When analyzing five or more replicates, standard statistical tests using *vast-tools tidy* output tables may be advisable.

vast-tools compare also provides various options to facilitate downstream analyses of differentially spliced events. Using the option *--GO*, it will provide lists of gene IDs for genes with differentially spliced events (for skipped exons and introns separately) as well as a list of background genes, based on the same read coverage filters. In addition, activating the *--print_sets* option, four additional control lists of events are printed for subsequent use in *Matt* (*see* Subheading 3) or other custom analyses. These events are (a) all AS events with sufficient read support, (b) events that are alternatively spliced but not differentially regulated, (c) constitutive events (high inclusion), and (d) cryptic events (low inclusion).

Finally, *vast-tools plot* (originally developed by Kevin Ha and inspired by Nuno Barbosa-Morais) can be used to create graphics with the PSI values for each differentially spliced AS event. *vast-tools plot* can either be called by *vast-tools compare* (*--plot_PSI* flag) or run as an independent module. It takes as arguments a subset of events from INCLUSION_LEVELS_FULL-species-N.tab and an optional config file with information about the colors and display order for each sample, and it generates plots with the PSI for each sample and the error bar related to the reads supporting each PSI measure (Fig. 1).

2.6 Quantification of Gene Expression and vast-tools compare_expr

vast-tools also provides quantification of steady-state mRNA levels, commonly referred to as gene expression levels. For this purpose, *vast-tools align* maps only the first 50 nts of each read (and only of the first read of the pair, if paired-end reads are provided) to a library with one representative transcript per gene. This minimizes the effect that using different read lengths might have on gene expression quantification. Subsequently, *vast-tools combine* generates tables with three possible measures of gene expression: raw read counts per gene, transcripts per million mapped reads (TPMs) and corrected-per-mappability reads per kilobase and million mapped reads (cRPKMs). Both TPMs and cRPKMs are computed by using the effective length of the transcript, which corresponds to the number of positions to which a 50-nt read can be mapped uniquely (see [33] for details).

In addition, using the TPM or cRPKM tables with raw read counts, *vast-tools compare_expr* allows the user to perform differential gene expression analyses using a qualitative approach similar to that applied by *vast-tools compare* for AS events. *vast-tools compare_expr* can also produce files for Gene Ontology enrichment analyses (foreground and background lists of gene IDs), as well as for Gene Set Enrichment Analysis (GSEA) (a ranked list of genes based on fold change, ready to import into GSEA). Although this module is useful to have an overall comparison with AS, for a proper study of differential gene expression, other widely used tools tailored for this type of analyses are recommended.

3 *Matt*: Differential Splicing Feature Analyses

3.1 Introduction

Matt is a Unix command-line toolkit for analyzing genomic sequences that was specifically designed for downstream analysis of AS events of interest. *Matt* is modular, comprising numerous commands, each designed for a specific task, and which can be piped together with other Unix utilities. The main goal of *Matt* is to provide a unique toolkit to perform many of the standard downstream computational analyses of AS studies, making investigation straightforward for researchers with little computational background.

Matt operates on tables as the basic data structure, where rows correspond to entities (e.g., exons, introns, genes) and columns to features of these entities (e.g., 5′ splice site strength, upstream intron length, PSI). *Matt* commands can be divided into two main categories (Fig. 1): commands for data preparation and table manipulation, and commands for data analysis. Extensive information on each command, including the specification of input and output, can be found on https://gitlab.com/aghr/matt and in [35]. In the following section, we will describe a few high-level commands for data analysis that are particularly useful.

3.2 matt cmpr_exons *and* matt cmpr_introns

matt cmpr_exons and *matt cmpr_introns* are commands to extract genomic features associated with provided groups of AS exons or introns, respectively. These features (around 75 for exons and 50 for introns; described on https://gitlab.com/aghr/matt/-/wikis/o5_analyze_seqs/get_efeatures) are related to exon and intron length and GC content, splice site, branch point and poly-pyrimidine tract strength and location, position within the transcript, and so on. By providing a table of exons or introns belonging to different groups (e.g., up- or downregulated in a given condition, constitutively spliced) *matt cmpr_exons* and *matt cmpr_introns*, respectively, can automatically extract those features and perform a statistical comparison of their distributions across defined groups. These commands output tables with all extracted features and determined statistics associated with each comparison (descriptive values and statistical significance) as well as a PDF with plots with all distributions and comparisons (e.g., for 3′ splice site strength in Fig. 3a, and *see also* Subheading 6). Therefore, with a single command, *Matt* generates a complete overview of the genomic and sequence characteristics of a set of exons or introns of interest. Moreover, among its multiple commands for data preparation, *Matt* has specific utility commands to define control groups of matched size by (a) stratifying by a specific feature (e.g., PSI in the control condition) with *matt stratify* or (b) randomly down-sampling with *matt rand_rows.*

3.3 matt cmpr_features

matt cmpr_features offers a complementary view to *matt cmpr_exons* and *matt cmpr_introns.* Given a list of exons (or any other entity) and at least two quantitative features (extracted using *Matt* or any other tool), it plots and statistically compares the distribution of the query feature (e.g., PSI) across bins of equal size based on the target feature (e.g., 5′ splice site strength) (Fig. 3b). These analyses allow the user to identify subtle as well as nonlinear associations between two quantitative variables.

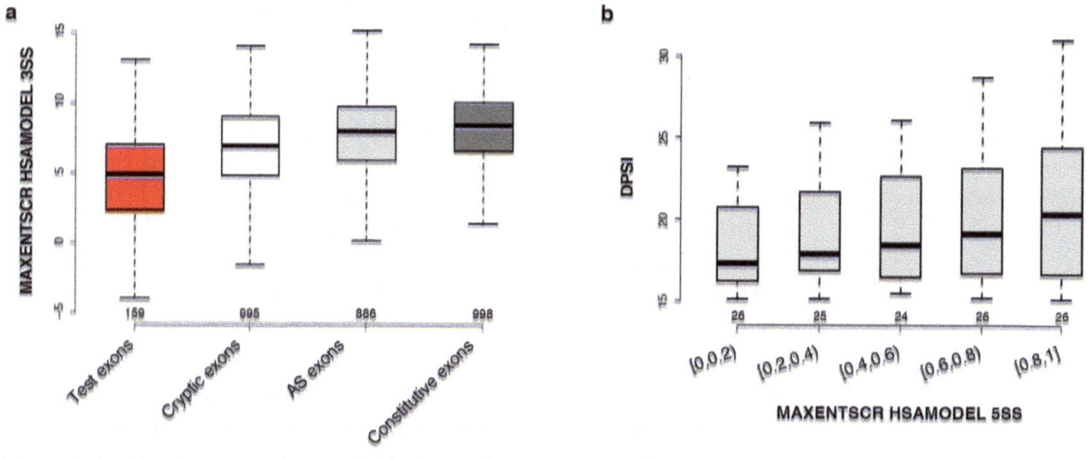

Fig. 3 Examples of plots from matt cmpr_exons and matt cmpr_features. (**a**) Output plot from matt cmpr_exons showing the distributions for four groups of exons of 3′ splice site strength values, using the maximum entropy score. (**b**) Output plot from *matt cmpr_features* showing the ΔPSI distributions of a set of exons split into five equal-sized bins based on their 5′ splice site strength values

3.4 Motif Enrichment Analyses and matt rna_maps

One key aspect when analyzing RNA-seq data from an experimental manipulation of a splicing factor is to be able to infer whether or not the effects are likely direct or indirect. In the absence of direct binding data, such as CLIP-seq, the most common approach is to search for enrichment of specific motifs in the sequences of and around regulated AS events. For this purpose, *Matt* has different commands to perform motif enrichment analyses using either (a) known binding motifs of some RNA binding proteins (from cisBP-RNA [36]) with *matt test_cisbp_enrich*, (b) position weight matrices (PWMs) with *matt test_pwm_enrich*, and (c) all possible k-mers of certain size or any custom regular expression with *matt test_regexp_enrich*. *Matt* performs its enrichment motif analyses for the entire length of the sequences provided, as well as for three segments independently (proximal, central and distal), to identify potential positional biases. Moreover, since *Matt* detects motif enrichment in any given set of sequences, it can also be used in other contexts, for example, to identify enriched motifs in CLIP-seq clusters.

RNA maps are an alternative way to study and visualize motif enrichment that is specific to the investigation of AS events. Using *matt rna_maps*, the user can produce plots with the distribution of specific motifs inside or around a set of exons or introns (*see* Subheading 6), which are often associated with the positional effects of most RBPs [37]. To the best of our knowledge, *Matt* is the only command-line tool with the capability to generate RNA maps.

4 *ExOrthist*: Evolutionary Conservation of Exons

4.1 Introduction

As with any other molecular process or genomic element, evolutionary conservation of AS events has long been considered a proxy for their functional relevance [38]. A key step to assess the conservation of AS events and their regulation is to infer homology relationships among exons from different species. Various tools exist to infer such relationships at the gene level (e.g., *Orthofinder* [39], *Broccoli* [40]). However, although different methodologies have been implemented in the past [3, 24, 34, 41, 42], no specific software existed so far to obtain exon homology relationships. To fulfill this need, we developed the Nextflow-based package *ExOrthist* [43], which infers exon homology relationships for multiple species at any evolutionary distance. In addition to inferring homology relationships, *ExOrthist* also offers two additional modules to perform evolutionary comparisons of subsets of exons of interest from two species (*compare_exon_sets*) and to plot intron-exon structures (*exint_plotter*). The algorithms as well as the input and output formats are described in detail on [43] and in the README file at https://github.com/biocorecrg/ExOrthist.

4.2 main.nf

The main module of *ExOrthist* allows the user to infer orthology and paralogy relationships among exons genome-wide. *ExOrthist* takes as input genome sequences and annotations (in FASTA and GTF formats, respectively) for a set of species, a file with the pairwise evolutionary distance ranges between those species (short, medium or long), and a file with gene orthogroups (i.e., gene homology relationships) among those species (Fig. 1). In addition, the user can provide a set of custom non-annotated exons, for example, those quantified by *vast-tools* or any other AS quantification tool. *ExOrthist* then generates Intron Position Aware (IPA) alignments for all pairs of isoforms from each pair of homologous genes. Next, for each exon in each species, *ExOrthist* parses those IPAs to identify the best matching exon in each homologous gene of the other species. Exon homology relationships are then inferred based on sequence similarity of the exon and its upstream and downstream exons, similarity of the length of the exon, and conservation of the flanking intron positions and phases. The cutoffs for each of these requirements depend on the evolutionary distance of the species being compared (short, medium, or long) and can be modified by the user. Finally, once pairwise exon homology relationships have been inferred, *ExOrthist* clusters them to create exon orthogroups across the set of compared species.

4.3 compare_exon_*sets and* exint_plotter

Once the exon orthogroups are defined by *main.nf*, *ExOrthist* offers two modules for the extraction of relevant information from them. The *compare_exon_sets* module allows the user to easily perform evolutionary analyses for subsets of exons of interest between two species. If one list of exons from one query species is provided (e.g., exons misregulated in a given disease), *ExOrthist* will provide statistics on the gene- and exon-level genomic conservation in the target species. The user can also retrieve the orthologous exons using the *--print_out* option. If two lists of exons are provided, one for each species (e.g., exons found to be regulated by the same splicing factor in human and in mouse), *ExOrthist* will provide gene- and exon-level statistics of genomic conservation, but also of regulatory conservation, both in text and graphical formats (see examples below). This enables the identification of orthologous exons regulated by the same splicing factor in both species (and thus more likely to be functionally relevant), but also of orthologous genes harboring nonhomologous exons that are regulated by the same splicing factor. Finally, the *exint_plotter* module generates PDF plots to visualize the evolution of exon-intron structures of homologous genes across multiple species.

5 *VastDB*: Looking at the Big Picture

5.1 Introduction

VastDB is a web resource containing multiple layers of information on AS and gene expression for, currently, seven animal species: human, mouse, rat, cow, chicken, zebrafish, and fruitfly. Arguably, the most relevant information is the RNA-seq-based genome-wide quantification of AS and gene expression across dozens of cell/tissue types and developmental stages. *VastDB* comprises a main panel of samples with an interactive display for all available species. The description and Short Read Archive (SRA) IDs for each of these samples are available in the "Samples" section. In addition, for a subset of species, *VastDB* displays various extra sets of samples of particular interest (e.g., GTEx, TCGA, neuronal diversity or differentiation, early embryonic development) in the form of static PNG plots under the section "Special datasets."

The information contained in *VastDB* is accessible in two main ways: by downloading genome-wide data from https://vastdb.crg.eu/wiki/Downloads, or through direct browsing for individual genes or AS events. For the latter, various search functions are available: by gene ID or name, VastID, coordinate, or by general semantic search (top right box). In addition, the user can browse a dedicated UCSC Browser instance (https://ucsc.crg.eu), hosting the *VastDB* event tracks and genome annotations for many other model and nonmodel species.

Currently, *VastDB* hosts data for seven species, but it is expected to grow over the next two years up to 30 species. Moreover, a sister database with the same structure contains information for the model plant *Arabidopsis thaliana* [27] (http://pastdb.crg.eu/). Answers to frequently asked questions and other additional information are available in the FAQ section and in the original publication [3].

5.2 Gene View Pages

Each gene from each species has its own *VastDB* page that can be browsed to search for AS events of potential interest (e.g., https://vastdb.crg.eu/gene/ENSG00000004487@hg19). The data displayed in the Gene view pages include basic information about the gene (coordinates, name, description, etc.), its expression using the cRPKM metric across different samples and special datasets (if available) and a list of orthologs in the other *VastDB* species. More importantly, for each gene, *VastDB* automatically displays its most potentially interesting AS events. These correspond to AS events with wide regulation (i.e., large PSI variation across samples) and/or that have annotated functions (*see* Subheading 5.4). These selected AS events are shown in a UCSC Genome Browser shot and as a list with *VastID* links to the corresponding Event view pages, together with various summary statistics (average PSI, PSI range, and number of annotated functions). Finally, the inclusion levels of those selected AS events can be displayed together in an interactive manner in a single PSI plot.

5.3 Event View Pages

The main page type of *VastDB* is the Event view, which contains all the information available for each VastID (e.g., https://vastdb.crg.eu/event/HsaEX0033219@hg19). Although the Gene view displays only a few selected AS events per gene, *VastDB* has an Event view for every AS event quantified by *vast-tools*. Each Event view offers rich information about the AS event, including coordinates and sequences for the alternative sequence and the upstream and downstream (C1 and C2) exons, length of the alternative sequence, associated 5′ and 3′ splice site strength (measured using the maximum entropy score [44]), technical *vast-tools*-related information, and functions of events (*see* Subheading 5.4). In addition, it has other layers of information that are particularly useful.

1. ORF impact predictions: for each AS event, *VastDB* provides a prediction of the impact that the inclusion or exclusion of the alternative sequence has on the main ORF. These predictions are based on features of the alternative sequence (i.e., whether or not the length is multiple of three or whether or not it encodes in-frame stop codons) and on the available transcript annotations. There are four major types of predicted impacts: (a) alternative protein isoforms both upon inclusion and exclusion, (b) ORF disruption upon sequence inclusion, (c) ORF disruption upon sequence exclusion, and (d) sequences in the untranslated regions (UTRs). These major types may have additional information, including whether they encode alternative N or C termini and so on. It should be kept in mind that these predictions are done for each AS event in isolation, but AS events may occur in combinations and/or together with alternative promoters and polyadenylation sites, which may affect the predicted ORF impact.

2. Overlap with protein disordered regions and domains: the fraction of disordered residues calculated using *disopred3* [45] and the overlap with PFAM domains is shown for the alternative sequence as well as for the upstream and downstream exons. In addition, *VastDB* displays domain plots for a representative inclusion and exclusion protein isoform (when available), highlighting the sequences of the AS event (Fig. 1 and Subheading 6).
3. Evolutionary conservation: for each of the other *VastDB* species, the orthologous AS event (if available) is shown with a link to the corresponding Event page, together with the coordinate and whether the event is alternatively, constitutively or lowly included. This greatly facilitates the exploration of evolutionary conservation and the interconnection among *VastDB* species. Currently, AS event-level orthologies were obtained using both *ExOrthist* and independent *liftOver*-based approaches [3], but they will exclusively rely on *ExOrthist* in future *VastDB* versions.
4. Suggestion of RT-PCR primers: primer sequences and the expected band sizes are provided for validation of the AS events through RT-PCR. These primer sequences are generated automatically expecting two bands (inclusion and exclusion) and optimizing the sizes of the PCR amplicons to allow good visualization in agarose gels and at the same time reducing amplification biases toward the shorter exclusion isoform.
5. Plots of inclusion levels: as per gene expression, PSI values are displayed in the main sample panel for all events, as well as in the special datasets, when available. In the case of PSIs, it is common that a given sample does not have sufficient read coverage (VLOW or higher) to confidently estimate its PSI in some samples, and therefore no value is displayed for those samples. The minimum coverage can be adjusted interactively, as well as the samples to be plotted.

5.4 Functions of Events

As part of the Event view, *VastDB* displays the annotated functions for each event, when available. Annotations of event functions correspond to individual publications (i.e., each has a PubMed link; a total of 1,991 so far) assessing the function of AS events experimentally. The majority of these annotations come from parsing studies on individual AS events. Other sources include experimentally validated Eukaryotic Linear Motifs (ELMs) [46], with associated PubMed IDs, which often map to constitutive exons. In addition, there are currently two studies of exon-level CRISPR-based screens [47, 48], whose number is expected to increase substantially over the following years. The ultimate goal of *VastDB* is to collect every study assessing the function of an AS event ever published. For this purpose, we put in place a collective effort, in

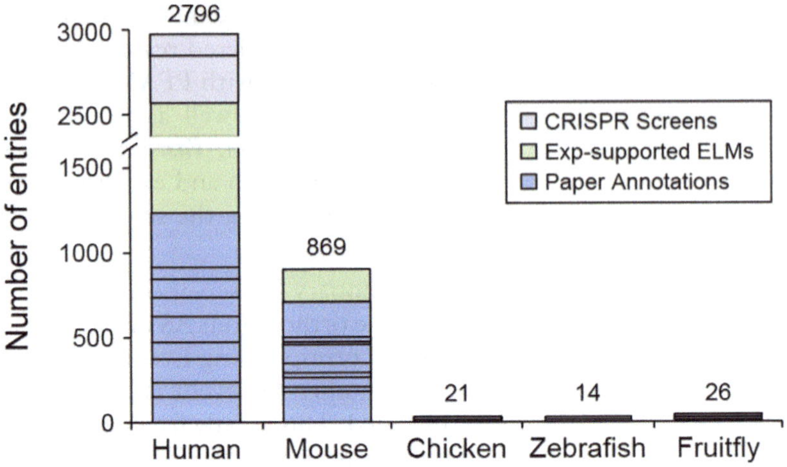

Fig. 4 Descriptive summary of the number of annotations of event functions per type and species. Paper annotations are normally single event case studies; the blue blocks in each stacked bar correspond to individual releases (currently version 9)

which users can submit annotations from their studies or any other publication through https://vastdb.crg.eu/wiki/EventFunction.

Currently, in its release 9, there are a total of 3,722 annotations (i.e., a pair of VastID and PubMed ID), covering 2,524 unique events. Therefore, most VastIDs with annotations have one associated study, although some have multiple. In particular, alternative exons in *Fn1* (MmuEX0019422), *INSR* (HsaEX0031843), and *TP73* (HsaEX0066617) are currently the top VastIDs in terms of associated functions, with 15, 13, and 13 associated studies, respectively. With regard to the distribution of annotations per species, most come from human and mouse (currently also including rat annotations) (Fig. 4). Nonetheless, the functional information for each event is displayed also in the Event pages of all orthologs of that event in the different *VastDB* species.

6 A Case Example: Identification and Characterization of Conserved *Srrm4* Targets in Mammals

To exemplify the use of the different tools and resources within the *VastDB* framework, we will describe a well-characterized case: the identification of microexons as the main targets of the splicing factor *Srrm4*. The commands, code and input/output files from this analysis, together with a markdown explanation of each step, are available on https://github.com/vastgroup/vastdb_framework_code_example.

First, in the Gene view of *VastDB* we can confirm the neural-specific expression of *Srrm4* across mouse cell and tissue types (Fig. 5).

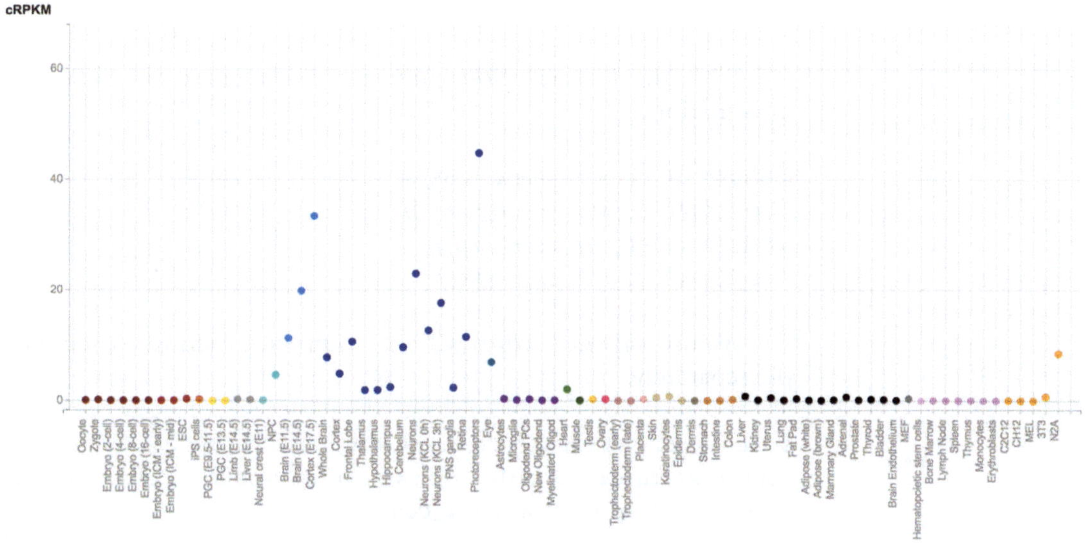

Fig. 5 Expression of mouse *Srrm4* across cell and tissue types form the main *VastDB* Gene view sample panel plot, using the cRPKM metrics

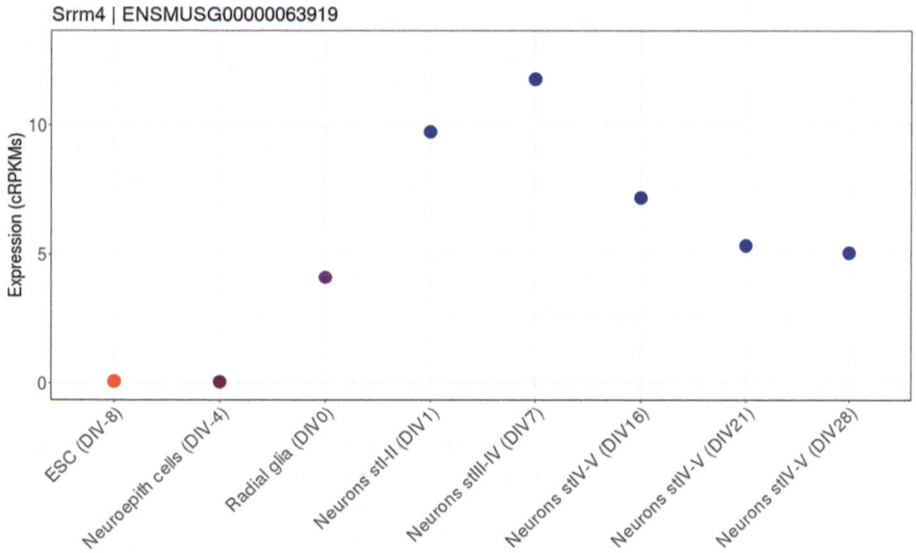

Fig. 6 Expression of mouse *Srrm4* during differentiation of embryonic stem cells (ESC) to glutamatergic neurons, using the cRPKM metrics

In addition, we can see other profiles of interest in the Special datasets section, for example the temporal dynamics of expression during neuronal differentiation (Fig. 6).

Next, we use *matt retr_rnaseq* to download RNA-seq data for a knockdown (KD) of *Srrm4* in neuroblastoma N2A cells (SRP041656), including two replicates for KD and control.

```
$ matt retr_rnaseq accessions_mouse.txt
```

With the file accessions_mouse.txt containing the following information.

```
SRR1269746 CL_N2A_Srrm4_Cont_a
SRR1269748 CL_N2A_Srrm4_Cont_b
SRR1269747 CL_N2A_Srrm4_KD_a
SRR1269749 CL_N2A_Srrm4_KD_b
```

We use *vast-tools align* (v2.5.1) to process each RNA-seq sample separately.

```
$ vast-tools align CL_N2A_Srrm4_Cont_a_R1.fq.gz CL_N2A_Srrm4_-
Cont_a_R2.fq.gz -sp mm10 -o vast_out/mm10/MMB --expr --IR_ver-
sion 2 -c 8 -n CL_N2A_Srrm4_Cont_a

$ vast-tools align CL_N2A_Srrm4_Cont_b_R1.fq.gz CL_N2A_Srrm4_-
Cont_b_R2.fq.gz -sp mm10 -o vast_out/mm10/MMB --expr --IR_ver-
sion 2 -c 8 -n CL_N2A_Srrm4_Cont_b

$ vast-tools align CL_N2A_Srrm4_KD_a_R1.fq.gz CL_N2A_Srrm4_K-
D_a_R2.fq.gz -sp mm10 -o vast_out/mm10/MMB --expr --IR_version
2 -c 8 -n CL_N2A_Srrm4_KD_a

$ vast-tools align CL_N2A_Srrm4_KD_b_R1.fq.gz
CL_N2A_Srrm4_KD_b_R2.fq.gz -sp mm10 -o vast_out/mm10/MMB --
expr --IR_version 2 -c 8 -n CL_N2A_Srrm4_KD_b
```

Afterward, *vast-tools combine* generates a single table with PSIs for each sample:

```
$ vast-tools combine -sp mm10 -o vast_out/mm10/MMB
```

After changing the working directory (cd vast_out/mm10/MMB), we employ *vast-tools compare* on the INCLUSION table to extract the differentially regulated AS events, here with thresholds $|\Delta PSI| \geq 25$ and margin ≥ 5, as well as lists of gene IDs to perform Gene Ontology analyses and of control AS events for subsequent *Matt* analyses:

```
$ vast-tools compare INCLUSION_LEVELS_FULL-mm10-4.tab
-a CL_N2A_Srrm4_Cont_a,CL_N2A_Srrm4_Cont_b
-b CL_N2A_Srrm4_KD_a,CL_N2A_Srrm4_KD_b
--print_dPSI --GO -sp mm10 --print_sets
--min_dPSI 25 --min_range 5
-name_A Control -name_B Srrm4_KD
```

Table 2
Summary output of *vast-tools compare*

AS_TYPE	Higher_in_Control	Higher_in_Srrm4_KD	TOTAL_EV	TOTAL_AS (10<PSI<90)
Microexons	108	0	2065	297
Long_AltEx	52	9	110,247	4786
Intron_ret	23	6	111,423	9969
Alt_3ss	NA	15	95,664	4298
Alt_5ss	NA	8	67,039	3140

Table 3
Summary of event types from *vast-tools compare*

AS_TYPE	Constitutive	Cryptic	AS_non_change
Microexons	286	1363	64
Long_AltEx	88,361	11,856	2627
Intron_ret	85,949	541	6715
Alt_3ss	34,192	51,950	2800
Alt_5ss	24,449	36,117	1865

The summary output from *vast-tools compare* already reveals a clear bias: most misregulated AS events upon *Srrm4* knockdown are exons, especially microexons of length ≤ 27 nt, which show higher inclusion in the control (108 vs. 0 for microexons and 52 vs. 9 for longer exons), consistent with the known role of *Srrm4* enhancing inclusion of very short exons [49, 50].

*** Options: dPSI = 25, range_dif = 5, min_ALT_use = 25.

*** Summary statistics (*see* Table 2).

*** Summary statistics for extra event sets (Max_dPSI = 5) (*see* Table 3).

In addition to the descriptive tables, the run of *vast-tools compare* generates several output files, whose names are automatically generated to include all the key parameters used for the analysis.

- The set of differentially regulated AS events (DiffAS), with average ΔPSI in the last column:

```
DiffAS-mm10-4-dPSI25-range5-min_ALT_use25-upreg_ALT_Control-
vs-Srrm4_KD-with_dPSI.tab
```

- Other sets of events: AS events ($10 < PSI < 90$ in at least one group) that do not change ($|\Delta PSI| \leq 5$; AS_NC), constitutive exons ($PSI > 95$ in both groups; CS), and cryptic exons ($PSI < 5$ in both groups; CR).

```
AS_NC-mm10-4-dPSI25-range5-min_ALT_use25-upreg_ALT_Control-
vs-Srrm4_KD-with_dPSI-Max_dPSI5.tab
CS-mm10-4-dPSI25-range5-min_ALT_use25-upreg_ALT_Control-vs-
Srrm4_KD-with_dPSI.tab
CR-mm10-4-dPSI25-range5-min_ALT_use25-upreg_ALT_Control-vs-
Srrm4_KD-with_dPSI.tab
```

- GeneIDs of cassette exons (AltEx) and background (BG) genes for GO analysis:

```
AltEx-mm10-4-dPSI25-range5-min_ALT_use25-upreg_ALT_Control-
vs-Srrm4_KD-with_dPSI.txt
BG-mm10-4-dPSI25-range5-min_ALT_use25-upreg_ALT_Control-vs-
Srrm4_KD-with_dPSI.txt
```

To perform a GO enrichment analysis, we next upload the two Gene ID lists in DAVID (https://david.ncifcrf.gov), download the Chart results and plot the -log10 of the p-values of the resulting categories as bar plots (Fig. 7).

These results reveal enrichment in gene functions associated with GTPase regulation, synaptic organization, and cytoskeleton, as previously described [51, 52].

Next, we aim at identifying potential genomic and sequence features associated with *Srrm4*-regulated exons with *Matt* v1.4. We use *Matt*'s table manipulation commands (*add_val*, *rand_rows*, *add_rows*, and *get_rows*) to prepare the necessary input table from the *vast-tools compare* outputs. The overall goal is to merge the four exon sets (differentially spliced [DiffAS], nonchanging [AS_NC], constitutive [CS], and cryptic [CR]) into one table, and simultaneously add a column GROUP with the regulatory group to which each exon belongs. To keep this dataset at a reasonable size, we randomly downsample 1000 exons from each control group (AS_NC, CR, and CR).

```
$ matt add_val AS_NC-mm10-4-dPSI25-range5-min_ALT_use25-upre-
g_ALT_Control-vs-Srrm4_KD-with_dPSI-Max_dPSI5.tab GROUP AS_NC
| grep -P "(MmuEX|EVENT)" | matt rand_rows - 1000 > tmp.tab

$ matt add_val CR-mm10-4-dPSI25-range5-min_ALT_use25-upre-
g_ALT_Control-vs-Srrm4_KD-with_dPSI.tab GROUP CR | grep -P
"(MmuEX|EVENT)" | matt rand_rows - 1000 | matt add_rows tmp.
tab -
```

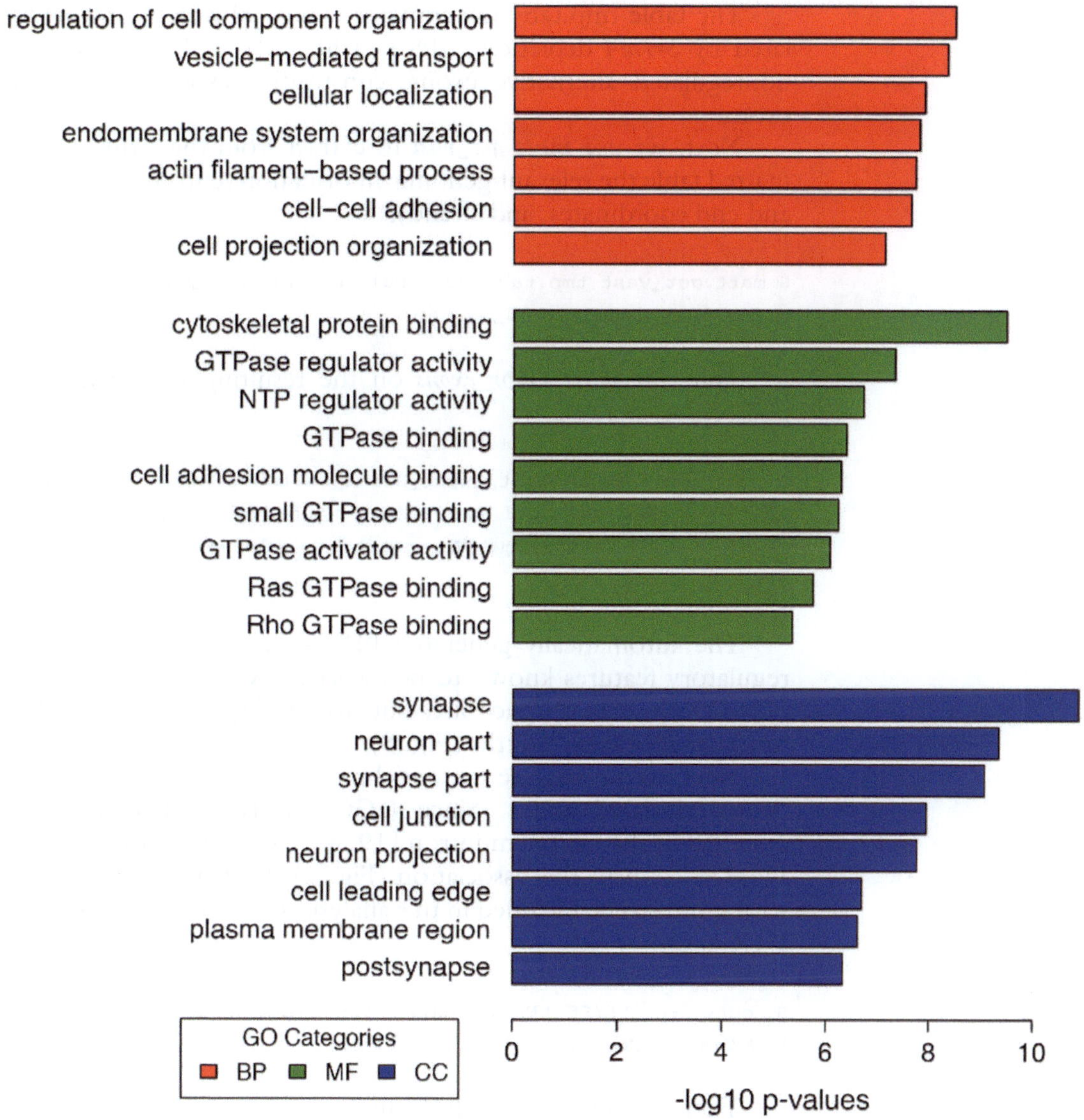

Fig. 7 Gene Ontology enrichment analyses using DAVID for genes containing exons that are differentially regulated upon *Srrm4* KD

```
$ matt add_val CS-mm10-4-dPSI25-range5-min_ALT_use25-upreg_ALT_Control-vs-Srrm4_KD-with_dPSI.tab GROUP CS | grep -P "(MmuEX|EVENT)" | matt rand_rows - 1000 | matt add_rows tmp.tab -

$ matt get_rows DiffAS-mm10-4-dPSI25-range5-min_ALT_use25-upreg_ALT_Control-vs-Srrm4_KD-with_dPSI.tab dPSI[-100,-15] | matt add_val - GROUP Srrm4_DOWN | grep -P "(MmuEX|EVENT)" | matt add_rows tmp.tab -
```

The table tmp.tab now contains exons (MmuEX) downregulated by *Srrm4* depletion (dPSI[−100,-15]), and 1000 random nonregulated alternative exons, constitutive exons, and cryptic exons.

Next, we use *matt get_vast* to extract from the *vast-tools* formatted table the relevant genomic information (chromosome, start and end coordinates, and strand).

```
$ matt get_vast tmp.tab COORD FullCO COMPLEX LENGTH -gtf mm10.
gtf > Matt_input_Srrm4_ex.tab
```

And run *matt cmpr_exons* on the resulting table (Matt_input_Srrm4_ex.tab):

```
$ matt cmpr_exons Matt_input_Srrm4_ex.tab START END SCAFFOLD
STRAND GENEID mm10.gtf mm10.fasta Mmus 150 GROUP[Srrm4_DOWN,
CR,AS_NC,CS] Matt_Srrm4_KD -notrbts -colors:red,white,light-
gray,darkgray
```

The automatically generated PDF summary shows the main regulatory features known to be associated with *Srrm4*-regulated exons [53]: weak 3′ splice sites but strong 5′ splice sites, as well as much smaller exon lengths (Fig. 8).

Next, we investigate the enrichment of motifs associated with *Srrm4*-regulated exons. Since UGC motifs are known to be enriched in the upstream intron [19, 49, 50], we used *matt rna_-maps* to confirm this association (Fig. 9). First, we create a table with motifs to be included in this analysis, here containing only the UGC motif.

```
$ echo -e 'TYPE\tNAME\tEXPR_FILE\tTHRESH\tBGMODEL\nREGEXP
\tUGC\tTGC\tNA\tNA' > ugc_motif.tab
```

Then, we generate the RNA map.

```
$ matt rna_maps Matt_input_Srrm4_ex.tab UPSTRM_EX_BORDER START
END DOSTRM_EX_BORDER SCAFFOLD STRAND GROUP 15 50 150 mm10.
fasta ugc_motif.tab TYPE NAME EXPR_FILE THRESH BGMODEL -d
UGC_map_Matt_Srrm4_KD
```

Where 15 is the length of the sliding window for the RNA motif map, 50 and 150 are, respectively, the number of exon and intron positions to be considered.

To exemplify the use of *VastDB* resources in the context of a downstream analysis of AS events of interest, we next download two files: PSI_TABLE-mm10.tab.gz (containing the PSIs across the main sample panel for each AS event) and PROT_DISORDER-mm10.tab.gz (fraction of disordered regions overlapped by each AS event).

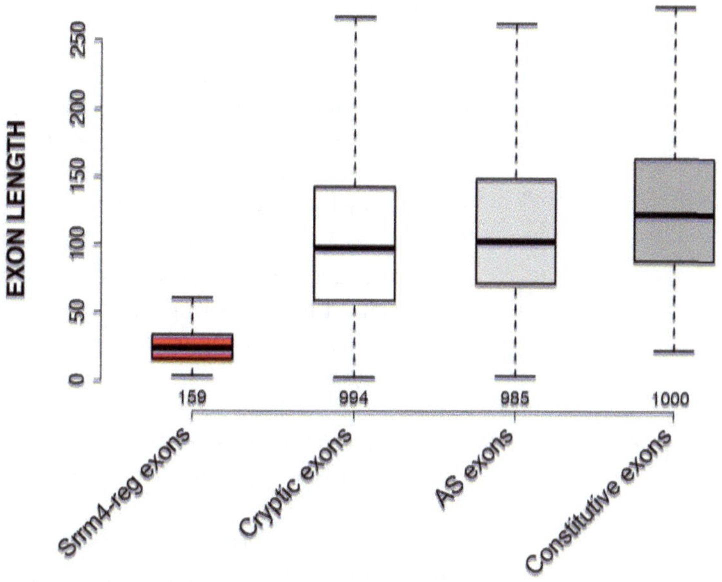

Fig. 8 Example of a boxplot from matt cmpr_exons, visualizing the drastic bias toward shorter lengths among *Srrm4*-regulated exons compared to other exon sets

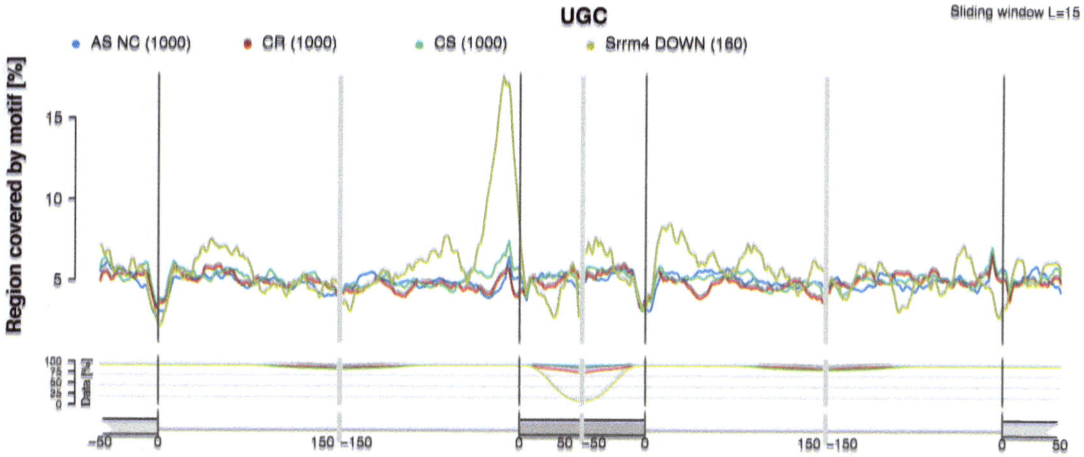

Fig. 9 RNA map for the UGC motif. The motif is strongly enriched in a window from approximately −30 to −5 upstream of *Srrm4*-regulated exons, but not in other exons

```
$ wget https://vastdb.crg.eu/downloads/mm10/EVENT_METRICS-
mm10.tab.gz

$ wget https://vastdb.crg.eu/downloads/mm10/PROT_DISORDER-
mm10.tab.gz
```

First, we calculate the ΔPSI between neural and nonneural samples. For this purpose, we use the utility script *Get_Tissue_Specific_AS.pl* [3, 27], also available for download in the GitHub repo https://github.com/vastgroup/vastdb_framework_code_example. This script takes as input a *vast-tools* INCLUSION table and a config file (a tab-separated table with the tissue groups for each sample, either "Neural" or "Other") and identifies tissue-specific AS events. With the option "*--test_tis* Neural", it generates a table with the average ΔPSI between neural and nonneural samples (as specified in the config file) for all AS events with sufficient read coverage for the comparison (at least five replicates in each group; *−min_rep* 5).

```
$ perl Get_Tissue_Specific_AS.pl PSI_TABLE-mm10.tab.gz -g Con-
fig_Neural.txt -min_N 2 -test_tis Neural -min_rep 5
```

With this information, we plotted the ΔPSI per type of exon, which shows a very strong bias for neurally upregulated exons among *Srrm4*-dependent exons (Fig. 10a).

Next, we made use of the disorder information to plot the average percentage of disorder residues for the alternative as well as the upstream (C1) and downstream (C2) exons for each exon set (Fig. 10b). As expected for tissue-specific exons, *Srrm4*-dependent exons more often encode disorder regions [54]. However, some *Srrm4*-dependent microexons are also known to be inserted within structured domains, as exemplified by a 15-nt exon in *Vav2* (MmuEX0051282; Fig. 10c).

Finally, we perform a conservation analysis using *ExOrthist* v0.1.0. We run *ExOrthist main.nf* for mouse and human, using default parameters for the short evolutionary distance range, adding all *vast-tools* exons, and using Ensembl 1-to-1 orthologs as gene orthogroups.

```
$ nextflow main.nf
```

main.nf needs two config files in the pipeline execution directory: first, a nextflow.config file, automatically downloaded during *ExOrthist* installation and containing the pipeline configuration properties; second, a params.config file specifying the required parameters for the run, the location of the input files as well as of the output folder (hg38_mm10_output, for this example). Among other files of interest, the output folder will include the EX_clusters.tab file, which will contain all exon orthogroups within the provided human and mouse gene orthogroups.

Next, we perform two types of analyses with *compare_exon_sets.pl*: (a) genome conservation of each mouse exon set in human, and (b) evolutionary conservation of human and mouse *Srrm4*-regulated exons.

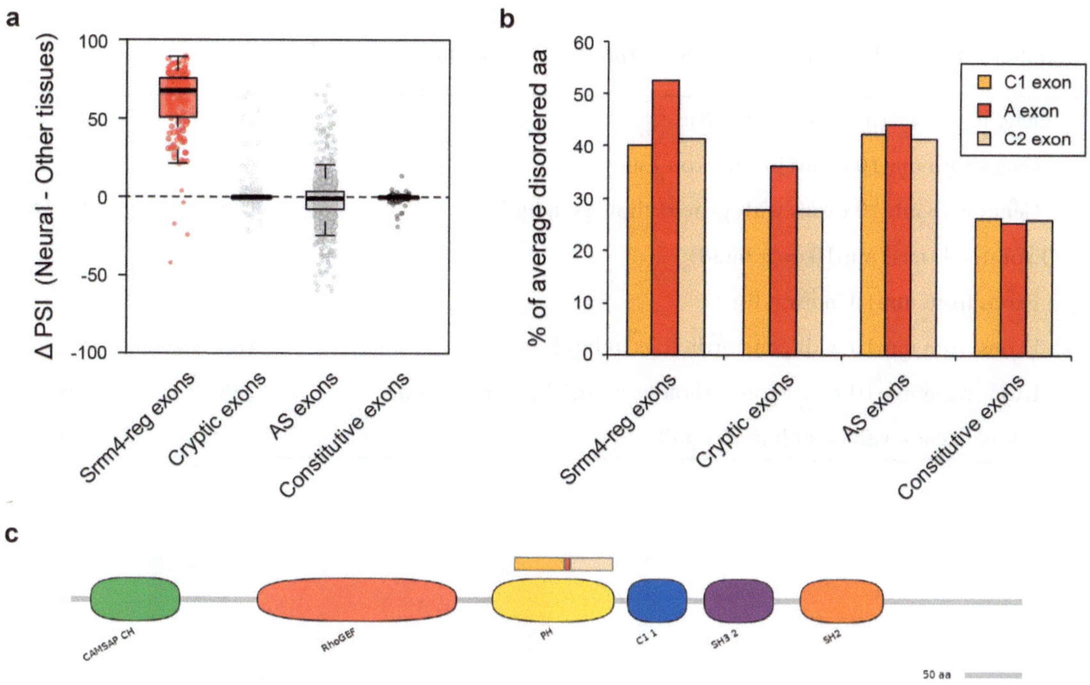

Fig. 10 Example of the use of VastDB resources for the downstream characterization of AS events of interest. (**a**) Difference in PSI between neural and nonneural samples (ΔPSI) for *Srrm4* downregulated exons. (**b**) Average percentage of residues overlapping disordered regions for each exon set. (**c**) Example of a microexon (MmuEX0051282) overlapping a structured domain in Vav2

From the first comparison, using a single list (Table 4), we obtain the following statistics.

```
$ perl ~/ExOrthist/bin/compare_exon_sets.pl -sp1 mm10 -sp2 hg38 -exon_list_sp1 Exons_mm10-4-dPSI25.txt -main_folder hg38_mm10_output/
```

Among other statistics, *ExOrthist* highlights that the percentage of mouse *Srrm4*-regulated exons hosted by genes with 1:1 human orthologs with an exon ortholog in human is 89%. We then run the same analysis for each control exon set (AS_NC, CR, CS), and plot the percentage of genome conservation in human, showing higher conservation of Srrm4-regulated exons compared to nonregulated alternative and cryptic exons (Fig. 11).

```
$ perl ~/ExOrthist/bin/compare_exon_sets.pl -sp1 mm10 -sp2 hg38 -exon_list_sp1 Exons_mm10-CR.txt -main_folder hg38_mm10_output/
```

```
$ perl ~/ExOrthist/bin/compare_exon_sets.pl -sp1 mm10 -sp2 hg38 -exon_list_sp1 Exons_mm10-AS_NC.txt -main_folder hg38_mm10_output/
```

Table 4
Summary output of *ExOrthist compare_exon_sets* for a single exon set

- Gene-level stats (mm10 = > hg38):		
Genes with mm10 exons in the exon lists	141	
Genes with mm10 exons with gene orthologs in hg38	137	97.16%
- Exon-level stats (mm10 = > hg38):		
Exons from mm10 in exon list	159	
Exons from mm10 with gene orthologs in hg38	155	97.48%
Exons from mm10 with exon orthologs in hg38 (G-conserved)	138	86.79%
Only genes with orthologs in hg38		89.03%

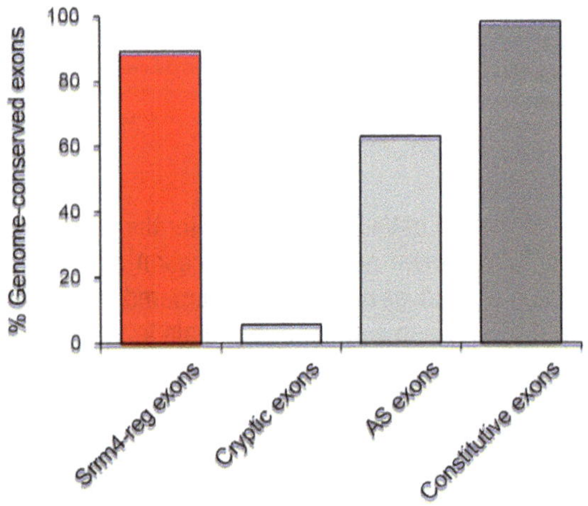

Fig. 11 Percent of genome conservation between human and mouse for various sets of exons with different regulatory patterns

```
$ perl ~/ExOrthist/bin/compare_exon_sets.pl -sp1 mm10 -sp2
hg38 -exon_list_sp1 Exons_mm10-CS.txt -main_folder
hg38_mm10_output/
```

To assess the regulatory conservation of *Srrm4*-regulated exons between mouse and human, and identify orthologs exons regulated in both species, we utilized *compare_exon_sets.pl* for two exon lists. The second list, *SRRM4*-regulated exons in human, was obtained by analyzing with *vast-tools* a RNA-seq dataset (SRP149913) in which SRRM4 or GFP (as control) were ectopically expressed in HEK293 cells.

```
$ perl ~/ExOrthist/bin/compare_exon_sets.pl -sp1 mm10 -sp2
hg38 -exon_list_sp1 Exons_mm10-4-dPSI25.txt -exon_list_sp2
Exons_hg38-2-dPSI50.txt -main_folder hg38_mm10_output/
-print_out
```

This provides a richer output, with the percent of conservation at the genomic and regulatory levels for both directional comparisons (*see* Table 5).

In addition, it also provides a graphical summary from species 1 to species 2 (mm10 to hg38, in this case) (Fig. 12).

Moreover, with the option -print_out, *compare_exon_sets.pl* generates a text file with all the exon orthogroups (i.e., clusters of conserved exons between human and mouse) containing *Srrm4*-regulated exons for either one or both species.

...					
Exon_cluster	**Species**	**GeneID**	**Ex_Coord**	**Length**	**INFO**
GF_000036.007	mm10	ENSMUSG00000021969	chr14:57850145-57850180	36	DOWN
GF_000036.007	hg38	ENSG00000180776	chr13:21392959-21392994	36	UP
GF_000127.026	mm10	ENSMUSG00000027799	chr3:55994048-55994056	9	DOWN
GF_000127.026	hg38	ENSG00000172915	chr13:35168987-35168995	9	NO_INFO
GF_002697.009	mm10	ENSMUSG00000031558	chr5:48191003-48191014	12	DOWN
GF_002697.009	hg38	ENSG00000145147	chr4:20490803-20490814	12	UP
GF_003628.012	mm10	ENSMUSG00000070733	chr5:73050867-73050875	9	DOWN
GF_003628.012	hg38	ENSG00000075539	chr4:48523488-48523496	9	UP
...					

In summary, starting from RNA-seq data for a single gene KD, *Srrm4*, we combined the different tools and resources provided in the *VastDB* framework to (a) characterize the neural-specific expression of this gene; (b) identify its exon targets; (c) perform functional enrichment studies on the genes containing the exon targets; (d) discover the major genomic features defining these targets, including their extremely short length, weaker 3′ splice site and strong 5′ splice site; (e) assess the positional enrichment of a specific *cis*-regulatory motifs (UGC) around those targets; (f) study their tissue-specific regulation; (g) characterize the overlap of exon targets with disordered regions and structured domains; (h) evaluate the evolutionary conservation of the exon targets; and (i) identify the targets with conserved regulation, and thus more likely relevant.

Table 5
Summary output of *ExOrthist compare_exon_sets* for two exon sets

- Gene-level stats:		
- mm10 = > hg38		
Genes with mm10 exons in the exon list	141	
Genes with mm10 exons with gene orthologs in hg38	137	97.16%
Genes with mm10 exons with gene orthologs in hg38 with regulated exons	83	58.87%
- hg38 = > mm10		
Genes with hg38 exon in the exon list	223	
Genes with hg38 exons with gene orthologs in mm10	209	93.72%
Genes with hg38 exons with gene orthologs in mm10 with regulated exons	83	37.22%
- Exon-level stats:		
- mm10 = > hg38		
Exons from mm10 in exon list	159	
Exons from mm10 with gene orthologs in hg38	155	97.48%
Exons from mm10 with exon orthologs in hg38 (G-conserved)	138	86.79%
Out of genes with orthologs		89.03%
Exons from mm10 with regulated exon orthologs in hg38 (R-conserved)	81	50.94%
Out of genes with orthologs		52.26%
Percent of R-conserved / G-conserved exons in mm10		58.70%
Exons from mm10 with gene orthologs with regulated exons in hg38	97	61.01%
Exon orthologous	81	83.51%
Best hit exon	1	1.03%
Unclear case	5	5.15%
Not conserved	10	10.31%
- hg38 = > mm10		
Exons from hg38 in exon list	252	
Exons from hg38 with gene orthologs in mm10	238	94.44%
Exons from hg38 with exon orthologs in mm10 (G-conserved)	173	68.65%
Out of genes with orthologs		72.69%
Exons from hg38 with regulated exon orthologs in mm10 (R-conserved)	81	32.14%
Out of genes with orthologs		34.03%
Percent of R-conserved / G-conserved exons in hg38		46.82%
Exons from hg38 with gene orthologs with regulated exons in mm10	98	38.89%

(continued)

Table 5 (continued)

Exon orthologous	81	82.65%
Best hit exon	1	1.02%
Unclear case	5	5.10%
Not conserved	11	11.22%

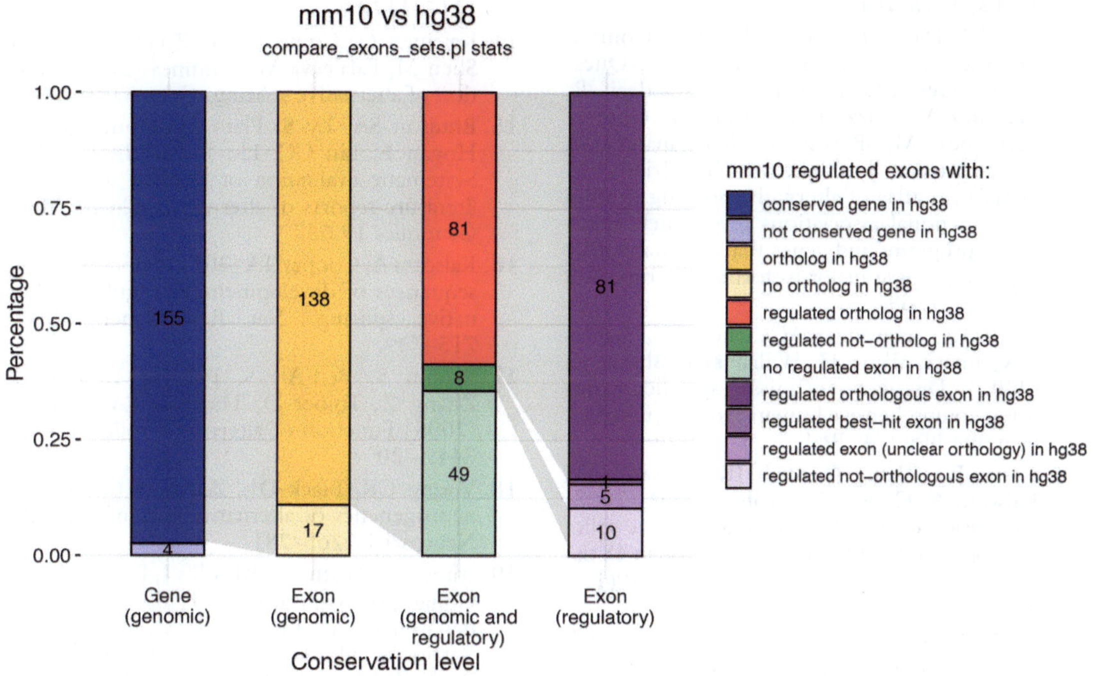

Fig. 12 Graphical summary of the comparison between two exon sets (mm10 against hg38) as output by *ExOrthist compare_exon_sets*

Acknowledgments

We thank Peter Kjer and Sophie Bonnal for their contribution to event function annotations in *VastDB*, as well as current and former members of the Irimia and Valcárcel labs at CRG for their constant feedback. The research associated with the development of the *VastDB* framework has been funded mainly by the European Research Council (ERC) under the European Union's Horizon 2020 research and innovation program (ERC-StG-LS2-637591) and the Spanish Ministerio de Ciencia (BFU2014-55076-P and BFU2017-89201-P). We acknowledge the support of the CERCA Programme/Generalitat de Catalunya and of the Spanish Ministry of Economy, Industry and Competitiveness (MEIC) to the EMBL partnership.

References

1. Wang ET, Sandberg R, Luo S, Khrebtukova I, Zhang L, Mayr C, Kingsmore SF, Schroth GP, Burge CB (2008) Alternative isoform regulation in human tissue transcriptomes. Nature 456(7221):470–476
2. Pan Q, Shai O, Lee LJ, Frey BJ, Blencowe BJ (2008) Deep surveying of alternative splicing complexity in the human transcriptome by high-throughput sequencing. Nat Genet 40(12):1413–1415
3. Tapial J, Ha KCH, Sterne-Weiler T, Gohr A, Braunschweig U, Hermoso-Pulido A, Quesnel-Vallières M, Permanyer J, Sodaei R, Marquez Y, Cozzuto L, Wang X, Gomez-Velazquez M, Rayon T, Manzanares M, Ponomarenko J, Blencowe BJ, Irimia M (2017) An atlas of alternative splicing profiles and functional associations reveals new regulatory programs and genes that simultaneously express multiple major isoforms. Genome Res 27:1759–1768
4. Clark TA, Schweitzer AC, Chen TX, Staples MK, Lu G, Wang H, Williams A, Blume JE (2007) Discovery of tissue-specific exons using comprehensive human exon microarrays. Genome Biol 8(4):R64
5. Castle JC, Zhang C, Shah JK, Kulkarni AV, Kalsotra A, Cooper TA, Johnson JM (2008) Expression of 24,426 human alternative splicing events and predicted cis regulation in 48 tissues and cell lines. Nat Genet 40(12): 1416–1425
6. Irimia M, Denuc A, Burguera D, Somorjai I, Martín-Durán JM, Genikhovich G, Jimenez-Delgado S, Technau U, Roy SW, Marfany G, Garcia-Fernàndez J (2011) Stepwise assembly of the nova-regulated alternative splicing network in the vertebrate brain. Proc Natl Acad Sci U S A 108(13):5319–5324
7. Porter RS, Jaamour F, Iwase S (2018) Neuron-specific alternative splicing of transcriptional machineries: implications for neurodevelopmental disorders. Mol Cell Neurosci 87:35–45
8. Raj B, Blencowe BJ (2015) Alternative splicing in the mammalian nervous system: recent insights into mechanisms and functional roles. Neuron 87:14–27
9. Nakka K, Ghigna C, Gabellini D, Dilworth FJ (2018) Diversification of the muscle proteome through alternative splicing. Skelet Muscle 8:8
10. Vernia S, Edwards YJ, Han MS, Cavanagh-Kyros J, Barrett T, Kim JK, Davis RJ (2016) An alternative splicing program promotes adipose tissue thermogenesis. elife 5:e17672
11. Tress ML, Abascal F, Valencia A (2017) Alternative splicing may not be the key to proteome complexity. Trends Biochem Sci 42(2):98–110
12. Blencowe BJ (2017) The relationship between alternative splicing and proteomic complexity. Trends Biochem Sci 42:407–408
13. Weatheritt RJ, Sterne-Weiler T, Blencowe BJ (2016) The ribosome-engaged landscape of alternative splicing. Nat Struct Mol Biol 23: 1117–1123
14. Kelemen O, Convertini P, Zhang Z, Wen Y, Shen M, Falaleeva M, Stamm S (2013) Function of alternative splicing. Gene 514:1–30
15. Bhuiyan SA, Ly S, Phan M, Huntington B, Hogan E, Liu CC, Liu J, Pavlidis P (2018) Systematic evaluation of isoform function in literature reports of alternative splicing. BMC Genomics 19:637
16. Kalsotra A, Cooper TA (2011) Functional consequences of developmentally regulated alternative splicing. Nat Rev Genet 12(10): 715–729
17. Stamm S, Ben-Ari S, Rafalska I, Tang Y, Zhang Z, Toiber D, Thanaraj TA, Soreq H (2005) Function of alternative splicing. Gene 344:1–20
18. Vuong CK, Black DL, Zheng S (2016) The neurogenetics of alternative splicing. Nat Rev Neurosci 17:265–281
19. Irimia M, Weatheritt RJ, Ellis J, Parikshak NN, Gonatopoulos-Pournatzis T, Babor M, Quesnel-Vallières M, Tapial J, Raj B, O'Hanlon D, Barrios-Rodiles M, Sternberg MJE, Cordes SP, Roth FP, Wrana JL, Geschwind DH, Blencowe BJ (2014) A highly conserved program of neuronal microexons is misregulated in autistic brains. Cell 159:1511–1523
20. Torres-Méndez A, Bonnal S, Marquez Y, Roth J, Iglesias M, Permanyer J, Almudí I, O'Hanlon D, Guitart T, Soller M, Gingras AC, Gebauer F, Rentzsch F, Blencowe BJ, Valcárcel J, Irimia M (2019) A novel protein domain in an ancestral splicing factor drove the evolution of neural microexons. Nat Ecol Evol 3:691–701
21. Braunschweig U, Barbosa-Morais NL, Pan Q, Nachman EN, Alipanahi B, Frey BJ, Irimia M, Blencowe BJ (2014) Widespread intron retention in mammals functionally tunes transcriptomes. Genome Res 24:1774–1786
22. Gueroussov S, Gonatopoulos-Pournatzis T, Irimia M, Raj B, Lin ZY, Gingras AC, Blencowe BJ (2015) An alternative splicing event amplifies evolutionary differences between vertebrates. Science 349:868–873

23. Solana J, Irimia M, Ayoub S, Orejuela MR, Zywitza V, Jens M, Tapial J, Ray D, Morris Q, Hughes TR, Blencowe BJ, Rajewsky N (2016) Conserved functional antagonism of CELF and MBNL proteins controls stem cell-specific alternative splicing in planarians. elife 5:e16797
24. Burguera D, Marquez Y, Racioppi C, Permanyer J, Torres-Mendez A, Esposito R, Albuixech-Crespo B, Fanlo L, D'Agostino Y, Gohr A, Navas-Perez E, Riesgo A, Cuomo C, Benvenuto G, Christiaen LA, Marti E, D'Aniello S, Spagnuolo A, Ristoratore F, Arnone MI, Garcia-Fernandez J, Irimia M (2017) Evolutionary recruitment of flexible Esrp-dependent splicing programs into diverse embryonic morphogenetic processes. Nat Commun 8:1799
25. de Mendoza A, Suga H, Permanyer J, Irimia M, Ruiz-Trillo I (2015) Complex transcriptional regulation and independent evolution of fungal-like traits in a relative of animals. elife 4:e08904
26. Sebe-Pedros A, Irimia M, Del Campo J, Parra-Acero H, Russ C, Nusbaum C, Blencowe BJ, Ruiz-Trillo I (2013) Regulated aggregative multicellularity in a close unicellular relative of metazoa. elife 2:e01287
27. Martín G, Márquez Y, Mantica F, Duque P, Irimia M (2021) Alternative splicing landscapes in Arabidopsis thaliana across tissues and stress conditions highlight major functional differences with animals. Genome Biol 22:35
28. Langmead B, Trapnell C, Pop M, Salzberg SL (2009) Ultrafast and memory-efficient alignment of short DNA sequences to the human genome. Genome Biol 10:R25
29. Shen S, Park JW, Lu ZX, Lin L, Henry MD, Wu YN, Zhou Q, Xing Y (2014) rMATS: robust and flexible detection of differential alternative splicing from replicate RNA-Seq data. Proc Natl Acad Sci U S A 111: E5593–E5601
30. Vaquero-Garcia J, Barrera A, Gazzara MR, Gonzalez-Vallinas J, Lahens NF, Hogenesch JB, Lynch KW, Barash Y (2016) A new view of transcriptome complexity and regulation through the lens of local splicing variations. elife 5:e11752
31. Trincado JL, Entizne JC, Hysenaj G, Singh B, Skalic M, Elliott DJ, Eyras E (2018) SUPPA2: fast, accurate, and uncertainty-aware differential splicing analysis across multiple conditions. Genome Biol 19:40
32. Sterne-Weiler T, Weatheritt RJ, Best AJ, Ha KCH, Blencowe BJ (2018) Efficient and accurate quantitative profiling of alternative splicing patterns of any complexity on a laptop. Mol Cell 72:187–200
33. Labbe RM, Irimia M, Currie KW, Lin A, Zhu SJ, Brown DD, Ross EJ, Voisin V, Bader GD, Blencowe BJ, Pearson BJ (2012) A comparative transcriptomic analysis reveals conserved features of stem cell pluripotency in planarians and mammals. Stem Cells 30:1734–1745
34. Barbosa-Morais NL, Irimia M, Pan Q, Xiong HY, Gueroussov S, Lee LJ, Slobodeniuc V, Kutter C, Watt S, Colak R, Kim T, Misquitta-Ali CM, Wilson MD, Kim PM, Odom DT, Frey BJ, Blencowe BJ (2012) The evolutionary landscape of alternative splicing in vertebrate species. Science 338:1587–1593
35. Gohr A, Irimia M (2019) Matt: Unix tools for alternative splicing analysis. Bioinformatics 35: 130–132
36. Ray D, Kazan H, Cook KB, Weirauch MT, Najafabadi HS, Li X, Gueroussov S, Albu M, Zheng H, Yang A, Na H, Irimia M, Matzat LH, Dale RK, Smith SA, Yarosh CA, Kelly SM, Nabet B, Mecenas D, Li W, Laishram RS, Qiao M, Lipshitz HD, Piano F, Corbett AH, Carstens RP, Frey BJ, Anderson RA, Lynch KW, Penalva LO, Lei EP, Fraser AG, Blencowe BJ, Morris QD, Hughes TR (2013) A compendium of RNA-binding motifs for decoding gene regulation. Nature 499:172–177
37. Witten JT, Ule J (2011) Understanding splicing regulation through RNA splicing maps. Trends Genet 27:89–97
38. Irimia M, Rukov JL, Roy SW, Vinther J, Garcia-Fernandez J (2009) Quantitative regulation of alternative splicing in evolution and development. BioEssays 31:40–50
39. Emms DM, Kelly S (2019) OrthoFinder: phylogenetic orthology inference for comparative genomics. Genome Biol 20:238
40. Derelle R, Philippe H, Colbourne JK (2020) Broccoli: combining phylogenetic and network analyses for orthology assignment. Mol Biol Evol 37:3389–3396
41. Merkin J, Russell CB, Chen P, Burge CB (2012) Evolutionary dynamics of gene and isoform regulation in mammalian tissues. Science 338:1593–1599
42. Alekseyenko AV, Kim N, Lee CJ (2007) Global analysis of exon creation versus loss and the role of alternative splicing in 17 vertebrate genomes. RNA 13(5):661–670
43. Marquez Y, Mantica F, Cozzuto L, Burguera D, Hermoso-Pulido A, Ponomarenko J, Roy SW, Irimia M (2021) ExOrthist: a tool to infer exon orthologies at any evolutionary distance. Genome Biol 22: 239. https://doi.org/10.1186/s13059-021-02441-9

44. Yeo GW, Burge CB (2004) Maximum entropy modeling of short sequence motifs with applications to RNA splicing signals. J Comput Biol 11:377–394
45. Jones DT, Cozzetto D (2012) DISOPRED3: precise disordered region predictions with annotated protein-binding activity. Bioinformatics 31:857–863
46. Kumar M, Gouw M, Michael S, Sámano-Sánchez H, Pancsa R, Glavina J, Diakogianni A, Valverde JA, Bukirova D, Čalyševa J, Palopoli N, Davey NE, Chemes LB, Gibson TJ (2020) ELM-the eukaryotic linear motif resource in 2020. Nucleic Acids Res 48:D296–D306
47. Gonatopoulos-Pournatzis T, Aregger M, Brown KR, Farhangmehr S, Braunschweig U, Ward HN, Ha KCH, Weiss A, Billmann M, Durbic T, Myers CL, Blencowe BJ, Moffat J (2020) Genetic interaction mapping and exon-resolution functional genomics with a hybrid Cas9-Cas12a platform. Nat Biotechnol 38: 638–648
48. Thomas JD, Polaski JT, Feng Q, De Neef EJ, Hoppe ER, McSharry MV, Pangallo J, Gabel AM, Belleville AE, Watson J, Nkinsi NT, Berger AH, Bradley RK (2020) RNA isoform screens uncover the essentiality and tumor-suppressor activity of ultraconserved poison exons. Nat Genet 52:84–94
49. Raj B, O'Hanlon D, Vessey JP, Pan Q, Ray D, Buckley NJ, Miller FD, Blencowe BJ (2011) Cross-regulation between an alternative splicing activator and a transcription repressor controls neurogenesis. Mol Cell 43(5): 843–850
50. Nakano Y, Jahan I, Bonde G, Sun X, Hildebrand MS, Engelhardt JF, Smith RJ, Cornell RA, Fritzsch B, Bánfi B (2012) A mutation in the Srrm4 gene causes alternative splicing defects and deafness in the Bronx waltzer mouse. PLoS Genet 8:e1002966
51. Calarco JA, Superina S, O'Hanlon D, Gabut M, Raj B, Pan Q, Skalska U, Clarke L, Gelinas D, van der Kooy D, Zhen M, Ciruna B, Blencowe BJ (2009) Regulation of vertebrate nervous system alternative splicing and development by an SR-related protein. Cell 138(5): 898–910
52. Quesnel-Vallières M, Irimia M, Cordes SP, Blencowe BJ (2015) Essential roles for the splicing regulator nSR100/SRRM4 during nervous system development. Genes Dev 29: 746–759
53. Raj B, Irimia M, Braunschweig U, Sterne-Weiler T, O'Hanlon D, Yuan-Lin Z, Chen IG, Easton L, Ule J, Gingras AC, Eyras E, Blencowe BJ (2014) Global regulatory mechanism underlying the activation of an exon network required for neurogenesis. Mol Cell 56: 90–103
54. Ellis JD, Barrios-Rodiles M, Colak R, Irimia M, Kim T, Calarco JA, Wang X, Pan Q, O'Hanlon D, Kim PM, Wrana JL, Blencowe BJ (2012) Tissue-specific alternative splicing remodels protein-protein interaction networks. Mol Cell 46:884–892

Chapter 8

Identification and Quantification of Microexons Using Bulk and Single-Cell RNA-Seq Data

Guillermo E. Parada and Martin Hemberg

Abstract

The analysis of RNA-seq has greatly improved the characterization and understanding of the transcriptome. In particular, RNA-seq experiments have extended catalogs of alternative splicing events. However, the analysis of RNAs-seq data for detection and quantification of microexons, extremely short exons of length up to 30 nt, require specialized computational workflows. Here, we describe MicroExonator, a reproducible computational workflow for microexon splicing analysis using bulk or single-cell RNA-seq data.

Key words Microexons, Alternative splicing, Single-cell RNAseq

1 Introduction

Alternative splicing (AS) is a pivotal step for RNA processing in higher eukaryotes and it significantly expands the complexity of the transcriptome [1]. In metazoans, AS events are found for most genes, and they are regulated across developmental stages, tissues, and cell types [2, 3]. The most extensive AS programs are associated with neurons, and an incompletely studied subset of AS events are those involving microexons [4–7]. Microexons are defined as exons of length $\leq$ 30 nts and they constitute the most evolutionarily conserved network of AS events described to date, suggesting a central role in neuronal development [8, 9].

Today, one of the most powerful and widely used assays for studying AS events is RNA-seq, and these experiments have greatly expanded the catalog of AS events [3, 10–12]. Since RNA-seq aligners often fail to detect microexon splice junctions while mapping reads to the genome, the development of custom computational strategies are required to analyze microexon AS events. To facilitate the discovery and quantification of microexons, we recently developed MicroExonator [13] which is the first publicly available computational workflow which is purpose-built for these

Peter Scheiffele and Oriane Mauger (eds.), *Alternative Splicing: Methods and Protocols*,
Methods in Molecular Biology, vol. 2537, https://doi.org/10.1007/978-1-0716-2521-7_8,

tasks. To ensure reproducibility and compatibility on high performance computing (HPC) clusters, MicroExonator was implemented on Snakemake [14] and it is freely available under MIT licence at https://github.com/hemberg-lab/MicroExonator. Here, we provide a detailed description of how to run MicroExonator.

2 Materials

MicroExonator can run on any unix based system as a Snakemake workflow, it is fully compatible with multiple job schedulers, including LSF and SLURM where it has been tested. MicroExonator's core software dependencies are automatically installed by conda through an integrated Snakemake virtual environment management. As an option, MicroExonator also carries out differential splicing analyses with the use of other downstream computational tools, and we provide a workflow for full integration with the widely used Whippet [15] software.

2.1 Hardware

MicroExonator can be run on standalone workstations, but if available we strongly recommend using a HPC cluster that operates a unix-based system such as Linux or Mac OSX. Unless mentioned otherwise, these instructions will assume that MicroExonator is run on a HPC cluster. The minimal amount of RAM is 16 GB and at least 4 CPUs is recommended.

2.2 Installation

MicroExonator's software dependencies are handled by conda, and it needs to be installed by the user. A minimal free installer of conda is available at https://docs.conda.io/en/latest/miniconda.html. Once conda is installed, it needs to be configured to incorporate the bioconda channel to allow access to a large collection of bioinformatics software [16], and it can be setup with the following commands.

```
conda config --add channels defaults
conda config --add channels bioconda
conda config --add channels conda-forge
```

This enables the creation of a virtual environment that contains Snakemake, from which MicroExonator can be executed. To fully take advantage of recent developments of package managing system, mamba should be installed in the base environment and it can then be used to install Snakemake with the following commands.

```
conda install -n base -c conda-forge mamba
mamba create --name snakemake_env snakemake
```

The above line creates a virtual environment called `snakemake_env` containing the latest stable version of Snakemake. The rest of MicroExonator's software dependencies will be installed by Snakemake through the dynamic creation of conda virtual environments during run time.

2.3 Setup

The most recent version of MicroExonator is available on GitHub and can be downloaded by cloning the repository.

```
git clone https://github.com/hemberg-lab/MicroExonator
```

This will create a folder called `MicroExonator` where the code required to execute MicroExonator will be downloaded, and it will store all output there. To execute MicroExonator, different files need to be created inside this folder. These files enable MicroExonator to work under a HPC cluster job scheduler and obtain the mandatory parameters for execution. For each different project or run, we recommend starting with a fresh copy of MicroExonator to avoid interference due to intermediary files or to accidentally overwrite final outcomes from previous runs. Additionally, this ensures that all the configuration files used to run MicroExonator are saved.

Since MicroExonator is still under development, to ensure the compatibility with the commands presented in this Chapter we recommend to access the MiMB branch.

```
cd MicroExonator
git checkout MiMB
```

2.3.1 Cluster Configuration

The execution of tasks on a HPC cluster is often organized by a job scheduler. For optimal execution of MicroExonator, users have the option to create a JavaScript Object Notation (JSON) file that instructs Snakemake how to assign the resources to each process and perform a direct job submission through the scheduler of choice. JSON is a standard file format that enables data interchange, and it is typically used to define a number of name/value pairs [17]. Throughout this chapter it is assumed that the file is placed inside MicroExonator's folder under the name of `cluster.json`.

Inside the `cluster.json` file all relevant parameters for job submission need to be defined and these will vary depending on which job scheduler is used. Each process of the MicroExonator pipeline is defined as a Snakemake rule (located in the `./rules` folder) and a default assignment of resources can be done under `"__default__"` inside the `cluster.json` file. For example in the case of the LSF scheduler, this entry may look like this.

Table 1
List of recommended CPUs and RAM memory for each rule

Rule name	#CPUs	RAM memory (GB)
Round1_bwa_mem_to_tags	5	Default
hisat2_genome_index	5	Default
Round2_bowtie_to_tags	5	Default
bowtie_genome_index	Default	30 for human genome
bowtie_to_genome	2	Default
total_hisat2_to_genome	5	Default
Output	2	30
whippet_quant	Default	2

```
"__default__" :
    {
        "queue"      : "normal",
        "nCPUs"      : "1",
        "memory"     : 10000,
        "resources"  : "\"select[mem>10000] rusage[mem=10000]
span[hosts=1]\"",
        "name"       : "JOBNAME.{rule}.{wildcards}",
        "output"     : "logs/{rule}.{wildcards}.out",
        "error"      : "logs/{rule}.{wildcards}.err",
        "Group"      : "team_hemberg",
        "tCPU"       : "99999"
    }
```

The above code ensures that 10 GB of RAM memory and one CPU are assigned to each rule. These are the recommended minimum hardware resources for the human and mouse genomes. Additional entries within the `cluster.json` are required to assign specific resources to certain rules. For example, the rule named "Round1_bwa_mem_to_tags" runs faster if additional CPUs are provided. To assign 5 CPUs instead of 1 CPU given by default one can define the following entry within `cluster.json`.

```
"Round1_bwa_mem_to_tags" :
{
    "nCPUs"    : 5
},
```

The full list of recommended resources for each rule can be found in Table 1 and an example of a valid `cluster.json` for running MicroExonator under LSF is located in `./Examples/Cluster_config`.

2.3.2 Listing Input Datasets

MicroExonator can process samples that are stored on a local hard drive or a remote server. In the latter case, files will be downloaded and temporarily stored to reduce the required disk space. Specific files need to be created inside MicroExonator's folder to enable the processing of the samples. The information about locally stored RNA-seq data needs to be provided as a tab-separated values (TSV) file named `local_samples.tsv`. This file should contain two columns named `path` and `sample`, which must indicate the path of fastq.gz files and their corresponding sample names. Example.

```
path    sample
/foo/condition_A.rep1.fastq.gz    A1
/foo/condition_A.rep2.fastq.gz    A2
/foo/condition_A.rep3.fastq.gz    A3
/foo/condition_B.rep1.fastq.gz    B1
/foo/condition_B.rep2.fastq.gz    B2
/foo/condition_B.rep3.fastq.gz    B3
```

Similarly, MicroExonator can take fastq files compressed using gzip from URLs if these are specified in a file named `sample_url.tsv`. This file should have the same structure as the file indicated above, but instead of `path`, a column named `url` must contain the direct URLs from which gzip compressed fastq files can be downloaded. Finally, samples stored in NCBI's Short Read Archive (SRA) can also be used as input if the corresponding accession numbers are listed (one per row) in a text file called `NCBI_accession_list.txt`. All these files need to be located inside the MicroExonator folder under these specific names. Note that all files do not need to be stored in the same location and it is possible to take files from all three types of locations as inputs to a single run.

2.3.3 Create a Configuration File

The parameters for MicroExonator need to be specified in a configuration file. This file must be located inside MicroExonator's folder under the name config.yaml. The values for each parameter should be provided as YAML format [18] and the set of parameters required to run MicroExonator will be discussed in the sections below.

3 Discovery and Quantification of Microexons

MicroExonator contains a collection of modules for processing bulk and single-cell RNA-seq data. The two main MicroExonator modules are discovery (Round1) and quantification (Round2) of microexons. Each module comprises a collection of Snakemake rules, and the code for each of these modules can be found in the `rules` folder and the temporary files generated are stored in the

Round1 and Round2 folders. Snakemake's rule definition syntax makes it possible to read each command directly from MicroExonator's code. Moreover, the execution of each command can be listed before running the pipeline through a dry-run and or visual representation of the directed acyclic graph that Snakemake builds to organize the job execution during runtime.

Both modules can take either bulk or single-cell data as input. During the discovery phase reads are first mapped to exon–exon junction tags using BWA-MEM [19]. Then MicroExonator processes the obtained alignments to generate a list of putative microexons, which is filtered and considered for quantification. During the quantification phase the reads are mapped to exon–microexon–exon tags that represent both annotated microexons and putative microexons identified during the discovery phase. The quantification phase uses bowtie and reads that span splice junctions are further filtered and quantified for percent spliced-in (PSI) calculations.

3.1 Minimal Configuration and Optional Parameters

To execute the discovery and quantification modules, minimal and optional parameters need to be defined inside the `config.yaml` file. The list of parameters can be found in Table 2.

Examples of valid `config.yaml` files can be found inside `./Examples/Runs` directory. Please note that the provided genome must have all the chromosomes present in the transcript annotation files (GTF/BED12), therefore it is recommended they are taken from the same source, for example, UCSC Genome database, ENSEMBL, or FlyBase. Processing of paired-end samples can be arranged by supplying an additional TSV file within the `config.yaml` file using the `paired_samples` parameter indicated in Table 2.

3.2 Execution

Once the discovery and quantification modules have been configured, MicroExonator can be run through a Snakemake call. Depending on the amount of data that is being analysed and the hardware resources, MicroExonator can take several hours or days to run. Thus, to ensure an interrupted execution of MicroExonator, we recommend to first create a screen session.

```
screen -S session_name
```

Screen is a GNU software application that can be used to multiplex virtual consoles which are administered within sessions. This command creates and activates a session named `session_name` (choose a name of preference), which enables the execution of any command while preventing the interruption of their execution after losing the connection with the terminal or accidentally closing the terminal. From a screen session, one needs to activate the conda virtual environment that contains Snakemake.

Table 2
Parameters required for configuring discovery and quantification modules

Parameter	Mandatory	Description
Genome_fasta	Yes	Path to multifasta file containing the genomic sequences
working_directory	Yes	Path to MicroExonator cloned folder
Gene_anontation_bed12 Gene_anontation_GTF	Yes (at least one of these)	Path to transcriptome annotation file formatted as 12-column BED file (BED12) Path to transcriptome annotation file formatted as gene transfer file (GTF). Note that GTF must have gene, transcripts and exons annotated as features. Examples of valid GTFs can be found at GENCODE or ENSEMBL
ME_len	No	Maximum microexon size (default: 30)
GT_AG_U2_5 GT_AG_U2_3	No	Path to file containing position weight matrices for U2 GT-AG 5′/3′ splice sites. These are available at SpliceRack [20] for some organisms. Example files can be found in the ./PWM folder
min_number_files_detected	No	The minimum number of files where a putative microexon needs to be detected in order to be considered (default: 3)
conservation_bigwig	No	Path to a bigwig file containing genome-wide conservation scores generated by phyloP or PhastCons. These scores are used to prevent highly conserved microexons from being discarded
min_conservation	No	If conservation_bigwig is provided, then this is the microexon mean conservation value used as threshold for the filtering steps
Optimize_hard_drive	No	Temporarily remove the internal copies of fastq.gz files between the discovery and quantification module. This reduces the total disk space that MicroExonator uses, but it will require data to be redownloaded
ME_DB	No	Path to a BED file of 6 or 12 columns that contains annotated microexons. For example, microexons annotated in VastDB can be included in this way
min_reads_PSI	No	Minimum number of reads required to compute a PSI value. NAs will be provided when this threshold is not meet (default = 3)
paired_samples	No	Path to a TSV file containing two (unnamed) columns that correspond to the paired samples. If this parameter is provided, MicroExonator will pool the counts obtained from both pairs into a single sample name (first TSV column)
Keep_fastq_gz	No	If this is set as T, then MicroExonator will retain all processed fastq.gz files at .FASTQ/ directory. Otherwise they will be deleted as soon as the run is completed
skip_mixture_model_filter	No	Enables skipping of the Gaussian mixture model filter that removes putative microexons based on splice site strength

```
conda activate snakemake_env
```

The discovery and quantification modules can then be run with the following command.

```
snakemake -s MicroExonator.smk --cluster-config cluster.
json --cluster {cluster system params} --use-conda -j {number
of parallel jobs}
```

where `-s` leads to the execution of the master MicroExonator's script (MicroExonator.smk). The communication between Snakemake and the job scheduler is specified by the cluster configuration file, `cluster.json`, with `--cluster-config` and the specific scheduler command should be declared within quotes after the `--cluster` flag. For example, for LSF the command to submit jobs is bsub, so `{cluster system params}` needs to be replaced by `"bsub -n {cluster.nCPUs} -R {cluster.resources} -c {cluster.tCPU} -G {cluster.Group} -q {cluster.queue} -o {cluster.output} -e {cluster.error} -M {cluster.memory}"`, where all the specific parameters that are used by bsub are obtained from the `cluster.json` file by the `cluster.*` operations. To directly run the MicroExonator commands on a workstation, both `--cluster-config` and `--cluster` should be omitted. The flag `--use-conda` enables the dynamic creation of conda virtual environments during run time, which will lead to the installation of all necessary dependencies for executing this module. Finally, `{number of parallel jobs}` needs to be replaced by an integer that indicates the number of parallel jobs the user aims to run simultaneously. For example, running the discovery and quantification modules under LSF while allowing 50 simultaneous jobs will require the following command.

```
snakemake -s MicroExonator.smk --cluster-config cluster.
json --cluster "bsub -n {cluster.nCPUs} -R {cluster.resources}
-c {cluster.tCPU} -G {cluster.Group} -q {cluster.queue} -o
{cluster.output} -e {cluster.error} -M {cluster.memory}" --
use-conda -j 50
```

3.2.1 Snakemake Additional Parameters

Snakemake provides several additional flags that allow fine control over the execution of MicroExonator. For example, as mentioned in Subheading 3, Snakemake can perform a dry-run where it will only check if all the correct parameters are in place to build the execution graph. The dry-run can be executed by including the `-n` flag in the Snakemake call to prevent jobs from being submitted. To visualize the commands that will be submitted, users can provide the -p flag, which will print over execution time (which could be a dry-run) the commands derived from applying the MicroExonator rules to the input samples. Any errors encountered during MicroExonator's execution will prevent other parallel jobs from being submitted. This default behavior can be changed by including the -k flag, which allows MicroExonator to keep running even after some jobs have failed. We recommend using the -k flag, as the failed steps can be fixed later and when rerunning MicroExonator files that have been successfully generated do not need to be recreated.

To optimize the hard drive space occupied by MicroExonator during run time, output files generated by the rules are defined as temporary files. All the temporary files can be kept by including the `--no-temp` flag. Although this is not recommended if disk space is limited, it can be useful for troubleshooting. To further limit the disk usage or Internet connection (in case the input files are not stored locally), MicroExonator defines the number of simultaneous downloads as a resource named `get_data`. To limit the number of files that are downloaded simultaneously by MicroExonator, users can include `--resources get_data = N` into the Snakemake call, where N is the number of simultaneous download processes allowed.

3.2.2 Using Target Rules for Workflow Control

A central concept when building a Snakemake workflow are targets. Targets are specified at the end of the command line call, and the choice determines the output. When no targets are defined, as in the examples shown in Subheading 3.2, then MicroExonator will use `quant` as the default target, which will result in the execution of the discovery and quantification modules. If one wants to execute only the discovery module, then `discovery` should be used as a target.

```
snakemake -s MicroExonator.smk  --cluster-config cluster.
json --cluster {cluster system params} --use-conda -j {number
of parallel jobs} discovery
```

This command produces `*.sam.row_ME.filter1` files in the Round1 folder, which contain all putative microexons that are discovered using this module. The list of putative microexons detected for each file constitutes a preliminary list of detected microexons, but further filtering steps are performed during the quantification module. To avoid accidental deletion or overwriting of these files, they are marked as protected files by Snakemake, and they can only be manually removed by the users. Moreover, note that if discovery is provided as a target, fastq files will be only temporarily stored inside MicroExonator's folder. Other modules from MicroExonator that can be run individually will be discussed in Subheadings 4 and 5.

Table 3
Description of MicroExonator highly quality list of detected microexons

Column name	Description
ME	Microexon coordinates
Transcript	Transcript where the microexon was detected
Total_coverage	Total coverage across all microexon splice junctions
Total_SJs	Splice junctions where the microexon was detected
ME_coverages	Comma-separated list of coverage values for each microexon splice junction
ME_length	Microexon length
ME_seq	Microexon sequence
ME_matches	Microexon number of matches inside the intron
U2_score	U2 splicing score
Mean_conservation	Mean conservation values (if phyloP score was provided)
P_MEs	Microexon confidence score
Total_ME	ME coordinates, U2 score and conservation for all microexon matches
ME_P_value	Value used for the final microexon filters
ME_type	Microexon type (IN, RESCUED, or OUT)

3.3 Output

The main results from the discovery and quantification modules can be found inside the `./Report` folder. Inside this folder, the `out.high_quality.txt` file contains information about the final list of microexons (Table 3). As separate files, the same type of information is reported for other microexons that are likely to be false positive in the `out_shorter_than_3_ME.txt` (<3 nt) and `out_low_scored_ME.txt` (≥ 3 nt) files. Additionally, quantitative results can be found in `Report/quant` as individual files that provide the estimated percentage of spliced in (PSI) values for microexons present in any of the output files (Table 4).

To calculate PSI values that takes into account the number of reads that fully span over both microexon splice sites, `use_corrected_PSI` should be set to `T` in config.yaml. This will generate additional quantification files located at `Report/quant/corrected/PSI_sparse/`, which will be considered for downstream quantitative analyses.

4 Detecting Differentially Included Microexons in Bulk RNA-Seq

MicroExonator can generate a GTF to incorporate detected microexons into gene transcripts models. This file allows the new microexons to be included in downstream alternative splicing analysis, for

Table 4
Description of MicroExonator quantification output

Column name	Description
File	Sample name
ME_coords	Microexon coordinates
SJ_coords	Splice junctions where the microexon was detected
ME_coverages	Comma-separated list of values corresponding to the number of reads that evidence microexon inclusion
SJ_coverages	Comma-separated list of values corresponding to the number of reads that evidence microexon exclusion
PSI	Microexon percent spliced-in values
CI_Lo	Lower bound of the PSI confidence interval
CI_Hi	Upper bound of the PSI confidence interval
Alt5	Alternative microexon 5′ splice sites
Alt3	Alternative microexon 3′ splice sites
Alt5_coverages	Alternative microexon 5′ splice site supporting reads
Alt3_coverages	Alternative microexon 3′ splice site supporting reads
Unique_ME_reads	Unique read sequences that support a microexon inclusion
sum_ME_SJ_coverage_up	Number of splice reads covering upstream microexon splice junction
sum_ME_SJ_coverage_down	Number of splice reads covering downstream microexon splice junction

example, using DEXseq [21], rMATS [22], or Whippet [15], where they otherwise would have been missed. To generate this file one need to run MicroExonator specifying it as a target.

```
    snakemake -s MicroExonator.smk  --cluster-config cluster.
json --cluster {cluster system params} --use-conda -j {number
of parallel jobs} Report/out.high_quality.gtf
```

4.1 Using Whippet to Assess Alternative Splicing of Microexons

Here, we present MicroExonator's downstream module for integrating Whippet to assess microexon differential inclusion between sample conditions. Whippet is a fast, lightweight and accurate method for quantifying alternative splicing [15]. Given an input gene annotation file, Whippet builds contiguous splice graphs (CSGs) to represent each transcript. Each CSG node represents a nonoverlapping exonic sequence, while edges represent splice junctions or contiguous exonic regions. By mapping reads directly to the CSG, Whippet enables a fast annotation-oriented quantification of splicing events.

4.1.1 Installing Whippet

To enable Whippet to be executed as part of the MicroExonator workflow, this needs to be installed from its repository (`https://github.com/timbitz/Whippet.jl.git`) along with its compatible version of Julia. For example, the current version of Whippet runs with Julia v1.6.0, which can be installed by cloning Julia's repository hosted in GitHub.

```
git clone https://github.com/JuliaLang/julia
git checkout v1.6.0
make
```

This will provide an executable version of julia that needs to be used to install Whippet from its repository.

```
git clone https://github.com/timbitz/Whippet.jl.git
cd Whippet.jl
/path/to/julia --project -e 'using Pkg; Pkg.instantiate()'
```

where `/path/to/julia` correspond to the absolute path where the compatible version of Julia can be found.

4.1.2 Configure MicroExonator to Use Whippet

Before running Whippet through MicroExonator, additional terms need to be defined in the `config.yaml` file. These indicate the full path of Whippet and Julia. Moreover, a transcriptome annotation file needs to be provided as GTF format. Examples of these additional lines in the config.yaml are provided below:

```
whippet_bin_folder : /path/to/Whippet.jl/bin
julia : /path/to/julia/julia
Gene_anontation_GTF : /path/to/gencode.annotation.gtf
```

Furthermore, it is possible to skip the MicroExonator discovery and quantification modules by including the following row in the `config.yaml` file.

```
downstream_only : T
```

4.1.3 Running Pairwise Comparison between Experiments

To arrange pairwise comparisons between groups of RNA-seq samples, MicroExonator takes as input a YAML file in which users can specify the desired comparisons.

Example.

```
comparison_ID:
  A : sample1,sample2,sample3
  B : sample4,sample5,sample6
```

There is no upper limit on the number of comparisons that can be submitted inside the YAML file. Condition groups must always be defined as A and B, and a comma separated list of files provided as a value for the conditions. Where in the example above `sample {1.0.6}` are the sample names of the input fastq.gz files. If data is downloaded from NCBI's SRA, then these would correspond to SRA accession numbers. In addition to the parameters indicated in Subheading 4.1.2, the path to this YAML file needs to be included in the `config.yaml` file using a key named `whippet_delta`. Finally MicroExonator can be instructed to execute this comparisons by using `differential_inclusion` as a target rule.

```
snakemake -s MicroExonator.smk  --cluster-config cluster.
json --cluster {cluster system params} --use-conda -j {number
of parallel jobs} differential_inclusion
```

If the discovery and quantification modules were not run previously, this command will run these modules as well, and the results will be automatically coupled to the Whippet analysis (unless `downstream_only` is set as `T`). Advanced users may want to run multiple modules with a single command, which for a large number of input files may involve the execution of thousands of commands. The Snakemake dry-run mode and visualization options are particularly useful and recommended in these cases.

4.1.4 Description of Output

The results from this module can be found inside the `Whippet` folder. In particular, Whippet requantification results can be found in `Whippet/Quant`, where splicing node quantification for each sample is reported as `.psi.gz` files. Note that Whippet provides a full quantification for all alternative splicing events indexed using its CGS model. Therefore, these files also contain the quantification of splicing nodes involved in other types of AS events (for more information, please visit Whippet's online documentation https://github.com/timbitz/Whippet.jl). MicroExonator integrates the quantitative results obtained for microexons nodes, which can be found at `.psi.ME.gz` files. MicroExonator report the analysis of differentially included microexons in the Whippet/ Delta/ folder, where the *.diff.microexons files correspond to the result computed directly with Whippet node quantification, while the *.diff.ME.microexons files correspond to results derived from MicroExonator's initial quantification (omitted when `downstream_only` is set as `T`).

5 Using scRNA-Seq Data to Quantify Microexons

One of the main challenges when analyzing scRNA-seq is the increased demand on computational resources. Each cell can be thought of as a sample, and experiments today often profile tens of thousands of cells. Furthermore, in most scRNA-seq experiments, the sequencing depth is much lower than for bulk RNA-seq experiments and the sparse nature of the data presents additional challenges. Here we describe different objectives that can be achieved using MicroExonator to process scRNA-seq data. MicroExonator takes advantage of the fast and lightweight algorithms implemented in Whippet (installing instructions are described in Subheading 4.1.1). To activate these modules, the following line needs to be included in the YAML file:

```
Single_Cell : T
```

Additionally, some commands will require the definition of cell clusters, which is normally determined through unsupervised clustering of the gene expression matrix. The cluster assignments need to be provided in a TSV file, containing two columns indicating the name of the scRNA-seq files and the cluster which they were assigned to. The path to a TSV file containing clustering metadata should be provided with the `cluster_metadata` parameter in the `config.yaml` file. The names of the relevant columns should be indicated by the `cluster_name` and `file_basename` parameters, which should be also provided in the same file.

5.1 Quantifying Microexons at the Single-Cell Level

MicroExonator can take as input thousands of files resulting from a scRNA-seq experiment and coordinate the steps to provide AS quantification at single-cell level with the aid of Whippet. This can be achieved by defining `quant_unpool_single_cell` as a target.

```
snakemake -s MicroExonator.smk --cluster-config cluster.
json --cluster {cluster system params} --use-conda -k -j
{number of parallel jobs} quant_unpool_single_cell
```

This command will execute the splicing node quantification for every scRNA-seq file using `whippet-quant` and store the results inside the `./Whippet/Quant/Single_Cell/Unpooled/` directory. To avoid writing excessive amounts of data to files, users can provide clutestering metadata (as indicated in Subheading 5) to aggregate the results by cluster group. For this purpose one needs to use `collapse_whippet` as target instead and the results will be stored in the `./Whippet/Quant/Collapsed/` directory.

Table 5
Parameters required for pseudo bulk microexon quantification

Parameter	Description	Default
n_pseudo_bulks	Number of pseudobulks per cluster	5
min_number_of_samples_single_cell	Limits the number of pseudobulk per cluster so that each group is guaranteed to contain a minimum number of cells.	3
min_number_of_reads_single_cell	Minimum number of reads used to calculate PSI. This parameter is equivalent to min_reads_PSI, but it only affects this module.	5
seed	Seed to fix the pseudorandom number generation	123

5.2 Microexon Quantification Across Cell Types

ScRNA-seq experiments typically provide considerably fewer reads for each cell than bulk RNA-seq, and the splicing node quantification is adversely affected by the lack of coverage over splice junctions. Here we present strategies designed to overcome the technical challenges due to sparse read coverage in scRNA-seq data.

5.2.1 Quantifying Microexons as Pseudobulk

To effectively increase the number of splicing nodes that can be quantified using scRNA-seq data, MicroExonator can randomly assign the input .fastq.gz files, corresponding to each single cell, into pseudobulk groups. Each group contains files from the same cell-type (defined by the provided clustering metadata). To run this module additional parameters need to be defined inside `config.yaml` (Table 5) and then this module can be executed using `collapse_pseudo_pool` as a target.

```
snakemake -s MicroExonator.smk --cluster-config cluster.
json --cluster {cluster system params} --use-conda -k -j
{number of parallel jobs} collapse_pseudo_pools
```

This command executes whippet-quant for each of the pseudobulks, and the outputs are stored in the `Whippet/Quant/Single_Cell/Pseudo_bulks/` directory. The output includes `pseudo_bulk_membership.tsv` which reports how cells were assigned to pseudobulk groups, and other files containing the obtained whippet-quant output for every pseudobulk group.

5.2.2 Performing Pairwise Comparisons Between Cell Types

To detect alternatively included microexons across cell types, MicroExonator has a dedicated module named Snakepool. In this module, pseudobulks are defined according to user-defined comparisons which need to be provided as a TSV file (run metadata) with a format that is specified in Table 6. Snakepool can perform pairwise comparisons between two groups of cell clusters (defined by provided clustering metadata), for which pseudobulk

Table 6
Description of run metadata file

Column name	Description
Compare_ID	User-defined name for scheduled comparison
A.cluster_names	Comma-separated list of cell types to be considered as part of sample group A
A.number_of_pools	Number of pseudobulk to be generated for sample group A
B.cluster_names	Comma-separated list of cell types to be considered as part of sample group B
B.number_of_pools	Number of pseudobulk to be generated for sample group B
Repeat	Number of computational replicates

quantification is performed using whippet-quant. Then whippet-delta is used to assess differential inclusion of splicing nodes across each comparison. To increase the reproducibility of the results, this process is repeated a number of times for each comparison as defined inside the run metadata file. Finally, the posterior probabilities of differential inclusion for each splicing node are fitted using a Beta distribution to identify nodes that are consistently differentially included across repetitions. Optional parameters to adjust this process can be provided as part of the `config.yaml` file (Table 7).

To execute the defined comparisons between cell-type groups `snakepool` needs to be given as a target.

```
snakemake -s MicroExonator.smk --cluster-config cluster.
json --cluster {cluster system params} --use-conda -k -j
{number of parallel jobs} snakepool
```

Final results from these analyses are stored in the `Whippet/Delta/Single_Cell/Sig_nodes/` directory, where differentially included microexons detected for each defined comparison will be listed in the `.all_nodes.microexons.txt` files. The content of these files is described in Table 8.

6 Troubleshooting

When an error is encountered during a MicroExonator run we recommend paying close attention to the error message reported from the failed step. Most job schedulers can redirect the standard output and standard error output to files, for example using the flags `-o` and `-e` with the LSF bsub command. To name the output files according to the error, `{rule}` and `{wildcards}` can be used inside `cluster.json`. For the `cluster.json` example given at Subheading 2.3.1, these files will be written to the `./logs` folder, where other log files generated by MicroExonator are also placed.

Table 7
Parameters required to perform differential inclusion analyses between cell types

Parameter	Description	Default
run_medatada	Path to file described by Table 6	None
cdf_t	Threshold to run a cumulative distribution function over the obtained Beta distribution of posterior probability values calculated for each splicing node. In practical terms, it can be considered as a user-defined threshold (between 0.5 and 1) for calculating a p-value associated with the node's probability of differential inclusion	0.8
min_p_mean	Minimum mean probability value to be considered to correspond to a significant AS difference	0.9
min_delta	Minimum mean DeltaPSI value to be considered to correspond to a significant AS difference	0.1
min_rep	When nodes have limited read coverage across cells, only some computational replicates could enable quantitative alternative splicing analysis. This parameter sets a minimum number of successful computational replicates that needs to be considered in order to catalogue an splicing node as differentially included for a given comparison	25
snakepool_seed	Seed to fix the pseudorandom number generation	123
Only_snakepool	A Boolean variable that when set to T, MicroExonator only performs AS analysis for all splicing nodes without specifically considering the nodes that represent microexons	F

Note that when a small number of microexons are detected, the Gaussian mixture model that is used to compute the microexon confidence score may fail. This failure will induce an error when executing the `Output` rule. To overcome this issue, users can opt to skip this filtering step by setting `skip_mixture_model_filter` as `T` inside the `config.yaml` file. If one introduces this setting after encountering this error, MicroExonator should resume the analysis without rerunning any step. To avoid overwriting final outcomes, several files generated during the final stages are declared as "protected" and MicroExonator runs that would replace these files will trigger errors during Snakemake execution time. These errors need to be carefully resolved to avoid having to start all over again. Modifying file properties—for example, timestamp, write permissions and path—of these protected outputs are common solutions to these issues. To fully restart a MicroExonator run with other settings (alternative genomes or transcriptome annotation files) users need to start with a fresh MicroExonator git clone or remove folders that were generated during the rung, such as data, Round1, Round2, Report, and Whippet.

Table 8
Description of output obtained from differential inclusion analyses between cell types

Column name	Description
Gene	Gene ID
Node	Node number inside the gene
Coord	Node coordinate
Strand	Plus or minus strand
Type	Whippet's node type
Psi_A.mean	Mean PSI values for group cluster A across computational replicates
Psi_B.mean	Mean PSI values for group cluster B across computational replicates
DeltaPsi.mean	Mean DeltaPsi values obtained across computational replicates
DeltaPsi.sd	Standard deviation of DeltaPsi values obtained across computational replicates
Probability.mean	Mean probability of differential inclusion obtained across computational replicates.
Probability.var	Variance of probability across computational replicates.
N.detected.reps	Number of replicates in which the differential inclusion could be assessed.
cdf.beta	p-value of being above the used defined probability threshold cdf_t
is.diff	Boolean variable defining whether the node was differentially included according to the user-defined criteria (min_rep, min_p_mean, and min_delta)
microexon_ID	Microexon ID based on its genomic coordinates

Acknowledgments

This work was supported by a core grant from the Wellcome Trust. We thank Dr. John Calarco and Dr. Gonzalo Riadi for useful comments over this chapter and Camilo Fuentes Beals for software testing.

References

1. Kornblihtt AR, Schor IE, Alló M, Dujardin G, Petrillo E, Muñoz MJ (2013) Alternative splicing: a pivotal step between eukaryotic transcription and translation. Nat Rev Mol Cell Biol 14:153–165
2. Maniatis T, Tasic B (2002) Alternative pre-mRNA splicing and proteome expansion in metazoans. Nature 418:236–243
3. Pan Q, Shai O, Lee LJ, Frey BJ, Blencowe BJ (2008) Deep surveying of alternative splicing complexity in the human transcriptome by high-throughput sequencing. Nat Genet 40: 1413–1415
4. Vuong CK, Black DL, Zheng S (2016) The neurogenetics of alternative splicing. Nat Rev Neurosci 17:265–281
5. Tapial J, Ha KCH, Sterne-Weiler T, Gohr A, Braunschweig U, Hermoso-Pulido A, Quesnel-Vallières M, Permanyer J, Sodaei R, Marquez Y, Cozzuto L, Wang X, Gómez-Velázquez M, Rayon T, Manzanares M, Ponomarenko J, Blencowe BJ, Irimia M (2017) An atlas of alternative splicing profiles

and functional associations reveals new regulatory programs and genes that simultaneously express multiple major isoforms. Genome Res 27:1759–1768

6. Ustianenko D, Weyn-Vanhentenryck SM, Zhang C (2017) Microexons: discovery, regulation, and function. Wiley Interdiscip Rev RNA 8. https://doi.org/10.1002/wrna.1418
7. Curry-Hyde A, Chen BJ, Mills JD, Janitz M (2018) Microexons: novel regulators of the transcriptome. J Hum Transcript 2:1–6
8. Irimia M, Weatheritt RJ, Ellis JD, Parikshak NN, Gonatopoulos-Pournatzis T, Babor M, Quesnel-Vallières M, Tapial J, Raj B, O'Hanlon D, Barrios-Rodiles M, Sternberg MJE, Cordes SP, Roth FP, Wrana JL, Geschwind DH, Blencowe BJ (2014) A highly conserved program of neuronal microexons is misregulated in autistic brains. Cell 159: 1511–1523
9. Li YI, Sanchez-Pulido L, Haerty W, Ponting CP (2015) RBFOX and PTBP1 proteins regulate the alternative splicing of micro-exons in human brain transcripts. Genome Res 25:1–13
10. Sultan M, Schulz MH, Richard H, Magen A, Klingenhoff A, Scherf M, Seifert M, Borodina T, Soldatov A, Parkhomchuk D, Schmidt D, O'Keeffe S, Haas S, Vingron M, Lehrach H, Yaspo M-L (2008) A global view of gene activity and alternative splicing by deep sequencing of the human transcriptome. Science 321:956–960
11. Wang ET, Sandberg R, Luo S, Khrebtukova I, Zhang L, Mayr C, Kingsmore SF, Schroth GP, Burge CB (2008) Alternative isoform regulation in human tissue transcriptomes. Nature 456:470–476
12. Haas BJ, Zody MC (2010) Advancing RNA-Seq analysis. Nat Biotechnol 28: 421–423
13. Parada GE, Munita R, Georgakopoulos-Soares I, Fernandes HJR, Kedlian VR, Metzakopian E, Andres ME, Miska EA, Hemberg M (2021) MicroExonator enables systematic discovery and quantification of microexons across mouse embryonic development. Genome Biol 22:43
14. Köster J, Rahmann S (2012) Snakemake--a scalable bioinformatics workflow engine. Bioinformatics 28:2520–2522
15. Sterne-Weiler T, Weatheritt RJ, Best AJ, Ha KCH, Blencowe BJ (2018) Efficient and accurate quantitative profiling of alternative splicing patterns of any complexity on a laptop. Mol Cell 72:187–200.e6
16. Grüning B, Dale R, Sjödin A, Chapman BA, Rowe J, Tomkins-Tinch CH, Valieris R, Köster J, Bioconda Team (2018) Bioconda: sustainable and comprehensive software distribution for the life sciences. Nat Methods 15: 475–476
17. Petković D (2017) SQL/JSON standard: properties and deficiencies. Datenbank-Spektrum 17:277–287
18. Ben-Kiki O, Evans C, Ingerson B (2009) Yaml ain't markup language (yaml™) version 1.1. Working Draft 2008-05 11
19. Li H, Durbin R (2009) Fast and accurate short read alignment with burrows-wheeler transform. Bioinformatics 25:1754–1760
20. Sheth N, Roca X, Hastings ML, Roeder T, Krainer AR, Sachidanandam R (2006) Comprehensive splice-site analysis using comparative genomics. Nucleic Acids Res 34: 3955–3967
21. Anders S, Reyes A, Huber W (2012) Detecting differential usage of exons from RNA-seq data. Genome Res 22:2008–2017
22. Shen S, Park JW, Lu Z-X, Lin L, Henry MD, Wu YN, Zhou Q, Xing Y (2014) rMATS: robust and flexible detection of differential alternative splicing from replicate RNA-Seq data. Proc Natl Acad Sci U S A 111: E5593–E5601

Chapter 9

Functional Annotation of Custom Transcriptomes

Fursham Hamid, Kaur Alasoo, Jaak Vilo, and Eugene Makeyev

Abstract

Many eukaryotic genes can give rise to different alternative transcripts depending on stage of development, cell type, and physiological cues. Current transcriptome-wide sequencing technologies highlight the remarkable extent of this regulation in metazoans and allow for RNA isoforms to be profiled in increasingly small biological samples and with a growing confidence. Understanding biological functions of sample-specific transcripts is a major challenge in genomics and RNA processing fields. Here we describe simple bioinformatics workflows that facilitate this task by streamlining reference-guided annotation of novel transcripts. A key part of our protocol is the R package *factR* that rapidly matches custom-assembled transcripts to their likely host genes, deduces the sequence and domain structure of novel protein products, and predicts sensitivity of newly identified RNA isoforms to nonsense-mediated decay.

Key words Custom transcriptome, RNA sequencing, Single-cell RNA-seq, Long-read RNA-seq, Alternative pre-mRNA processing, Protein domains, Nonsense-mediated decay

1 Introduction

The structure of metazoan mRNAs is extensively regulated by the choice of the transcription start site, alternative splicing and 3′-terminal cleavage/polyadenylation events [1–4]. These mechanisms often depend on the tissue and cell type and are additionally modulated by various external cues [5–7]. In addition to diversifying the proteome [8, 9], alternative mRNA isoforms can also contribute to gene expression dynamics, for example, by giving rise to premature translation termination codons (PTCs) and subjecting PTC-containing transcripts to nonsense-mediated decay (NMD) [10, 11].

Rapidly developing high-throughput transcriptome sequencing approaches have bolstered our ability to analyze gene expression at the isoform level. Initial short-read RNA sequencing (RNA-seq) studies suggested that most mammalian genes produce more than one alternatively spliced product [12, 13]. Similarly, analyses of poly(A)-proximal mRNA sequences pointed at a

Peter Scheiffele and Oriane Mauger (eds.), *Alternative Splicing: Methods and Protocols*,
Methods in Molecular Biology, vol. 2537, https://doi.org/10.1007/978-1-0716-2521-7_9,

widespread nature of alternative cleavage/polyadenylation [14–16]. Using continuously improving single-cell RNA sequencing (scRNA-seq) approaches it is now possible to tease apart transcriptomes of different cells types in heterogeneous populations, organs and entire organisms [17–21]. Long-read, or third-generation, sequencing is becoming increasingly popular in transcriptomics studies [22–24], and further development of this technology will likely lead to an explosion of high-quality custom transcriptome data in the nearest future.

Elucidating functional consequences of alternative splicing and other mechanisms diversifying mRNA sequence has been a long-standing challenge in the field. The first step in this direction often involves bioinformatics reconstruction of sample-specific transcriptomes based on high-throughput sequencing data. This can be achieved with some success in an entirely de novo mode, that is, using RNA-seq data with no additional inputs (e.g., Trans-ABySS [25], Trinity [26], Oases [27], rnaSPAdes [28], and RNA-Bloom [29]). More reliable results can be obtained when genomic and ideally some cDNA sequences are available to guide the assembly process as a reference. This is the preferred approach for human samples and model organisms with sequenced genomes. Several bioinformatics tools have been developed to align RNA-seq reads to reference sequences (e.g., HISAT2 [30], STAR [31], GMAP [32], and minimap2 [33]) and then assemble the alignments into a likely set of "parental" transcripts (e.g., Cufflinks [34], StringTie2 [35, 36], CLASS2 [37], Scallop [38], scRNAss [39], and FLAIR [40]).

Transcriptome assemblers typically report their predictions as a set of genomic coordinates. Unlike the reference transcriptome data that use a General Feature Format (e.g., GTF or GFF3) to specify RNA coding potential, position of the protein-coding sequence (CDS) and the gene of origin, newly assembled transcripts often lack this information. Attempts to predict their possible biological functions currently require the use of an unwieldy collection of computer programs and time-consuming manual inspection of sequences of interest. We therefore believe that a more streamlined and integrated approach to functional annotation of novel transcripts would be beneficial for the field.

Here we describe how custom transcriptomes reconstructed from bulk, single-cell or long-read RNA-seq data can be rapidly annotated using our newly developed R package *factR*. *factR* assigns newly discovered mRNA isoforms to their genes of origin and uses the reference information to map ORFs, deduce domain architecture of the protein products, predict mRNA susceptibility to NMD, and pinpoint alternative splicing events underlying the difference between sample-specific and reference transcripts.

2 Materials

2.1 Computer

Parts of the code described in this chapter are designed for a shell command-line interpreter. We therefore recommend a Mac or a Linux-based computer with at least 8 Gb of RAM, 4 CPU cores and 80 Gb of free disk storage space to run our example workflows. More space may be required to assemble custom transcriptomes starting form user-specific deep sequencing data.

2.2 R, RStudio and factR

Several key steps of our annotation workflow are implemented in the R programming language (https://www.r-project.org/). The latest software to run R code can be downloaded from https://cloud.r-project.org/. Optionally, the R scripts can be executed using the Integrated Development Environment (IDE) RStudio downloadable from https://rstudio.com/products/rstudio/. We use our newly developed R package, *factR*, to perform the functional annotation steps. Detailed instructions on installing *factR* and its dependencies can be found at https://fursham-h.github.io/factR/.

2.3 Preassembled Transcriptome Files

Our workflow requires a custom-assembled transcriptome in the GTF format as an input. Users may provide their own custom GTF files, or use our sample GTF files containing preassembled transcriptomes from bulk, single-cell and long-read RNA-sequencing experiments. These files are available in the folder named "Custom transcriptome" of our GitHub repository (https://github.com/fursham-h/MiMB-factR-supplement). Users may generate their own custom transcriptomes by following and modifying the workflows described in Subheadings 3.1–3.3.

2.4 Shell Scripts

We provide shell script files for three custom transcriptome assembly examples in (download_assemble_bulk.sh, download_assemble_sc.sh, download_assemble_lr.sh). These are available in the "Workflow" folder of our GitHub repository (https://github.com/fursham-h/MiMB-factR-supplement). These scripts require the following additional programs to be installed and added to computer's PATH.

2.5 Additional Programs

1. SRA Toolkit. Unprocessed RNA-seq data stored at NCBI Sequence Read Archive database can be easily downloadable using the fastq-dump program from this toolkit (https://github.com/ncbi/sra-tools/wiki/01.-Downloading-SRA-Toolkit).
2. HISAT2. This program [30] aligns next-generation sequencing reads to a reference genome. Other alignment programs may be used depending on user's preferences (e.g., STAR [31]).

3. Minimap2. This aligner [33], downloadable from https://github.com/lh3/minimap2, is used to process long-read transcriptome sequencing data in one of the workflows described in this chapter.
4. Samtools. This is a suite of programs to read and process aligned sequence files (SAM/BAM/CRAM) [41]. Information on downloading and installing Samtools can be found at Samtools home page (www.htslib.org).
5. StringTie2. We use this tool [35, 36] to assemble aligned RNA-seq reads from bulk, single-cell and long-read sequencing experiments into a likely set of "parental" transcripts. A step-by-step guide on installing StringTie2 can be found at https://ccb.jhu.edu/software/stringtie/#install.

3 Methods

Here, we describe assembly of custom transcriptome files (GTFs) from bulk (*see* Subheading 3.1), single-cell (*see* Subheading 3.2) or long-read RNA-seq data (*see* Subheading 3.3) and functional annotation of newly identified transcripts using *factR* (*see* Subheading 3.4) (Fig. 1). Examples of biological insights that can be obtained using these workflows are summarized in Subheading 3.5.

3.1 Custom Transcriptome Assembly Using Bulk RNA-Seq Data

Unprocessed RNA-seq data are available from public repositories including Sequence Read Archive (SRA), European Nucleotide Archive (ENA), Neuroscience Multi-omic (NeMO), and so on. In this section, RNA-seq data for mouse embryonic stem cells treated with the NMD inhibitor cycloheximide (CHX) or untreated control [42] are downloaded from SRA and processed by HISAT2 and StringTie2 to produce a custom GTF file. For further details on this dataset, see https://www.ncbi.nlm.nih.gov/Traces/study/?acc=PRJNA178171.

1. There are two ways to implement this workflow. Users may simply download the shell script containing the entire code for this section (download_assemble_bulk.sh; folder "Workflow" of GitHub repository; https://github.com/fursham-h/MiMB-factR-supplement) and run it by issuing the following commands in the Terminal window.

```
cd <path to the Workflow directory containing the script>
bash ./download_assemble_bulk.sh
```

2. Alternatively, the process can be followed step by step by copying and pasting relevant parts of the code to the Terminal window. In this case, we begin by downloading the raw sequence files (FASTQs) from the SRA database.

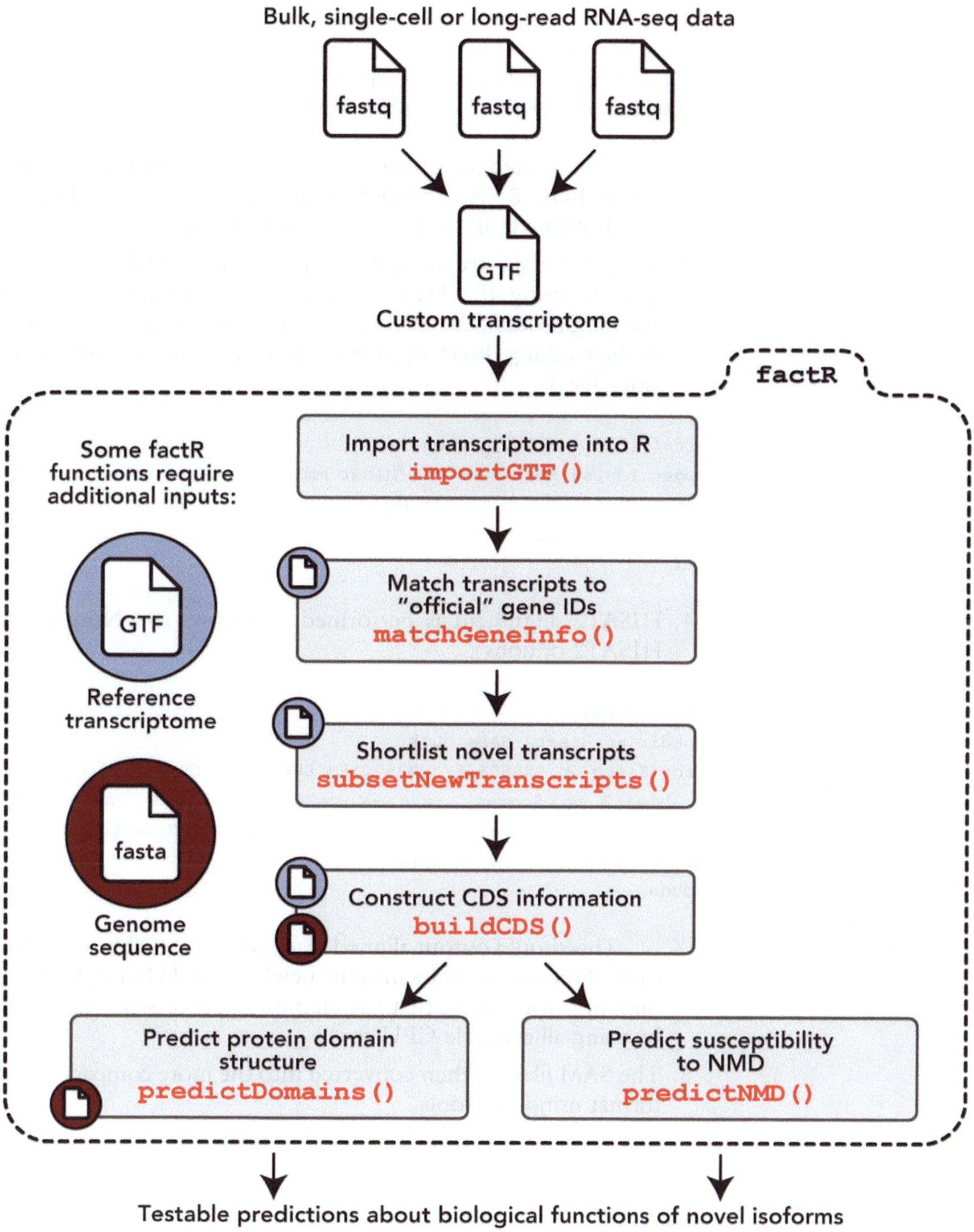

Fig. 1 Functional annotation process described in this chapter

```
cd <path to the Workflow directory>
mkdir -p FASTQ_bulk
fastq-dump -O FASTQ_bulk --split-files -gzip \
SRR606187 SRR606188 SRR606189 SRR606190
```

This should download 8 compressed FASTQ files for 4 paired-end RNA-seq experiments to the newly created directory FASTQ_bulk using the *fastq-dump* program.

3. Next, the data are mapped to the mouse mm10 reference genome using the HISAT2 aligner. In preparation for the mapping process, the HISAT2 mm10 index file is downloaded to the working directory, if not already present (*see* **Note 1** on index files):

```
if [ ! -d "mm10" ]; then
wget idx.s3.amazonaws.com/hisat/mm10_genome.tar.gz
tar -xf "mm10_genome.tar.gz"
rm mm10_genome.tar.gz
fi
```

4. HISAT2 alignment is performed as follows (*see* **Note 2** on HISAT2 options).

```
cores='nproc'
mkdir -p Hisat2_SAMs_bulk
for fastq in SRR606187 SRR606188 SRR606189 SRR606190;do
  hisat2 -p $cores --dta -x mm10/genome -1 \ FASTQ_bulk/
"$fastq"_1.fastq.gz -2 \ FASTQ_bulk/"$fastq"_2.fastq.gz -S \
Hisat2_SAMs_bulk/$fastq.sam
done
```

This should output aligned read files in the SAM format for each of the four samples into the newly created Hisat2_SAMs_-bulk directory. Note that the alignment process is accelerated by using all available CPU cores.

5. The SAM files are then converted into the more compact BAM format using Samtools.

```
mkdir -p Hisat2_sorted_BAMs_bulk
for file in SRR606187 SRR606188 SRR606189 SRR606190; do
 samtools view -@ $cores -Su Hisat2_SAMs_bulk/$file | \
samtools sort -@ $cores -o Hisat2_sorted_BAMs_bulk/$file.bam
done
```

The resultant BAM files are saved in the new Hisat2_sorted_BAMs_bulk directory.

6. We can now assemble the aligned reads into custom transcripts using the reference transcriptome-guided mode of StringTie2. A mouse reference transcriptome file can be downloaded, for example, from GENCODE as follows (version M25).

```
url="ftp://ftp.ebi.ac.uk/pub/databases/gencode/Gencode_-
mouse/release_M25/gencode.vM25.annotation.gtf.gz"
if [ ! -f "gencode.vM25.annotation.gtf" ]; then
 wget $url
 gzip -d gencode.vM25.annotation.gtf.gz
fi
```

7. To assemble a custom transcriptome using default settings of StringTie2 (*see* **Note 3** on running StringTie2 with additional parameters), run the following:

```
mkdir -p Stringtie_gtf_bulk
for file in SRR606187 SRR606188 SRR606189 SRR606190; do
  stringtie Hisat2_sorted_BAMs_bulk/$file.bam -p $cores -o \
Stringtie_gtf_bulk/$file.gtf -G \ gencode.vM25.annotation.gtf
done
```

The above code should output a GTF file for each sample into the Stringtie_gtf_bulk directory.

8. The final set of commands will merge the four GTFs into a nonredundant set of transcripts and compress the output as bulk_merged.gtf.gz file in the current directory.

```
stringtie --merge -G gencode.vM25.annotation.gtf \
Stringtie_gtf_bulk/SRR606187.gtf \
Stringtie_gtf_bulk/SRR606188.gtf \
Stringtie_gtf_bulk/SRR606189.gtf \
Stringtie_gtf_bulk/SRR606190.gtf | gzip > \ bulk_merged.gtf.gz
```

3.2 Custom Transcriptome Assembly Using Single-Cell RNA-Seq Data

In this section, we describe a workflow for downloading single-cell mouse neocortex RNA-seq data [43], combining FASTQ files based on apparent cell identity, aligning reads using HISAT2 and finally constructing the custom transcriptome using StringTie2. The script to execute the code (download_assemble_sc.sh) for this section can be found in our GitHub repository (in subfolder "Workflow" https://github.com/fursham-h/MiMB-factR-supplement).

1. Users may run the entire script.

```
cd <path to the Workflow directory containing the script>
bash ./download_assemble_sc.sh
```

Alternatively, follow the walk-through below.

2. We begin by retrieving paired-end FASTQ files from 506 cells extracted from mouse neocortex from NeMO data archive (https://nemoarchive.org/). This is a subset of cells isolated from the L1-L2/3 and L4-L6 layers of the visual cortex (VISp) and assigned to the glutamatergic neuron, GABAergic neuron, astrocyte or endothelial cell types with a high confidence [43]. A tab-delimited text file (sc_fastq_curated.txt) containing sample names, apparent cell identity and download hyperlink has been created to facilitate with the large data download (from https://github.com/fursham-h/MiMB-factR-supplement). The first 3 entries of this file.

```
cd <path to the Workflow directory>
head -n 3 sc_fastq_curated.txt
```

appear as follows.

LS-15552_S28_E1-50 Endothelial http://data.nemoarchive.org/biccn/lab/zeng/transcriptome/scell/SSv4/mouse/raw/VISp/LS-15552_S28_E1–50.fastq.tar.

LS-15552_S31_E1-50 Endothelial http://data.nemoarchive.org/biccn/lab/zeng/transcriptome/scell/SSv4/mouse/raw/VISp/LS-15552_S31_E1–50.fastq.tar.

LS-15552_S44_E1-50 Endothelial http://data.nemoarchive.org/biccn/lab/zeng/transcriptome/scell/SSv4/mouse/raw/VISp/LS-15552_S44_E1–50.fastq.tar.

3. To download all 506 single-cell FASTQ files, run the following.

```
mkdir -p FASTQ_sc
while IFS=$'\t' read -r NAME ID LINK;do
 wget $LINK -P FASTQ_sc
 tar -xvf "FASTQ/$NAME.fastq.tar" -C FASTQ_sc
 rm FASTQ_sc /$NAME.fastq.tar
done < sc_fastq_curated.txt
```

4. Since the sequencing depth for individual cells is relatively low, we merge FASTQ files for glutamatergic neurons, GABAergic neurons, astrocytes or endothelial cells according to their predicted identities to increase the chance of discovering novel cell type-specific RNA isoforms. Pairs of merged FASTQ files are generated for each cell type using the shell cat function and sc_fastq_curated.txt as a guide.

```
# Create arrays to store filenames and group names
 declare -A fastqArray
 declare -A fastqNames

 while IFS=$'\t' read -r NAME CLASS LINK;do
    fastqArray["$CLASS.1"]="${fastqArray["$CLASS.1"]} \
FASTQ_sc/${NAME}_R1.fastq.gz"
    fastqArray["$CLASS.2"]="${fastqArray["$CLASS.2"]} \
FASTQ_sc/${NAME}_R2.fastq.gz"
   fastqNames["$CLASS.1"]=${CLASS}_R1
   fastqNames["$CLASS.2"]=${CLASS}_R2
 done < sc_fastq_curated.txt

 # Loop groups and concatenate FASTQ files
 for n in Glutamatergic GABAergic Endothelial Astrocyte;do
    cat ${fastqArray["$n.1"]} > \ FASTQ_sc/${fastqNames
["$n.1"]}.fastq.gz
    cat ${fastqArray["$n.2"]} > \ FASTQ_sc/${fastqNames
["$n.2"]}.fastq.gz
 done
```

5. As in the bulk RNA-seq workflow above, the data are aligned to the mouse mm10 reference genome by HISAT2. If you do not have the mm10 index file, follow **step 3** of Subheading 3.1.
6. Run HISAT2 alignment (*see* **Note 2** on running HISAT2 with additional parameters such as --known-splicesite-infile).

```
 cores='nproc'
 mkdir -p Hisat2_SAMs_sc
 for fastq in Glutamatergic GABAergic Endothelial Astrocyte;do
  hisat2 -p $cores --dta -x mm10/genome -1 \ FASTQ_sc/
"$fastq"_R1.fastq.gz -2 \ FASTQ_sc/"$fastq"_R2.fastq.gz -S
Hisat2_SAMs_sc/$fastq.sam
 done
```

7. Convert SAM files to BAM format.

```
 mkdir -p Hisat2_sorted_BAMs_sc
 for file in Glutamatergic GABAergic Endothelial Astrocyte; do
   samtools view -@ $cores -Su Hisat2_SAMs_sc/$file.sam | \
samtools sort -@ $cores -o Hisat2_sorted_BAMs_sc/$file.bam
 done
```

8. You will need a reference transcriptome file for the subsequent StringTie2 steps. This file can be downloaded from GENCODE as described in **step 6** of Subheading 3.1 above.

9. Perform StringTie2 transcript assembly:

```
mkdir -p Stringtie_gtf_sc
 for file in Glutamatergic GABAergic Endothelial Astrocyte; do
  stringtie Hisat2_sorted_BAMs_sc/$file.bam -p $cores -o \
Stringtie_gtf_sc/$file.gtf -G gencode.vM25.annotation.gtf
done
```

10. Finally, merge the four GTF files to generate a single unified compressed transcriptome file named sc_merged.gtf.gz:

```
stringtie --merge -G gencode.vM25.annotation.gtf \
 spiepr A3B2 riStringtie_gtf_sc/Glutamatergic.gtf \ spiepr6-
Stringtie_gtf_sc/GABAergic.gtf \ spiepr6Stringtie_gtf_sc/En-
dothelial.gtf \ spiepr6Stringtie_gtf_sc/Astrocyte.gtf | gzip
> spiepr6sc_merged.gtf.gz
```

3.3 Custom Transcriptome Assembly Using Long-Read RNA-Seq Data

In this example, we download brain-specific long-read Oxford Nanopore RNA sequencing data [44], align the reads with minimap2, and assemble a custom GTF file using StringTie2. The script to execute the code (download_assemble_lr.sh) for this section can be found in the "Workflow" folder of our GitHub repository (https://github.com/fursham-h/MiMB-factR-supplement).

1. To run the entire script:

```
cd <path to the Workflow directory containing the script>
bash ./download_assemble_lr.sh
```

2. Alternatively, start by downloading three single-end FASTQ files from the ENA database using *wget*. The URLs can be found in a tab-delimited text file (lr_fastq.txt) available from our GitHub repository (in subfolder "Workflow"; https://github.com/fursham-h/MiMB-factR-supplement). This file.

```
cd <path to the Workflow directory>
cat lr_fastq.txt
```

contains the following information.

```
ERR2680377 ftp.sra.ebi.ac.uk/vol1/fastq/ERR268/007/
ERR2680377/ERR2680377.fastq.gz brain_C1
ERR3363658 ftp.sra.ebi.ac.uk/vol1/fastq/ERR336/008/
ERR3363658/ERR3363658_1.fastq.gz brain_C2R1
ERR3363660 ftp.sra.ebi.ac.uk/vol1/fastq/ERR336/000/
ERR3363660/ERR3363660_1.fastq.gz brain_C2R2
```

3. Download the three long-read FASTQ files to the FASTQ_lr folder as follows.

```
mkdir -p FASTQ_lr
while IFS=$'\t' read -r ACC LINK NAME;do
 wget $LINK -P FASTQ_lr
done < lr_fastq.txt
```

4. To map the long-read data to mouse mm10 reference genome using minimap2, first download mouse mm10 genomic fasta file.

```
 url="ftp://ftp.ebi.ac.uk/pub/databases/gencode/Gencode_-
mouse/
 release_M25/GRCm38.primary_assembly.genome.fa.gz"
 if [ ! -f "GRCm38.primary_assembly.genome.fa" ]; then
  wget $url
  gzip -d GRCm38.primary_assembly.genome.fa.gz
 fi
```

5. The actual minimap2 alignment using recommended spliced alignment settings and all available CPU cores is performed as follows.

```
cores='nproc'
mkdir -p minimap2_SAMs_lr
for fastq in ERR2680377 ERR3363658_1 ERR3363660_1; do
 minimap2 -t $cores -ax splice -o \
 minimap2_SAMs_lr/$fastq.sam \
GRCm38.primary_assembly.genome.fa
FASTQ_lr/$fastq.fastq.gz
done
```

6. Convert SAM files to BAM format.

```
 mkdir -p minimap2_sorted_BAMs_lr
 for file in ERR2680377 ERR3363658_1 ERR3363660_1; do
  samtools view -@ $cores -Su minimap2_SAMs_lr/$file.sam | \
samtools sort -@ $cores -o \ minimap2_sorted_BAMs_lr/$file.bam
 done
```

7. Similar to the above workflows, we will use StringTie2 for reference-guided custom transcriptome assembly. If not done already, download the reference gencode.vM25.annotation.gtf file required for this purpose as described in **step 6** of Subheading 3.1.

8. Assemble custom transcriptomes for each long-read RNA-seq sample.

```
mkdir -p Stringtie_gtf_lr
for file in ERR2680377 ERR3363658_1 ERR3363660_1; do
  stringtie minimap2_sorted_BAMs_lr/$file.bam -p $cores -o \
Stringtie_gtf_lr/$file.gtf -G gencode.vM25.annotation.gtf
done
```

9. Finally merge individual transcriptomes and compress it into a single nonredundant lr_merged.gtf.gz file.

```
stringtie --merge -G gencode.vM25.annotation.gtf \
Stringtie_gtf_lr/ERR2680377.gtf \
Stringtie_gtf_lr/ERR3363658_1.gtf \
Stringtie_gtf_lr/ERR3363660_1.gtf | gzip > lr_merged.gtf.gz
```

3.4 Functional Annotation of Custom Transcriptomes Using factR

Here, we describe how custom GTF files can be rapidly annotated to deduce possible biological functions of newly predicted sample-specific RNA isoforms. We will use our merged transcriptome file for the four neural cell types (sc_merged.gtf.gz described in Subheading 3.2) for demonstration. Annotation procedures for the bulk and long-read RNA-seq transcriptomes described in Subheadings 3.1 and 3.3 are described on GitHub (https://fursham-h.github.io/factR/articles/factR.html).

1. Enter the R environment and load the *factR* package.

```
library(factR)
```

2. *factR* interacts with GTF data stored as a GenomicRanges object with genomic interval data and relevant metadata. To create such an object from the sc_merged.gtf.gz file, we use the importGTF() function:

```
custom.gtf <- importGTF("path/to/sc_merged.gtf.gz")
```

3. The first 3 entries of the newly created *GenomicRanges* object can be inspected using the base R function head().

```
head(custom.gtf, n = 3)
```

which should output the following in the R console.

```
GRanges object with 3 ranges and 9 metadata columns:
 seqnames ranges strand | source type score phase gene_id
transcript_id exon_number gene_name ref_gene_id
  <Rle> <IRanges> <Rle> | <factor> <factor> <numeric>
```

```
<integer> <character> <character> <character> <character>
<character>
 [1] chr1 3188551-3672854 - | StringTie transcript 1000 <NA>
MSTRG.1 MSTRG.1.1 <NA> <NA> <NA>
 [2] chr1 3188551-3207317 - | StringTie exon 1000 <NA> MSTRG.1
MSTRG.1.1 1 <NA> <NA>
 [3] chr1 3213439-3216968 - | StringTie exon 1000 <NA> MSTRG.1
MSTRG.1.1 2 <NA> <NA>
 -------
 seqinfo: 39 sequences from an unspecified genome; no seq-
lengths
```

Note that in addition to genomic coordinates (seqnames and ranges), a typical imported GTF file will contain metadata describing the feature type (e.g., transcript or exon), its parental gene information and its assigned transcript ID.

4. Users may inspect architectures of specific sets of transcripts using viewTranscripts() function (*see* **Note 4** for our syntax recommendations). For example, the following will plot all transcripts for the *Zfr* gene encoding a conserved zinc finger-containing RNA-binding protein with known neuronal functions [45, 46] (Fig. 2a):

```
viewTranscripts(custom.gtf, gene_name == "Zfr")
```

5. StringTie2, the transcript assembler used in Subheadings 3.1 and 3.2, typically assigns arbitrary names to newly identified transcripts (e.g., "MSTRG.x.y") and uses the same prefix for their gene IDs (e.g., "MSTRG.x"). Incidentally, this is why the output in Fig. 2a lacks any novel Zfr transcripts. To match newly identified transcripts to their "official" host genes, *factR* should be provided with the reference transcriptome data, for example, as follows.

```
download.file("ftp://ftp.ebi.ac.uk/pub/databases/gencode/
 Gencode_mouse/release_M25/gencode.vM25.annotation.gtf.gz",
destfile = "gencode.vM25.annotation.gtf.gz")
ref.gtf <- importGTF("gencode.vM25.annotation.gtf.gz")
```

6. Genes found in the custom custom.gtf can now be compared with the reference ref.gtf object using the matchGeneInfo() function and novel transcripts assigned to the "official" genes, whenever possible. By default, this tool matches the query (custom.gtf) to the reference (ref.gtf) by finding overlapping coordinates. To tune the performance of this tool, we provide additional arguments specifying the name of columns containing "primary" and potentially "secondary" gene IDs from the

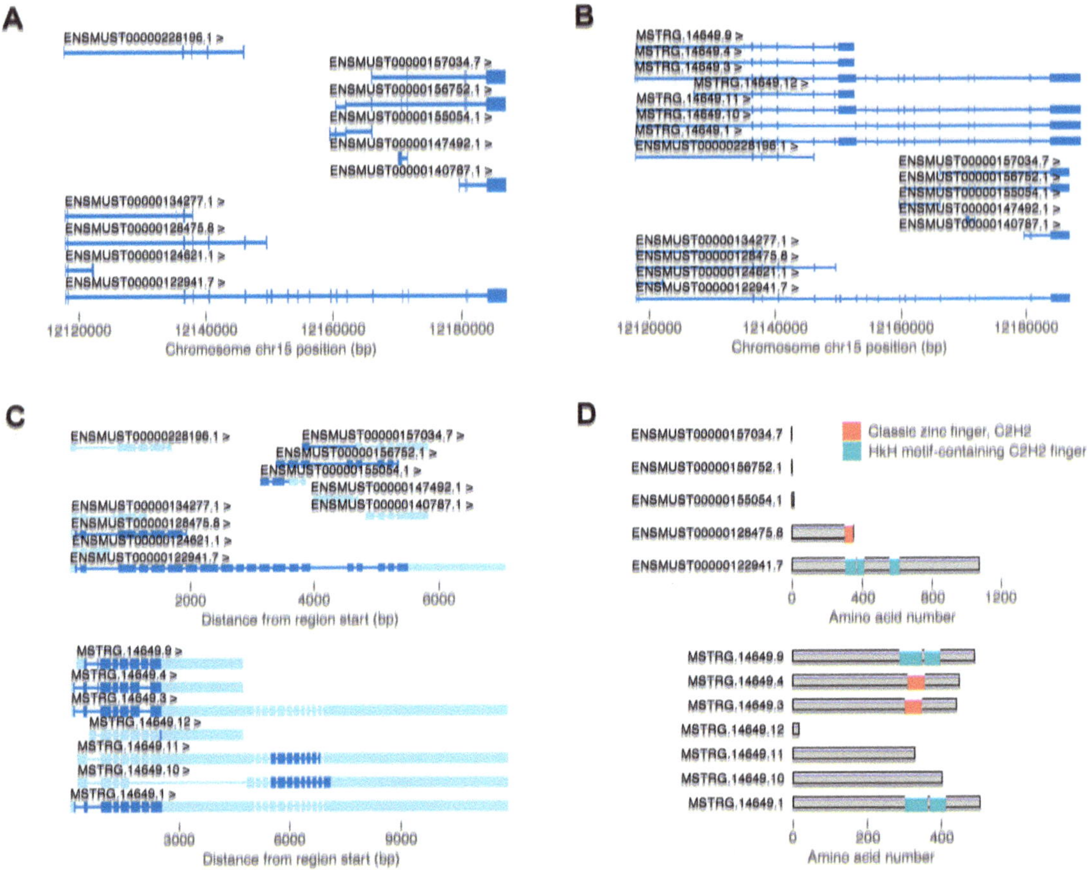

Fig. 2 Graphical outputs of factR. (**A-B**) Exon-intron architectures of known (**A**) and new (**B**) transcripts expressed from the *Zfr* gene. (**C**) Prediction of coding (dark blue) and non-coding (light blue) segments of *Zfr* transcripts. (**D**) Domain architectures of predicted protein-coding *Zfr* transcripts

query (custom.gtf) (*see* **Note 5** for details on matching process).

```
custom.gtf <- matchGeneInfo(custom.gtf, ref.gtf,
  primary_gene_id = "gene_id", secondary_gene_id = "ref_ge-
ne_id")
```

7. The output plot of viewTranscripts(custom.gtf, gene_name == "Zfr") should now show seven newly found transcripts (Fig. 2b).
8. Since mammalian introns are often much longer than exons, the overall exon structure of a transcript may be difficult to inspect when all its elements are shown to scale. *factR* can "shrink" introns for visualization purposes by switching the "rescale_introns" argument to TRUE, a modification inspired by the *wiggleplotr* package [47].

```
viewTranscripts(custom.gtf, gene_name == "Zfr",
 rescale_introns = TRUE)
```

9. Custom transcriptomes files typically combine both new and previously annotated transcripts. To select only the newly assembled transcripts, run the following.

```
custom_new.gtf <- subsetNewTranscripts(custom.gtf, ref.gtf)
```

   This command will subset custom.gtf transcripts with distinct exonic coordinates compared to ref.gtf and store these transcripts in the custom_new.gtf object.

10. Some custom-built transcripts may differ from their reference counterparts by having different start or/and end coordinates, with otherwise similar exon-intron structure. To shortlist only those novel transcripts that have distinct intronic coordinates, simply set the "refine.by" argument to "intron."

```
custom_new.gtf <- subsetNewTranscripts(custom.gtf, ref.gtf,
 refine.by = "intron")
```

11. Functional annotation of newly assembled transcripts in *factR* begins by deducing their protein-coding sequences (CDSs). To search for putative CDSs, *factR* requires a genome sequence file, which can be obtained, for example, from the AnnotationHub [48] database.

```
library(AnnotationHub)
ah <- AnnotationHub()
query(ah, c("mm10","2bit"))
```

   The above code should output the following information.

```
AnnotationHub with 1 record
# snapshotDate(): 2020-10-27
# names(): AH14005
# $dataprovider: UCSC
# $species: Mus musculus
# $rdataclass: TwoBitFile
# $rdatadateadded: 2014-12-15
# $title: mm10.2bit
# $description: UCSC 2 bit file for mm10
# $taxonomyid: 10090
# $genome: mm10
# $sourcetype: TwoBit
```

```
# $sourceurl: http://hgdownload.cse.ucsc.edu/goldenpath/mm10/
bigZips/mm10.2bit
# $sourcesize: NA
# $tags: c("2bit", "UCSC", "genome")
# retrieve record with 'object[["AH14005"]]'
```

12. Download the mm10 2bit genome file and attach it to your R environment as follows.

```
Mmusculus <- ah[['AH14005']]
```

13. *factR*'s buildCDS() is a multistep function that constructs CDSs on query transcripts using reference annotation as a guide. First, a database of annotated ATGs is created and used to search for a potential translation start site on query transcripts. If successfully found, buildCDS() will look for an open-reading frame and append the CDS coordinates to the query object. This function can be run for our novel transcripts as follows.

```
custom_new_CDS.gtf <- buildCDS(custom_new.gtf, ref.gtf,
 Mmusculus)
```

14. To visualize predicted CDS regions for novel Zfr transcripts, simply run the following.

```
viewTranscripts(custom_new_CDS.gtf, gene_name == "Zfr",
 rescale_introns = TRUE)
```

Note that CDSs are now colored dark blue and untranslated sequences are shown in light blue (top part of Fig. 2c). Also note that by setting rescale_introns to TRUE we can see exonic regions more clearly but lose their genomic coordinates.

15. To compare the new CDSs with their reference counterparts, run the following.

```
viewTranscripts(ref.gtf, gene_name == "Zfr",
 rescale_introns = TRUE)
```

The output of this command is shown in the bottom part of Fig. 2c.

16. We can now examine possible functional consequences of alternative processing events affecting the newly identified transcripts. To test which transcripts may be susceptible to NMD (*see* **Note 6** for the NMD features examined), use the predictNMD() function.

```
NMDprediction.out <- predictNMD(custom_new_CDS.gtf)
```

predictNMD() outputs a data-frame containing information on features that may trigger NMD, and a predicted NMD sensitivity for each coding transcript.

17. To identify putative NMD targets for specific genes (e.g., *Zfr*), run the following.

```
NMDprediction.zfr <- predictNMD(custom_new_CDS.gtf,
 gene_name == "Zfr"))
```

To view the first 3 entries of the NMDprediction.zfr object, run the following.

```
head(NMDprediction.zfr, n = 3)
```

which should return the following in the R console:

```
# A tibble: 3 x 5
 transcript stop_to_lastEJ num_of_downEJs '3'UTR_length'
is_NMD
 <chr> <dbl> <int> <dbl> <lgl>
1 MSTRG.14649.1 3857 11 8774 TRUE
2 MSTRG.14649.10 -177 0 4740 FALSE
3 MSTRG.14649.11 41 1 4958 FALSE
```

18. *factR* can also deduce domain structure of the protein products encoded by newly identified mRNA isoforms using its predictDomains() tool.

```
domains.out <- predictDomains(custom_new_CDS.gtf,
 Mmusculus)
```

The above code returns a data-frame containing the name, expected value and position of predicted protein domain families within each transcript.

19. predictDomains() connects to the PFAM database over the internet and may require a substantial amount of time and stable internet connection to query multiple transcripts simultaneously. To quickly explore this functionality, users may prefer to begin with a small subset of transcripts specified as described in **Note 4**. For example, to predict the domain structure for all transcripts of the *Zfr* gene, run the following.

```
domains.zfr <- predictDomains(custom_new_CDS.gtf,
 Mmusculus, gene_name == "Zfr")
```

20. The contents of domains.zfr can be previewed using head (domains.zfr, $n = 3$), which should produce the following output.

```
# A tibble: 3 x 5
 transcript description eval begin end
 <chr> <chr> <chr> <dbl> <dbl>
1 MSTRG.14649.1 HkH motif-containing C2H2 finger 0.044 303 364
2 MSTRG.14649.1 HkH motif-containing C2H2 finger 0.065 371 414
3 MSTRG.14649.3 Classic zinc finger, C2H2 0.063 303 350
```

In addition, the domain architectures for the set of query proteins can be plotted by switching the argument "plot" to TRUE (Fig. 2d).

```
domains.zfr <- predictDomains(custom_new_CDS.gtf,
 Mmusculus, gene_name == "Zfr", plot = TRUE)
```

21. Domain structures of known proteins annotated in the reference transcriptome can be plotted for comparison.

```
domains.known.zfr <- predictDomains(ref.gtf, Mmusculus,
 gene_name == "Zfr", plot == TRUE)
```

22. At the end of the session, the annotated custom_new_CDS.gtf *GenomicRanges* object can be exported to a GTF file as follows.

```
export(custom_new_CDS.gtf, "Custom_new.gtf", format =
 "GTF")
```

23. Finally, to export NMDprediction.out and domains.out data frames as tab-delimited text files, run the following.

```
write.table(NMDprediction.out, "Custom_new_NMD.txt", sep =
 "\t", row.names = FALSE, quote = FALSE)
write.table(domains.out, "Custom_new_domains.txt", sep =
"\t", row.names = FALSE, quote = FALSE)
```

3.5 Biological Significance and Concluding Remarks

The above workflows can provide valuable insights into the extent of post-transcriptional regulation and biological functions of alternative mRNA isoforms in user-specific samples. For example, *factR* analysis of cell type-specific transcriptome files (Glutamatergic.gtf, GABAergic.gtf, Astrocyte.gtf, and Endothelial.gtf prepared as described in Subheading 3.2) shows that the number of novel mRNA isoforms in glutamatergic and GABAergic neurons exceeds that in astrocytes and endothelial cells (Fig. 3a). A similar trend is evident for novel isoforms with altered CDS and a subset of these transcripts that gain or lose sequences encoding specific protein domains (Fig. 3b, c). Systematic analysis of domain families commonly gained or/and lost in the two types neurons compared to the reference transcriptome (Fig. 3d, e) shows that many of them (e.g., C2H2 zinc finger, KRAB, and RBD) are found in important regulators of gene expression - including the C2H2 zinc finger protein Zfr showcased in Fig. 2. This effect is also detected in the by custom transcriptome assembled from brain-specific long-read sequencing data (*see* Subheading 3.3; Fig. 3f, g), warranting further analyses of possible functions of the regulated targets in this biological context.

factR also shows that many new transcripts in the custom-assembled glutamatergic and GABAergic transcriptomes might be controlled by NMD (Fig. 3h), consistent with diverse regulatory functions of this mRNA destabilization mechanism in the nervous system [49–51]. As a positive control, we analyzed the merged custom transcriptome for the CHX- or untreated mESCs (*see* Subheading 3.1). Compared to the reference transcriptome, a significantly larger fraction of newly identified transcripts in this dataset contained NMD-promoting features and was upregulated by the NMD inhibitor CHX (Fig. 3i, j). Thus, *factR* may facilitate further studies of NMD functions in various biological systems.

The protocol described in this chapter should allow researchers to make better use of their sample-specific RNA sequencing data by quickly going beyond reference transcriptome annotations. Care should be taken to reduce the incidence of false-positive predictions by selecting sufficiently stringent HISAT2 and StringTie2 settings (*see* **Notes 2** and **3**), and by inspecting aligned data visually, for example using IGV browser (https://software.broadinstitute.org/software/igv/). Ultimately all bioinformatics predictions should be validated experimentally, for example using reverse transcription-PCR and immunoblotting. With these caveats, we hope that our annotation tools will streamline the ongoing efforts to understand functional significance of the widespread regulation of eukaryotic genes at the level of pre-mRNA processing.

4 Notes

1. Prebuilt indices for some other organisms are available at http://daehwankimlab.github.io/hisat2/download/ and can be downloaded similarly using wget. Alternatively, a custom index can be built as described in http://daehwankimlab.github.io/hisat2/manual/.
2. The example code shown aligns reads from paired-end sequencing data. For instructions on aligning single-end

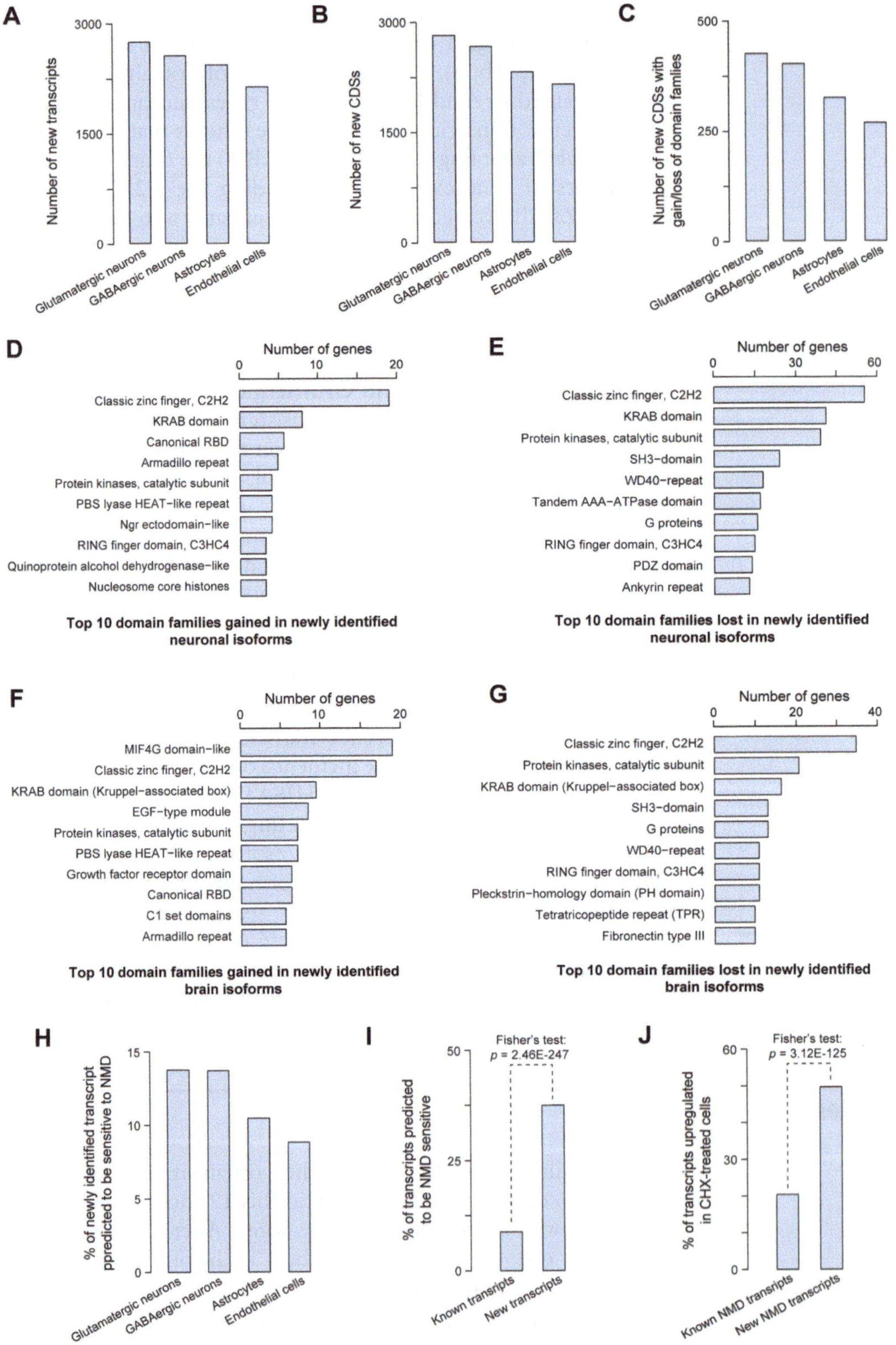

Fig. 3 Examples of biological insights obtained using our workflows. (**A-C**) New transcripts, CDSs and CDSs with distinct domain architectures identified in neural cell populations. (**D-G**) Functional domains involved in the regulation of gene expression are commonly gained or lost in neuronal cell populations and in the whole brain. (**H-J**) Many new transcripts identified by *factR* are predicted to be controlled by NMD

reads, as well as description of other options (including for example running HISAT2 with --known-splicesite-infile) please refer to http://daehwankimlab.github.io/hisat2/manual/.

3. The example in the text shows how to assemble transcripts using default StringTie2 parameters. Stringency of the assembly process can be tuned as described in https://ccb.jhu.edu/software/stringtie/index.shtml?t=manual.
4. Several *factR* tools including viewTranscripts(), predictNMD(), and predictDomains() can work with subsets of transcripts. These are selected by providing logical statements defined by the metadata found in input GenomicRanges object. In our examples, we showed how to select a list of transcripts from the *Zfr* gene, by picking entries containing the string "Zfr" in metadata column "gene_name". Other suggested ways to select transcripts include specifying gene identifier (e.g., gene_id == "ENSMUST00000157034.7") or transcript identifier (e.g., transcript_id == "ENSMUST00000122941.7").
5. matchGeneInfo() identifies query genes with inconsistent identifiers ("gene_id") from reference and attempts to match them using three distinct methods sequentially. Method (1) involves replacing identifiers in column "gene_id" with a secondary gene identifier occasionally found in custom GTF files, and rematching them to reference. StringTie2 transcript assembler outputs such secondary identifiers under the attribute "ref_gene_id". Method (2) attempts to correct Ensembl-style identifiers that contain version identifier (e.g., ENSMUSG00000006498.16). It drops the suffixes on these IDs and rematches them to reference. Method (3) will match the discordant query genes to reference by finding overlapping coordinates. Running matchGeneInfo() using default parameters will invoke only method (3). Methods (1) and (2) are activated by specifying "secondary_gene_id" and "primary_gene_id" arguments.
6. Nonsense-mediated decay can be triggered by several mechanisms. predictNMD() identifies three features potentially making a CDS-containing transcript sensitive to NMD: the position of translation termination site relative to the downstream-most exon-exon junction, the number exon junctions in the 3′ untranslated region (3′UTR) and the length of the 3′UTR. If translation of a transcript terminates >50 nt upstream of the last exon-exon junction, predictNMD() categorizes it as NMD sensitive.

Acknowledgments

This work has been supported by Biotechnology and Biological Sciences Research Council (BB/M007103/1 and BB/R001049/1), Estonian Research Council (PSG415 and PRG1095), and European Commission (H2020-MSCA-RISE-2016; Project ID 734791).

References

1. Landry JR, Mager DL, Wilhelm BT (2003) Complex controls: the role of alternative promoters in mammalian genomes. Trends Genet 19(11):640–648. https://doi.org/10.1016/j.tig.2003.09.014
2. Manning KS, Cooper TA (2017) The roles of RNA processing in translating genotype to phenotype. Nat Rev Mol Cell Biol 18(2):102–114. https://doi.org/10.1038/nrm.2016.139
3. Ule J, Blencowe BJ (2019) Alternative splicing regulatory networks: functions, mechanisms, and evolution. Mol Cell 76(2):329–345. https://doi.org/10.1016/j.molcel.2019.09.017
4. Tian B, Manley JL (2017) Alternative polyadenylation of mRNA precursors. Nat Rev Mol Cell Biol 18(1):18–30. https://doi.org/10.1038/nrm.2016.116
5. Vuong CK, Black DL, Zheng S (2016) The neurogenetics of alternative splicing. Nat Rev Neurosci 17(5):265–281. https://doi.org/10.1038/nrn.2016.27
6. Baralle FE, Giudice J (2017) Alternative splicing as a regulator of development and tissue identity. Nat Rev Mol Cell Biol 18(7):437–451. https://doi.org/10.1038/nrm.2017.27
7. Furlanis E, Scheiffele P (2018) Regulation of neuronal differentiation, function, and plasticity by alternative splicing. Annu Rev Cell Dev Biol 34:451–469. https://doi.org/10.1146/annurev-cellbio-100617-062826
8. Furlanis E, Traunmuller L, Fucile G, Scheiffele P (2019) Landscape of ribosome-engaged transcript isoforms reveals extensive neuronal-cell-class-specific alternative splicing programs. Nat Neurosci 22(10):1709–1717. https://doi.org/10.1038/s41593-019-0465-5
9. Tapial J, Ha KCH, Sterne-Weiler T, Gohr A, Braunschweig U, Hermoso-Pulido A, Quesnel-Vallieres M, Permanyer J, Sodaei R, Marquez Y, Cozzuto L, Wang X, Gomez-Velazquez M, Rayon T, Manzanares M, Ponomarenko J, Blencowe BJ, Irimia M (2017) An atlas of alternative splicing profiles and functional associations reveals new regulatory programs and genes that simultaneously express multiple major isoforms. Genome Res 27(10):1759–1768. https://doi.org/10.1101/gr.220962.117
10. Hamid FM, Makeyev EV (2014) Emerging functions of alternative splicing coupled with nonsense-mediated decay. Biochem Soc Trans 42(4):1168–1173. https://doi.org/10.1042/BST20140066
11. Hwang J, Maquat LE (2011) Nonsense-mediated mRNA decay (NMD) in animal embryogenesis: to die or not to die, that is the question. Curr Opin Genet Dev 21(4):422–430. https://doi.org/10.1016/j.gde.2011.03.008
12. Wang ET, Sandberg R, Luo S, Khrebtukova I, Zhang L, Mayr C, Kingsmore SF, Schroth GP, Burge CB (2008) Alternative isoform regulation in human tissue transcriptomes. Nature 456(7221):470–476. https://doi.org/10.1038/nature07509
13. Pan Q, Shai O, Lee LJ, Frey BJ, Blencowe BJ (2008) Deep surveying of alternative splicing complexity in the human transcriptome by high-throughput sequencing. Nat Genet 40(12):1413–1415. https://doi.org/10.1038/ng.259
14. Hoque M, Ji Z, Zheng D, Luo W, Li W, You B, Park JY, Yehia G, Tian B (2013) Analysis of alternative cleavage and polyadenylation by 3′ region extraction and deep sequencing. Nat Methods 10(2):133–139. https://doi.org/10.1038/nmeth.2288
15. Lianoglou S, Garg V, Yang JL, Leslie CS, Mayr C (2013) Ubiquitously transcribed genes use alternative polyadenylation to achieve tissue-specific expression. Genes Dev 27(21):2380–2396. https://doi.org/10.1101/gad.229328.113
16. Shepard PJ, Choi EA, Lu J, Flanagan LA, Hertel KJ, Shi Y (2011) Complex and dynamic landscape of RNA polyadenylation revealed by

PAS-Seq. RNA 17(4):761–772. https://doi.org/10.1261/rna.2581711

17. Arzalluz-Luque A, Conesa A (2018) Single-cell RNAseq for the study of isoforms-how is that possible? Genome Biol 19(1):110. https://doi.org/10.1186/s13059-018-1496-z
18. Packer JS, Zhu Q, Huynh C, Sivaramakrishnan P, Preston E, Dueck H, Stefanik D, Tan K, Trapnell C, Kim J, Waterston RH, Murray JI (2019) A lineage-resolved molecular atlas of C. elegans embryogenesis at single-cell resolution. Science 365(6459): eaax1971. https://doi.org/10.1126/science.aax1971
19. Fincher CT, Wurtzel O, de Hoog T, Kravarik KM, Reddien PW (2018) Cell type transcriptome atlas for the planarian Schmidtea mediterranea. Science 360(6391):eaaq1736. https://doi.org/10.1126/science.aaq1736
20. Tabula Muris Consortium, Overall coordination, Logistical coordination, Organ collection and processing, Library preparation and sequencing, Computational data analysis, Cell type annotation, Writing group, Supplemental text writing group, Principal investigators (2018) Single-cell transcriptomics of 20 mouse organs creates a Tabula Muris. Nature 562(7727):367–372. https://doi.org/10.1038/s41586-018-0590-4
21. Zeisel A, Hochgerner H, Lonnerberg P, Johnsson A, Memic F, van der Zwan J, Haring M, Braun E, Borm LE, La Manno G, Codeluppi S, Furlan A, Lee K, Skene N, Harris KD, Hjerling-Leffler J, Arenas E, Ernfors P, Marklund U, Linnarsson S (2018) Molecular architecture of the mouse nervous system. Cell 174(4):999–1014.e22. https://doi.org/10.1016/j.cell.2018.06.021
22. Hardwick SA, Joglekar A, Flicek P, Frankish A, Tilgner HU (2019) Getting the entire message: progress in isoform sequencing. Front Genet 10:709. https://doi.org/10.3389/fgene.2019.00709
23. Oikonomopoulos S, Bayega A, Fahiminiya S, Djambazian H, Berube P, Ragoussis J (2020) Methodologies for transcript profiling using long-read technologies. Front Genet 11:606. https://doi.org/10.3389/fgene.2020.00606
24. Stark R, Grzelak M, Hadfield J (2019) RNA sequencing: the teenage years. Nat Rev Genet 20(11):631–656. https://doi.org/10.1038/s41576-019-0150-2
25. Robertson G, Schein J, Chiu R, Corbett R, Field M, Jackman SD, Mungall K, Lee S, Okada HM, Qian JQ, Griffith M, Raymond A, Thiessen N, Cezard T, Butterfield YS, Newsome R, Chan SK, She R, Varhol R, Kamoh B, Prabhu AL, Tam A, Zhao Y, Moore RA, Hirst M, Marra MA, Jones SJ, Hoodless PA, Birol I (2010) De novo assembly and analysis of RNA-seq data. Nat Methods 7(11): 909–912. https://doi.org/10.1038/nmeth.1517
26. Grabherr MG, Haas BJ, Yassour M, Levin JZ, Thompson DA, Amit I, Adiconis X, Fan L, Raychowdhury R, Zeng Q, Chen Z, Mauceli E, Hacohen N, Gnirke A, Rhind N, di Palma F, Birren BW, Nusbaum C, Lindblad-Toh K, Friedman N, Regev A (2011) Full-length transcriptome assembly from RNA-Seq data without a reference genome. Nat Biotechnol 29(7):644–652. https://doi.org/10.1038/nbt.1883
27. Schulz MH, Zerbino DR, Vingron M, Birney E (2012) Oases: robust de novo RNA-seq assembly across the dynamic range of expression levels. Bioinformatics 28(8):1086–1092. https://doi.org/10.1093/bioinformatics/bts094
28. Bushmanova E, Antipov D, Lapidus A, Prjibelski AD (2019) rnaSPAdes: a de novo transcriptome assembler and its application to RNA-Seq data. Gigascience 8(9):giz100. https://doi.org/10.1093/gigascience/giz100
29. Nip KM, Chiu R, Yang C, Chu J, Mohamadi H, Warren RL, Birol I (2020) RNA-bloom enables reference-free and reference-guided sequence assembly for single-cell transcriptomes. Genome Res 30(8): 1191–1200. https://doi.org/10.1101/gr.260174.119
30. Kim D, Langmead B, Salzberg SL (2015) HISAT: a fast spliced aligner with low memory requirements. Nat Methods 12(4):357–360. https://doi.org/10.1038/nmeth.3317
31. Dobin A, Davis CA, Schlesinger F, Drenkow J, Zaleski C, Jha S, Batut P, Chaisson M, Gingeras TR (2013) STAR: ultrafast universal RNA-seq aligner. Bioinformatics 29(1):15–21. https://doi.org/10.1093/bioinformatics/bts635
32. Wu TD, Reeder J, Lawrence M, Becker G, Brauer MJ (2016) GMAP and GSNAP for genomic sequence alignment: enhancements to speed, accuracy, and functionality. Methods Mol Biol 1418:283–334. https://doi.org/10.1007/978-1-4939-3578-9_15
33. Li H (2018) Minimap2: pairwise alignment for nucleotide sequences. Bioinformatics 34(18): 3094–3100. https://doi.org/10.1093/bioinformatics/bty191
34. Trapnell C, Roberts A, Goff L, Pertea G, Kim D, Kelley DR, Pimentel H, Salzberg SL, Rinn JL, Pachter L (2012) Differential gene and transcript expression analysis of RNA-seq experiments with TopHat and cufflinks. Nat

Protoc 7(3):562–578. https://doi.org/10.1038/nprot.2012.016

35. Pertea M, Pertea GM, Antonescu CM, Chang TC, Mendell JT, Salzberg SL (2015) StringTie enables improved reconstruction of a transcriptome from RNA-seq reads. Nat Biotechnol 33(3):290–295. https://doi.org/10.1038/nbt.3122
36. Kovaka S, Zimin AV, Pertea GM, Razaghi R, Salzberg SL, Pertea M (2019) Transcriptome assembly from long-read RNA-seq alignments with StringTie2. Genome Biol 20(1):278. https://doi.org/10.1186/s13059-019-1910-1
37. Song L, Sabunciyan S, Florea L (2016) CLASS2: accurate and efficient splice variant annotation from RNA-seq reads. Nucleic Acids Res 44(10):e98. https://doi.org/10.1093/nar/gkw158
38. Shao M, Kingsford C (2017) Accurate assembly of transcripts through phase-preserving graph decomposition. Nat Biotechnol 35(12):1167–1169. https://doi.org/10.1038/nbt.4020
39. Liu J, Liu X, Ren X, Li G (2019) scRNAss: a single-cell RNA-seq assembler via imputing dropouts and combing junctions. Bioinformatics 35(21):4264–4271. https://doi.org/10.1093/bioinformatics/btz240
40. Tang AD, Soulette CM, van Baren MJ, Hart K, Hrabeta-Robinson E, Wu CJ, Brooks AN (2020) Full-length transcript characterization of SF3B1 mutation in chronic lymphocytic leukemia reveals downregulation of retained introns. Nat Commun 11(1):1438. https://doi.org/10.1038/s41467-020-15171-6
41. Li H, Handsaker B, Wysoker A, Fennell T, Ruan J, Homer N, Marth G, Abecasis G, Durbin R, Subgroup GPDP (2009) The sequence alignment/map format and SAMtools. Bioinformatics 25(16):2078–2079. https://doi.org/10.1093/bioinformatics/btp352
42. Hurt JA, Robertson AD, Burge CB (2013) Global analyses of UPF1 binding and function reveal expanded scope of nonsense-mediated mRNA decay. Genome Res 23(10):1636–1650. https://doi.org/10.1101/gr.157354.113
43. Tasic B, Yao Z, Graybuck LT, Smith KA, Nguyen TN, Bertagnolli D, Goldy J, Garren E, Economo MN, Viswanathan S, Penn O, Bakken T, Menon V, Miller J, Fong O, Hirokawa KE, Lathia K, Rimorin C, Tieu M, Larsen R, Casper T, Barkan E, Kroll M, Parry S, Shapovalova NV, Hirschstein D, Pendergraft J, Sullivan HA, Kim TK, Szafer A, Dee N, Groblewski P, Wickersham I, Cetin A, Harris JA, Levi BP, Sunkin SM, Madisen L, Daigle TL, Looger L, Bernard A, Phillips J, Lein E, Hawrylycz M, Svoboda K, Jones AR, Koch C, Zeng H (2018) Shared and distinct transcriptomic cell types across neocortical areas. Nature 563(7729):72–78. https://doi.org/10.1038/s41586-018-0654-5
44. Sessegolo C, Cruaud C, Da Silva C, Cologne A, Dubarry M, Derrien T, Lacroix V, Aury J-M (2019) Transcriptome profiling of mouse samples using nanopore sequencing of cDNA and RNA molecules. Sci Rep 9(1):14908. https://doi.org/10.1038/s41598-019-51470-9
45. Elvira G, Massie B, DesGroseillers L (2006) The zinc-finger protein ZFR is critical for Staufen 2 isoform specific nucleocytoplasmic shuttling in neurons. J Neurochem 96(1):105–117. https://doi.org/10.1111/j.1471-4159.2005.03523.x
46. Kjærgaard T, Desdorf R, Heuck A, Olsen A, Lykke-Hartmann K (2015) The zinc finger RNA binding protein, ZFR, contributes to axon guidance in Caenorhabditis elegans. Gene 557(1):11–18. https://doi.org/10.1016/j.gene.2014.11.063
47. Alasoo K (2020) wiggleplotr: make read coverage plots from BigWig files. R package version 1.14.0
48. Morgan M, Shepherd L (2020) AnnotationHub: client to access AnnotationHub resources. R package version 2220
49. Colak D, Ji SJ, Porse BT, Jaffrey SR (2013) Regulation of axon guidance by compartmentalized nonsense-mediated mRNA decay. Cell 153(6):1252–1265. https://doi.org/10.1016/j.cell.2013.04.056
50. Giorgi C, Yeo GW, Stone ME, Katz DB, Burge C, Turrigiano G, Moore MJ (2007) The EJC factor eIF4AIII modulates synaptic strength and neuronal protein expression. Cell 130(1):179–191. https://doi.org/10.1016/j.cell.2007.05.028
51. Makeyev EV, Zhang J, Carrasco MA, Maniatis T (2007) The MicroRNA miR-124 promotes neuronal differentiation by triggering brain-specific alternative pre-mRNA splicing. Mol Cell 27(3):435–448. https://doi.org/10.1016/j.molcel.2007.07.015

Chapter 10

Design, Labeling, and Application of Probes for RNA smFISH

Ewa Piskadlo, Bastian Th. Eichenberger, Luca Giorgetti, and Jeffrey A. Chao

Abstract

Visualization of single mRNA molecules in fixed cells can be achieved using single molecule fluorescent in situ hybridization (smFISH). This approach enables accurate quantification of mRNA numbers and localization at a single-cell level. To ensure reliable results using smFISH, it is critical to use fluorescent probes that are highly specific to their RNA target. To facilitate probe design, we have created anglerFISH, a user-friendly command-line based pipeline. In this chapter, we present how to perform a smFISH experiment using user-designed and labeled probes.

Key words Single-molecule fluorescence in situ hybridization (smFISH), Fluorescence labeling of oligonucleotides, FISH probe design, Command line interface (CLI), Nextflow workflow

1 Introduction

Single-molecule RNA fluorescence in situ hybridization (smFISH) enables the quantification of the number and location of individual transcripts within their native cellular environment. This has enabled the entire life cycle of mRNAs from transcription to degradation to be visualized in single cells [1]. Additionally, when imaged in combination with fluorescent antibodies, smFISH can also be used to characterize RNA-protein interactions [2].

smFISH can be performed on a variety of specimens ranging from bacteria to the entire Drosophila brain [3–5]. All of these specimens are fixed and subsequently permeabilized. This allows for RNA transcripts to become accessible to fluorescently labeled probes that hybridize to the target thus revealing its position. Probes are single-stranded DNA oligonucleotides that are typically between 20 and 50 nucleotides in length and are complementary to the target RNA [6, 7]. Alternative approaches using branched DNA oligonucleotide technology have also been used for smFISH [8]. Irrespective of the exact design of the probes, the hybridization

Peter Scheiffele and Oriane Mauger (eds.), *Alternative Splicing: Methods and Protocols*,
Methods in Molecular Biology, vol. 2537, https://doi.org/10.1007/978-1-0716-2521-7_10,

of multiple fluorescently labeled oligonucleotides to a single RNA amplifies the signal and enables distinguishing it from background signals.

While smFISH provides information that is not readily obtainable from ensemble or even single-cell RNA sequencing experiments, it is currently limited in the number of genes that can be visualized at one time. This limitation stems from the challenges in the specific detection of spectrally distinct fluorophores. Recent methods, however, employing barcoding techniques and sequential rounds of hybridization and bleaching have been able to visualize up to ~10,000 unique RNA transcripts within a single cell [9].

Sequences for smFISH probes have to be selected carefully taking into account binding specificity, probe length, spacing, and melting temperatures. Additionally, designing probes can be challenging as multiple bioinformatics tools are required. This chapter introduces anglerFISH, a workflow-based, command-line tool that allows nonexpert users to design smFISH probes for RNA targets of interest. Designing custom probes reduces experimental costs and allows for more flexibility in fluorescent dyes and hybridization conditions. In addition, this chapter explains how to fluorescently label these probes in the lab and contains a protocol for performing RNA smFISH.

2 Materials

2.1 Designing Custom Probes

1. A POSIX compatible operating system (Linux or macOS). Windows users are advised to install the free Linux operating system Ubuntu via a virtual machine software, such as Oracle VM VirtualBox.
2. Bash (version $\geq$3.2).
3. Java (version $\geq$8, 1.8).
4. Miniconda 3 or Anaconda 3.
5. Git.
6. GNU Make.

2.2 Probe Labeling

1. Silica gel in a container with a lid.
2. Anhydrous DMSO.
3. Nuclease-free water.
4. 100% ethanol.
5. Nanodrop.
6. Centrifuge and nuclease-free tubes (1.5 mL and PCR).
7. Thermocycler.

8. ddUTP-dye conjugation mix: 10 μL of 20 mM amino-11-ddUTP (*see* **Note 1**), 2 μL of 1 M $NaHCO_3$ (pH 8.4), 10 μL of 40 mM amine-dye (in DMSO).
9. ddUTP-dye-oligonucleotides labeling reaction per one labeling reaction: 10 μL of 100 μM oligonucleotide mix, 1 μL of 5 mM ddUTP–dye conjugation mix (keep in darkness), 1 μL of 20 U/μL of terminal deoxynucleotidyl transferase (TdT), 3 μL of 5× TdT reaction buffer.
10. Amber tubes.
11. 1 M sodium acetate, pH 5.5.
12. 5 mg/mL linear acrylamide.

2.3 RNA FISH Fixation and Hybridization (for 4 Coverslips)

1. Phosphate-buffered saline (PBS).
2. Incubator with shaker set to 37 °C.
3. Glass coverslips #1 thickness 18 mm diameter.
4. Prolong Gold Antifade Reagent with DAPI.
5. Transparent nail polish.
6. Fixation solution: 10 mL of 20% paraformaldehyde, 5 mL of 10× PBS, 35 mL of H_2O.
7. Hybridization buffer: Gently thaw probes on ice and prepare mixture on ice as well. Mix: 100 μL of 20% dextran sulfate (*see* **Note 2**), 20 μL of 20× SSC, 20 μL of formamide brought to room temperature (RT) before opening to reduce oxidation, 200 nM of FISH probes (approximately 1–2 μL). Fill H_2O up to 200 μL total volume. Gently pipette up and down to mix.
8. Wash buffer: 5 mL of 20× SSC, 5 mL of formamide brought to RT before opening, 40 mL of H_2O. Store in the dark by wrapping with aluminum.
9. Hybridization chamber: place a sheet of Parafilm on the bottom of a petri dish. Tape an inverted cap of a 50 mL conical tube to the edge of the chamber. Fill the inverted cap with PBS to create a humid environment. Prepare both a sheet of Parafilm to seal and aluminum foil to cover the chamber.
10. 70% ethanol.

3 Methods (*See* Fig. 1)

3.1 Designing Custom Probes

We have created anglerFISH [10], a custom Nextflow [11] workflow, based primarily on OligoMiner software [12], as well as several other common tools to design RNA smFISH probes [13–15]. This generates highly specific antisense probes against either endogenous or exogenous RNA targets in a reproducible and automated manner with a single command. Due to utilizing

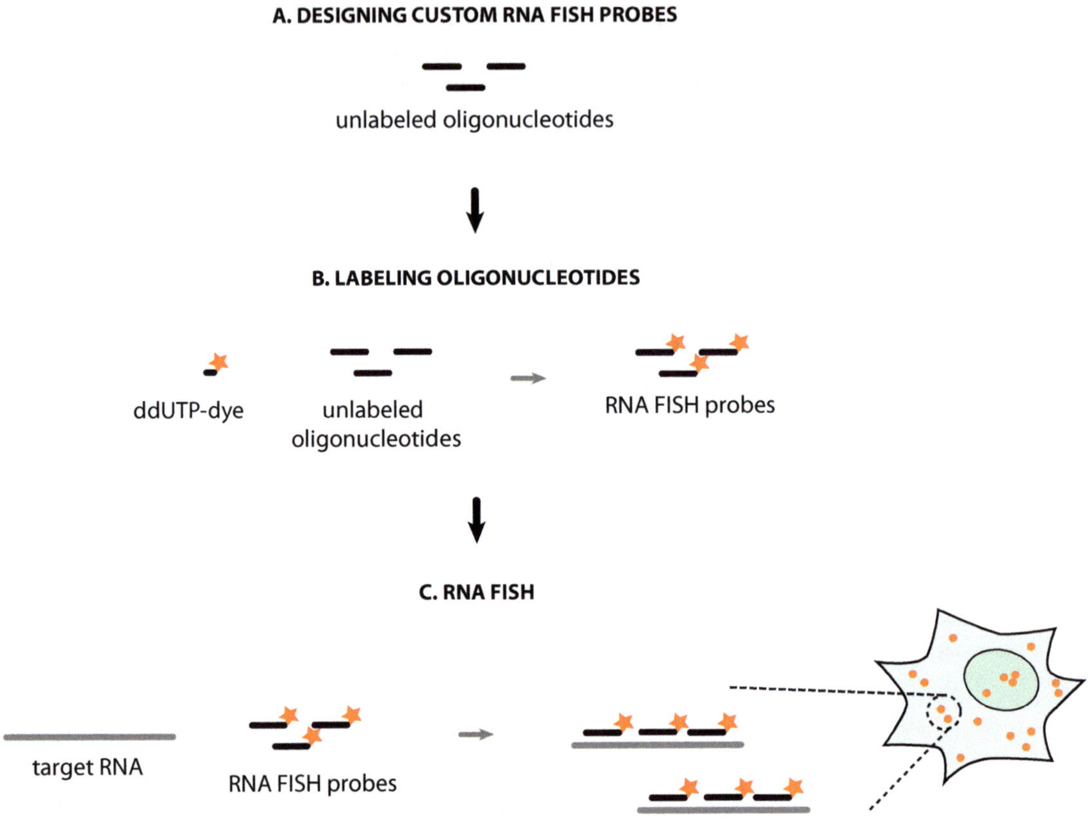

Fig. 1 RNA smFISH with custom-designed and labeled probes. (**a**) Design sets of custom probes (ssDNA oligonucleotides) against an RNA of interest. (**b**) The probes with sequences determined in the previous step are labeled with ddUTP-dyes to create fluorescently labeled RNA FISH probes. (**c**) Cells are fixed with paraformaldehyde and permeabilized to enable FISH probes to hybridize to their RNA targets, allowing visualization of single molecules of RNA

OligoMiner software, the workflow provides the researcher with control over probe parameters including probe length range, probe spacing and overlap, hybridizing temperature, buffer composition and GC content. Refer to the README.md file in the GitHub repository https://github.com/episkadlo/anglerFISH for more detailed instructions and examples, or consult the workflow's help message. Tips for probe design parameters can be found in **Note 3**.

3.1.1 Download and Configure the Workflow

1. Download the repository containing the workflow from GitHub: go to https://github.com/episkadlo/anglerFISH, click on the green "Code" button, and select the "Download ZIP" option, then unpack the file in the location of your choice. Rename the folder to "anglerFISH."
2. In the terminal, go inside the directory you just unpacked (anglerFISH), by typing the following path.

```
cd Path/To/Your/anglerFISH/
```

When inside the directory, install and configure all the necessary software and packages by typing the following command.

```
make
```

3. To test if the pipeline is ready to be used and to see the help message with instructions of use, run the following.

```
./nextflow main.nf --help
```

3.1.2 Prepare Genome Indexes for Specificity Filtering

1. Download a fasta (.fa) file for the genomic sequence of the organism you want to design probes against. Genomic sequences can be found at the UCSC Genome Browser website (https://hgdownload.soe.ucsc.edu/downloads.html). If the fasta is compressed (fa.gz), unpack it to a .fa manually (Linux: in the command line with gunzip <file>.fa.gz, Mac: unpacks by default). Place the unpacked genome fasta file in the directory "anglerFISH/UPLOAD_HERE/genome_raw/".
2. From the anglerFISH/ directory, build the genome indexes for HISAT2 and Jellyfish programs. *See* **Note 4** for more details on generating indexes and parameters.

```
./nextflow main.nf --genome_index <genome name> --createIn-
dexes true --l <min length> --L <max length>
```

The indexes will be automatically placed in a new directory "anglerFISH/UPLOAD_HERE/genome_index/<genome_index>". The indexes are kept in this directory and can be used repeatedly to design sets of probes. There is no need to repeat this step until new target organism must be used.

3.1.3 Run the Workflow to Generate RNA FISH Probes

1. Copy the target mRNA sequence in fasta format to the directory "anglerFISH/UPLOAD_HERE/fasta".
2. Once the indexes have been created (Subheading 3.1.2), run the pipeline to design probes, specifying the parameters of the desired probes.

```
./nextflow main.nf --genome_index <genome> --input_file
<RNA_sequence_base> <optional parameters>
```

Carefully read the help message of the workflow (./next-flow main.nf --help) to learn about all the possible options, modes, and default parameters of the probes.

After a successful run of the pipeline (*see* **Note 5**), a zip file with all the output files produced by the workflow can be found in the directory "anglerFISH/UPLOAD_HERE/results". Unpack the zip file, which should contain the following.

(a) Fasta and tab files with designed probes.

(b) Fasta file with reverse complementary sequences of the probes.

(c) Alignment file(s) of the final probes against the genome.

(d) File with a table of melting temperatures (Tm) of the probes.

(e) Log File with basic information about the run.

3. Inspect the number of probes the workflow returned. For reliable detection of single RNAs, we recommend using at least 40 probes made of 18–23 nucleotides or at least 25 probes made of 45–50 nucleotides. If your run returned fewer probes, try rerunning the workflow with broader parameters.
4. (Optional) Visualization of probes against endogenous RNAs.

   Go to a web application version of the Integrative Genomic Viewer at https://igv.org/app/ and click on the *Genome > selector* and select the same genome version that was used to filter and map your probes (*see* **Note 6**). Click on *Tracks > Local File* to load the probe alignment files created by the pipeline. Select both the .bam and .bam.bai file at the same time. Search for the name of your gene in the search box to zoom in to the gene and view the aligned probes.

3.2 Probe Labeling

A detailed description of the probe labeling can be found in the original publication describing TdT probe labeling [16].

3.2.1 Conjugating ddUTP and Dye

1. Assemble a moisture-free environment by enclosing silica beads in a box with a tightly closed lid for around an hour.
2. In the moisture-free environment, resuspend the amine-dye in anhydrous DMSO to a final concentration of 40 mM.
3. Assemble ddUTP-dye conjugation mix as described in Subheading 2 and incubate for 2 h at room temperature (RT) in the dark.
4. Add 18 μL of nuclease-free water to obtain 40 μL of 5 mM stock solution of ddUTP-dye mix. Aliquot to amber tubes and store at −20 °C.

3.2.2 Conjugating Oligonucleotides and ddUTP-Dye

1. Order the oligonucleotide ssDNAs with sequences designed by the workflow at a concentration of 100 μM and desalted.
2. In a PCR tube, assemble the ddUTP-dye-oligos labeling reaction described in Subheading 2. Vortex and briefly spin down the mixture in a minicentrifuge. Incubate the mixture for 16–24 h in a thermocycler set to 37 °C.
3. Add 60 μL 1 M sodium acetate (pH 5.5), 125 μL nuclease-free water, and 1.5 μL linear acrylamide to the PCR tube. Mix well and transfer to a 1.5 mL tube. Add 800 μL of 100% ethanol (−20 °C). Mix the contents of the tube by inverting several times. Transfer to −80 °C for 30–60 min.
4. Centrifuge for 30 min (16,000 × *g* at 4 °C) after which a colored pellet should be visible at the bottom of the tube. Keep the tubes on ice at all times.
5. Remove the supernatant, add 1 mL of 80% ethanol (4 °C) to the tube. Gently pipette the liquid up and down several times to detach the pellet from the bottom of the tube, trying to avoid fragmenting the pellet into smaller pieces. Centrifuge for 5 min (16,000 × *g* at 4 °C). Repeat this ethanol washing step three times in total, after the first wash transfer the entire ethanol with the floating pellet to a clean 1.5 mL tube.
6. Remove as much of the supernatant as possible and let the pellet dry for 5–10 min or until all of the liquid evaporated. Resuspend the pellet with 30 μL of nuclease-free water by pipetting up and down until completely dissolved (*see* **Note 7**).

3.2.3 Assessing the Labeling Efficiency

1. Measure the absorbance (optical density, OD) at 260 nm (OD_{260}) and at the maximum absorption of the dye (OD_{dye}) of the probe solution with Nanodrop. To calculate the degree of labeling, enter the values in the *DegreeOfLabeling.xlsx* spreadsheet attached in the anglerFISH directory or the anglerFISH GitHub repository. Alternatively, follow the step-by-step instructions below.
2. Calculate the mean molar extinction coefficient of the oligonucleotides mix (ε_{oligos}) by taking the mean of the molar extinction coefficient of each oligonucleotide. Add 9000 mol^{-1} cm^{-1} to correct for the extra UTP added to labeled oligonucleotide (ε_{corr}).
3. Calculate the concentration of the oligonucleotides.

$$c_{oligos} = \frac{OD_{260} - cf_{260} \times OD_{dye}}{\varepsilon_{corr}(mol^{-1}cm^{-1}) \times \text{light path (cm)}}$$

where cf_{260} is the contribution of the dye to OD at 260 nm (consult dye manual), ε_{dye} is the dye's extinction coefficient (consult dye manual). The standard light path length of the Nanodrop is 0.1 cm.

4. Calculate the concentration of the dye:

$$c_{\text{dye}} = \frac{\text{OD}_{\text{dye}}}{\varepsilon_{\text{dye}}\left(\text{mol}^{-1}\text{cm}^{-1}\right) \times \text{light path (cm)}}$$

5. Calculate the degree of labeling.

$$\text{degree of labeling} = \frac{c_{\text{dye}}}{c_{\text{oligo}}}$$

We consider a probe set to be sufficiently well-labeled for smFISH experiments when the degree of labeling is above 0.9 (*see* **Note 8**).

3.3 RNA smFISH

3.3.1 Fixation of Samples

1. Seed between 10,000 and 40,000 cells per well onto coverslips placed in a 12-well plate and return to the incubator to allow cells to attach and become well spread (~1–2 days). Avoid allowing cells to grow to confluence (*see* **Note 9**).
2. Aspirate the culture medium and wash twice with 1 mL of PBS.
3. Add 1 mL per well of fixation solution and incubate for 10 min at RT while gently rocking (*see* **Note 10**).
4. Wash thrice with PBS.
5. To permeabilize the cells, add 1 mL of fresh 70% w/v ethanol and incubate at 4 °C overnight.

3.3.2 Probe Hybridization

1. Aspirate ethanol and wash the wells once with PBS.
2. Add 1 mL per well of the wash buffer and incubate for 5 min at RT while gently rocking. Repeat the wash once more.
3. Pipette 45 μL of hybridization solution into a drop on a sheet of Parafilm in the hybridization chamber (*see* Subheading 2).
4. Remove the coverslip with tweezers from the wash buffer and gently dry it by tapping the side on absorbent tissue. Invert the coverslip (cell side down) onto the drop of hybridization solution.
5. Seal the chamber with Parafilm and cover with aluminum foil before hybridizing at 37 °C for 4 h up to overnight.
6. In a new 12-well plate, add 1 mL per well of prewarmed (37 °C) wash buffer.
7. Transfer the coverslips back to the 12-well plate using tweezers, placing them cell side up.
8. Incubate in the dark at 37 °C for 30 min, while gently rocking to wash out any unhybridized probes.
9. Repeat wash once more.

10. Wash twice with PBS.
11. Prepare the glass slides by centrally placing a drop of Prolong Gold Antifade Reagent with DAPI.
12. Use tweezers to remove coverslips from PBS and slightly dry by tapping the side on absorbent tissue.
13. Place the coverslip cell side down onto the drop.
14. Let the mounting medium solidify by storing the samples in a dark place at RT overnight.
15. Seal coverslips with nail polish around the edges and proceed to image or store for up to a week at 4 °C.

3.4 Image Acquisition

1. smFISH images can be collected on a wide-field epifluorescence microscope equipped with a metal-halide lamp, a high numerical aperture oil objective, and a CCD camera.
2. The DAPI channel should be used to identify regions for imaging where cells are well dispersed and to avoid any bias in data collection.
3. Acquire optical sections spanning 5–6 μm using a 100× objective with a z-step sampling close to Nyquist rate, which is typically 200–300 nm depending on objective's numerical aperture. Set exposure times resulting in a satisfactory signal histogram.

4 Notes

1. We have successfully generated working probes with ATTO 565 and ATTO 633 dyes.
2. Prepare the 20% dextran sulfate solution prior to use as it takes several hours to dissolve.
3. As a general starting point for probe design, we suggest designing 18–23 nucleotides long probes, with melting temperatures between 47–60 °C, in 10% formamide. We found those conditions to provide a broad selection of probes. In case the number of probes designed by the workflow is very high and only a subset of probes needs to be selected for synthesis, we suggest either narrowing the desired melting temperatures of probes closer to 57 °C and selecting probes with similar melting temperatures or alternatively, picking probes based on the sequence alignment. We suggest selecting probes targeting the 5′ end of the target RNA, or probes that target the suitable isoforms or other important features.
4. Genome indexes, built from reference genome files, allow mapping and processing sequences more efficiently. In the case of anglerfish, the hisat2 index is used for aligning probes

to the genome and the jellyfish index is used for finding and filtering out common k-mer sequences. The jellyfish index must be built for each k-mer length separately. In order to define k-mers (probe lengths) to build jellyfish indexes, select the range of probe lengths that you want to design and indicate the range by specifying a minimal (--l) and maximal (--L) length of the probes. For example, if you plan to design probes with lengths ranging from 18–23 nucleotides, specify --l 18 --L 23.

In case you have already built the indexes in the past and want to expand the range of k-mer jellyfish indexes (probe lengths), you can use the --jf-only flag to only make additional jellyfish indexes, without making the hisat2 indexes again. For example, if you decide later to also design probes between 24 and 25 nucleotides, to add jellyfish indexes for these k-mers run the following.

```
./nextflow main.nf --genome_index <genome name> --createIn-
dexes true --l 24 --L 25 --jf_only
```

5. To save disk space, regularly remove the content of the "anglerFISH/work/" directory, which contains intermediate files used in the process of designing probes and are obsolete after the workflow finishes designing a probe set. Alternatively, you can also run "make clean."
6. You can also upload any genome of interest.
7. If the pellet is difficult to resuspend at first, allow it to soak the water for a few minutes, then try to pipet again.
8. Occasionally, the labeling efficiency can be low (<90%). Usually, simply repeating the labeling of the unlabeled oligonucleotides solves the issue. The second major factor that we found affecting probe labeling is the batch of the TdT enzyme, which should be replaced if the degree of labeling is consistently low. If the probe labeling efficiency is still suboptimal, refer to the original labeling publication [16] for more details on troubleshooting this issue.
9. The number of cells and the time will depend on the experimental cell line.
10. Perform this step under the hood to prevent inhaling toxic paraformaldehyde fumes.

Acknowledgments

The authors would like to thank all the members of the Chao lab and the Giorgetti lab for their support in establishing the probe designing and experimental pipeline. We thank Daniel Mateju for critically reading the manuscript. In addition, we thank the Facility for Advanced Imaging and Microscopy at FMI for help in image acquisition. Research in the Chao lab is funded by the Novartis Research Foundation, the Swiss National Science Foundation (310031A_182314) and the SNF NCCR RNA & Disease network. Research in the Giorgetti lab is funded by the Novartis Research Foundation and the European Research Council (ERC) under the European Union's Horizon 2020 research and innovation (grant agreement no. 759366, 'BioMeTre').

References

1. Chao JA, Lionnet T (2018) Imaging the life and death of mRNAs in single cells. Cold Spring Harb Perspect Biol 10:a032086
2. Eliscovich C, Shenoy SM, Singer RH (2017) Imaging mRNA and protein interactions within neurons. Proc Natl Acad Sci 114
3. Skinner SO, Sepúlveda LA, Xu H et al (2013) Measuring mRNA copy number in individual Escherichia coli cells using single-molecule fluorescent in situ hybridization. Nat Protoc 8:1100–1113
4. Yang L, Titlow J, Ennis D et al (2017) Single molecule fluorescence in situ hybridisation for quantitating post-transcriptional regulation in Drosophila brains. Methods 126:166–176
5. Long X, Colonell J, Wong AM et al (2017) Quantitative mRNA imaging throughout the entire Drosophila brain. Nat Methods 14: 703–706
6. Femino AM et al (1998) Visualization of single RNA transcripts in situ. Science 280:585–590
7. Raj A, van den Bogaard P, Rifkin SA et al (2008) Imaging individual mRNA molecules using multiple singly labeled probes. Nat Methods 5:877–879
8. Battich N, Stoeger T, Pelkmans L (2013) Image-based transcriptomics in thousands of single human cells at single-molecule resolution. Nat Methods 10:1127–1133
9. Xia C, Fan J, Emanuel G et al (2019) Spatial transcriptome profiling by MERFISH reveals subcellular RNA compartmentalization and cell cycle-dependent gene expression. Proc Natl Acad Sci 116:19490–19499
10. Piskadlo E, Eichenberger B (2021) anglerFISH: initial anglerFISH release. Zenodo 4427720
11. Tommaso PD, Chatzou M, Floden EW et al (2017) Nextflow enables reproducible computational workflows. Nat Biotechnol 35: 316–319
12. Beliveau BJ, Kishi JY, Nir G et al (2018) OligoMiner provides a rapid, flexible environment for the design of genome-scale oligonucleotide in situ hybridization probes. Proc Natl Acad Sci 115
13. Marçais G, Kingsford C (2011) A fast, lock-free approach for efficient parallel counting of occurrences of k-mers. Bioinformatics 27: 764–770
14. Kim D, Paggi JM, Park C et al (2019) Graph-based genome alignment and genotyping with HISAT2 and HISAT-genotype. Nat Biotechnol 37:907–915
15. Zadeh JN, Steenberg CD, Bois JS et al (2010) NUPACK: analysis and design of nucleic acid systems. J Comput Chem 32:170–173
16. Gaspar I, Wippich F, Ephrussi A (2017) Enzymatic production of single-molecule FISH and RNA capture probes. RNA 23:1582–1591

Chapter 11

BaseScope™ Approach to Visualize Alternative Splice Variants in Tissue

Alexandra Bunda and Arturo Andrade

Abstract

Defining the cell-specific alternative splicing landscape in complex tissues is an important goal to gain functional insights. Deep-sequencing techniques coupled to genetic strategies for cell identification has provided important cues on cell-specific exon usage in complex tissues like the nervous system. BaseScope™ has emerged as a powerful and highly sensitive alternative to in situ hybridization to determine exon composition in tissue with spatial and morphological context. In this protocol, we will review how BaseScope was utilized to detect the e37a-*Cacna1b* splice variant of the presynaptic calcium channel $Ca_V2.2$ or N-type. This splice variant arises from a pair of mutually exclusive exons (e37a and e37b). E37a-*Cacna1b* is heavily underrepresented relative to e37b-*Cacna1b* and both exons share 60% of their sequence. By using BaseScope™, we were able to discover that e37a-*Cacna1b* is expressed in excitatory pyramidal neurons of hippocampus and cortex, as well as motor neurons of the ventral horn of the spinal cord.

Key words BaseScope™, *Cacna1b*, Alternative splicing, $Ca_V2.2$, N-type, In situ hybridization

1 Introduction

Alternative splicing of the pre-mRNA is a process by which cells increase the variety of transcripts without increasing their genetic load, because a single gene can originate multiple splice variants with distinct functions [1–3]. The exon selection that occurs during alternative splicing is under the fine control of RNA-binding proteins (splicing factors) and, in some cases, DNA-binding proteins [4, 5]. The expression and activity of regulators for alternative splicing greatly determines tissue- and cell-specific expression of splice variants [1]. Thus, alternative splicing is thought to add complexity to tissues by increasing the functional variety of cells within them [6].

Alternative splicing is abundant in the nervous system where more than 95% of multiexon genes undergo some form of splicing [7–9]. Given the complexity of the nervous system, it has been

Peter Scheiffele and Oriane Mauger (eds.), *Alternative Splicing: Methods and Protocols*,
Methods in Molecular Biology, vol. 2537, https://doi.org/10.1007/978-1-0716-2521-7_11,

challenging to identify the exon composition of genes in specific neurons. Recently, tremendous progress has been made at identifying splice variants on a global scale at the cell-specific level using multilevel approaches including genetic drivers, microdissection, and fluorescence activated cell-sorting coupled to next generation sequencing [10–13]. These methods have provided valuable insights into the cell-specific combination of exon usage in complex tissues like the nervous system.

To identify individual splicing events in complex tissues with spatial and morphological context, BaseScope®, a modality of in situ hybridization was developed [14–16]. This technique is extremely sensitive; thus, it is possible to detect transcripts with as few as one nucleotide difference. Here, we will describe how BaseScope™ was applied for tissue detection of splice variants of the gene that encodes for the α_1-pore forming subunit of the presynaptic calcium channel, $Ca_V2.2$ (*Cacna1b*). $Ca_V2.2$ channels control the influx of calcium that triggers transmitter release in central and peripheral synapses [17]. These channels are involved in pain, anxiety, and aggression [18–20]. The *Cacna1b* gene is a multiexon gene that undergoes extensive cell-specific alternative splicing [21]. We and others have shown that alternative splicing of the *Cacna1b* pre-mRNA has robust functional implications for $Ca_V2.2$ channels [6]. For example, we have shown that inclusion of exon 37a (e37a) enhances the analgesic response to morphine by producing $Ca_V2.2$ channels that are more sensitive to μ-opioid receptor signaling [22–25]. Furthermore, lack of splice variants containing e37a results in enhanced exploratory behavior under basal and stressful conditions [26]. These studies demonstrate the functional relevance of e37a-*Cacna1b* splice variants, thereby indicating an importance of gaining mechanistic insight into this splice variant and determining its pattern of expression.

Exon 37a together with 37b are a pair of mutually exclusive exons that encode for sequences in the C-terminus of $Ca_V2.2$ channels (Fig. 1a) [27]. E37a-*Cacna1b* was originally detected in nociceptors, but also in the brain where it is much less abundant [28]. In nociceptors, e37a-*Cacna1b* splice variants were detected using electrophysiology coupled to single-cell RT-PCR [27, 29]. The detection of e37a-*Cacna1b* splice variant in tissues posed two interesting challenges: e37a is disproportionally represented relative to e37b in total *Cacna1b* transcripts (~2% vs ~98% in the brain, respectively) [28], and ~60% of the sequence between e37a and e37b is shared (Fig. 1a). To overcome these challenges, we utilized BaseScope™, a technique that combines highly specific probes ("Z" probes) with a signal amplification system that allows for detection of low-abundance exons (Fig. 1a, b). "Z" probes contain an 18–25 base region complementary to the target mRNA sequence that in the case of BaseScope™ binds to exon junctions (in our case, between e36 and e37a in the *Cacna1b*

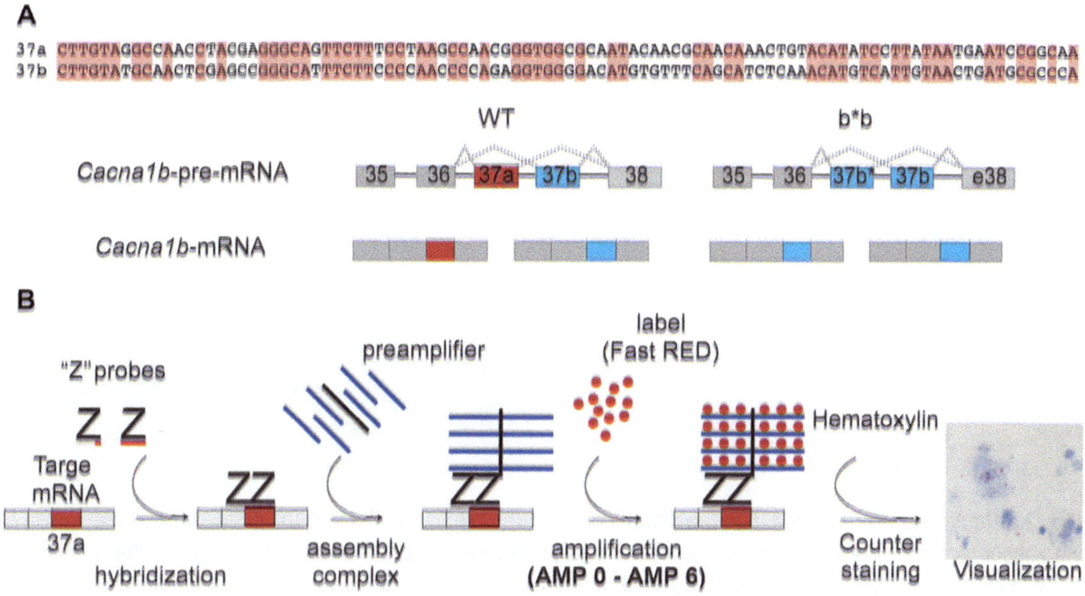

Fig. 1 Use of BaseScope™ for the detection of *e37a-Cacna1b* mRNA in tissue. (**a**) *Top*. Sequences for e37a and e37b. These two exons share ~60% of their pre-mRNA sequence. Red shades indicate the base pairs that are shared between the e37a and e37b. *Bottom*. Schematic representation of e37a/e37b mutually exclusive splicing in the WT mice. To eliminate *e37a-Cacna1b*, e37a was replaced with a second copy of e37b (e37b*), forcing the mice to express only splice variants with e37b. This mouse was used as a control for probe specificity. (**b**) Schematic representation of the workflow for BaseScope™, "Z" probes containing a complementary sequence to e37a hybridize, next a complex is an assembled using "Z" probes as a base, then a series of amplification steps lead to accumulation of dye specifically on the assembly complex. This accumulation forms a punctuated pattern where each dot represents a single mRNA molecule. In microscopy image, red dots can be seen in nuclei counterstained with hematoxylin

mRNA), a spacer sequence, and a 14-base tail sequence. A pair of "Z" probes (double Z) hybridize to the target mRNA, and the two 14-base tail sequences act as a platform for the assembly of the signal amplification complex. This amplification complex contains 20 binding sites for label compounds, which can be fluorescence and/or colorimetric based. Each puncta indicates a single mRNA molecule (Fig. 1b). Thanks to this technique, we determined new cell populations where e37a-*Cacna1b* splice variants are expressed, including excitatory neurons in the hippocampus and cortex [26], as well as motor neurons located in the ventral horn of the spinal cord (Fig. 2). We have also used BaseScope™ successfully to identify other Cacna1b splice variants (+18a-*Cacna1b*) in midbrain, cerebellum and spinal cord [30]. This new information is likely to unveil new functions for e37a-*Cacna1b* and +18a-*Cacna1b* splice variants.

We will describe a detailed protocol to perform BaseScope™ using a free-floating mounting technique of spinal cord and brain tissue sections. We will provide detailed list of materials, reagents, and notes that could be used to troubleshoot specific experimental needs.

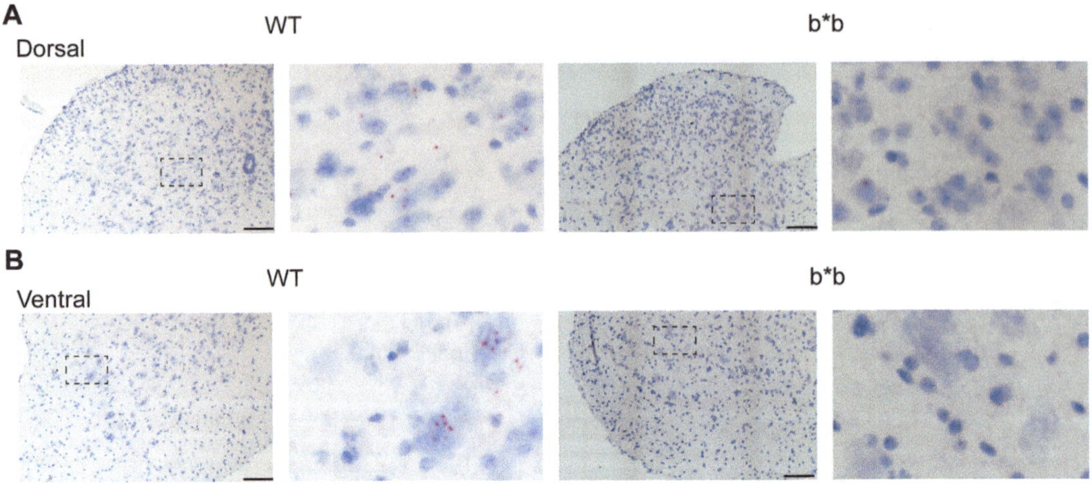

Fig. 2 Detection of *e37a-Cacna1b* in the spinal cord using BaseScope™. Little to no signal for *e37a-Cacna1b* was observed in the dorsal horn of the spinal cord (**a**), however this splice variant was found in some nuclei of the ventral areas of the spinal cord (**b**). Several puncta were localized in large nuclei of the ventral horn, likely motor neurons. Note that no signal was observed in similar areas of spinal cord sections of the b*b Mice (lack *e37a-Cacna1b* expression) demonstrating the specificity of the probes against *e37a*

2 Materials

1. Transgenic mice.

   Wild-type mice (WT) in C57BL/6 background (Charles River). *Cacna1b*$^{tm2.1Dili}$ (*b*b*) in a C57BL/6 background (Charles River) express only the *e37b-Cacna1b* splice variant, but not the e37a-*Cacna1b* variant. These mice were used as a negative control to assess the specificity of the probe against e37a (Fig. 1a and [23, 26]).
2. Isoflurane.
3. Euthasol.
4. Neutral buffered formalin fixative (NBF).
5. PBS.
6. Sucrose.
7. Optimal cutting temperature compound (OCT).
8. Isopentane.
9. 15 mm Netwell™ inserts.
10. SuperFrost® microscope slides or charged microscope slides.
11. RNAscope™ Hydrogen Peroxide and protease reagents (ACD). This kit contains RNAscope™ Hydrogen Peroxide, RNAscope™ Protease III, RNAscope Protease Plus, and RNAscope™ Protease IV.

12. RNAscope® Target Retrieval reagent (ACD) contains 10× target retrieval solution.
13. Molecular biology grade distilled water (RNase free) and tap water.
14. Ethanol.
15. Hydrophobic barrier pen.
16. BaseScope™ probe spanning exon junction 36:37a of *Cacna1b* mRNA (ACD).
17. BaseScope™ Detection Reagents. This contains BaseScope™ *AMP 0 to AMP 6* reagents and BaseScope™ Fast RED-A and BaseScope™ Fast RED-B (ACD).
18. RNAscope™ Wash Buffer (ACD).
19. HybEZ™ Hybridization oven (ACD) or HybEZ™ II Hybridization oven (ACD).
20. HybEZ™ Humidity Control Tray with lid (ACD).
21. HybEZ™ Slide Rack (ACD).
22. Gill's hematoxylin I (American Master Tech).
23. Ammonium hydroxide (Sigma-Aldrich).
24. VectaMount™ (Vector Laboratories).
25. Oster® Steamer 5712.
26. Paper towels.
27. Thermometer.
28. Tissue-Tek® Vertical Slide Rack (American Master Tech Scientific).
29. Tissue-Tek® Staining Dish (American Master Tech Scientific).
30. Cover Glass, 24 × 50 mm.
31. Cryostat.
32. Paintbrush.

3 Methods

3.1 Tissue Preparation

This procedure is key to obtain high-quality sections for BaseScope™. Improper preparation of tissue can lead to artifacts derived from the accumulation of dye (RNAscope Fast RED) in random places within the section.

1. Mice were bred and raised in a temperature-controlled pathogen-free room and moved into a separate temperature-controlled room prior to experimental use. Rooms were programmed on a 12-h light–dark cycle with water and food ad libitum. All experimental protocols were approved by the

Institutional Animal Care and Use Committee of University of New Hampshire.

2. Anesthetize mice in a chamber with a saturating dose of isoflurane until asleep and inject 150 mg/kg IP of Euthasol.
3. Perfuse mice transcardially with 1× PBS for 10 min, then NBF for 10 min, then 1× PBS for 10 min.
4. Remove tissue of interest (whole brain and/or spinal cord).
5. Dehydrate brain tissue in two sequential steps (20% and 30% sucrose 1× PBS at 4 °C). For spinal cord, three sequential steps are recommended (10%, 20%, and 30% sucrose 1× PBS at 4 °C). Change the sucrose solution until the tissue is seen at the bottom of the tube, this process takes 24–72 h.
6. Freeze tissue in OTC embedding media with dry ice or liquid nitrogen (tissue can be used immediately or kept in an airtight container at −80 °C for later use for up to 3 months).
7. Prior to sectioning, equilibrate the tissue blocks and 15 mm Netwell™ inserts in a 12-well plate at −20 °C in the cryostat for at least 1 h.
8. Cut tissue blocks in 12 μm sections (*see* **Note 1**).
9. Collect sections into prechilled 15 mm Netwell inserts. After collection, keep the Netwell™ inserts containing the tissue sections on dry ice until ready to mount.
10. Mount sections onto a glass slide (mounted sections can be stored at −80 °C for up to a week). To do this, transfer sections from Netwell™ inserts with a chilled paintbrush to a container with PBS (petri dish). Next, guide sections onto slides. Finally, remove excess of PBS with a Kimwipe®, and airdry (*see* **Note 2**).
11. Dry slides in a drying oven at 60 °C for 30 min. This step will help tissue to adhere to the glass to prevent random accumulation of dye.
12. Insert slides into the Tissue-Tek® vertical slide rack and fix slides in 10% NBF in a Tissue-Tek® staining dish for 15 min to ensure complete fixation.
13. Transfer slides to a Tissue-Tek® staining dish filled with 50% ethanol for 5 min at room temperature (RT).
14. Repeat the **step 13** with 70% ethanol, then again twice with 100% ethanol at RT.
15. Remove slides from the Tissue-Tek® vertical slide rack, place on absorbent paper, and air-dry for at least 5 min at RT, then proceed with target retrieval.

3.2 Target Retrieval

1. Before starting target retrieval, take a few minutes to prepare the Oster® Steamer. Fill the water reservoir with cold tap water

to the HI line, place a clear steaming bowl on the base, then place two slide holder in steam bowl. Fill one slide holder with 200 mL of 1× RNAscope® Target Retrieval solution, and another slide holder with 200 mL of distilled water. Set the Oster® Steamer to allow the temperature to rise to 99 °C. Monitor the temperature of the RNAscope 1× Target Retrieval solution by inserting a digital thermometer through the steamer holes of the lid of the container (*see* **Note 3**).

2. Lay slides on absorbent paper, after allowing RNAscope™ Hydrogen Peroxide to equilibrate at RT, add 1–2 drops (or enough to completely cover the section) and incubate for 10 min at RT. This step is important to block endogenous peroxidase activity and reduce nonspecific background staining (*see* **Note 4**).
3. Remove RNAscope™ Hydrogen Peroxide by tapping slide on absorbent paper and immediately insert the slide into the Tissue-Tek® Vertical Slide Rack and submerge in a Tissue-Tek® Staining Dish filled with distilled water.
4. Wash slides 3–5 times by moving the Tissue-Tek® Vertical Slide Rack up and down in distilled water.
5. Transfer slides to the slide holder with distilled water in the Oster® Steamer for 10 s to acclimate them, then immerse slides in the container with 1× RNAscope® Target Retrieval solution, cover steamer with lid and incubate for 15 min.
6. Remove slides from Oster® Steamer and transfer to a separate container with 200 mL of distilled water and rinse for 15 s at RT.
7. Transfer slides to 100% ethanol for 3 min.
8. Dry slides at 60 °C for 30 min. Proceed to protease reaction.

3.3 Protease Reaction

1. Draw a barrier around each section with a hydrophobic barrier pen and let dry for 1 min or overnight at RT protecting section from dust.
2. Place dried slides on the ACD EZ-Batch™ Slide Rack and add 1–2 drops (or enough to completely cover the section) (*see* **Note 5**) of RNAscope™ Protease III for brain tissue or RNAscope™ Protease IV for spinal cord (*see* **Note 6**).
3. Close the humidity tray and incubate at 40 °C in the HybEZ™ Hybridization oven for 15 min (*see* **Note 7**).
4. Take each slide one at a time from the slide rack. Then tap and/or flick to remove the excess liquid. Immediately place each slide in a Tissue-Tek® Vertical Slide Rack and submerge in a Tissue-Tek® Staining Dish filled with distilled water.

5. Wash slides 3–5 times by moving the Tissue-Tek® Vertical Slide Rack up and down.
6. Proceed to BaseScope™ protocol.

3.4 BaseScope™

In this part of the protocol a series of amplification steps will be performed (*AMP 0* to *AMP 6*) to allow amplifiers reagents like BaseScope™ Fast RED to form the amplification complex and thereby the puncta indicative of the presence of mRNA molecules. The following steps provide details on each amplification step.

1. Warm the 50× RNAscope™ Wash Buffer before use for 10–20 min at 40 °C and prepare 3 L of 1× RNAscope™ Wash Buffer with distilled water.
2. Prepare counterstaining reagents. In a fume hood, prepare 50% hematoxylin (1:1 Gill's hematoxylin and distilled water in a Tissue-Tek® Staining Dish), and ammonia water (1.43 mL of 1 N ammonium hydroxide to 250 mL of distilled water and mix well).
3. Equilibrate BaseScope™ Detection Reagents. Place *AMP 0* to *AMP 6* reagents at RT, preheat HybEZ™ Hybridization oven and HybEZ™ Humidity Control Tray at 40 °C. Equilibrate BaseScope™ probe for 30 min at RT.
4. To hybridize probes, remove excess of liquid from slides and add 1–2 drops (or enough to completely cover the section) of the probes. Place slides in HybEZ™ Slide Rack. Insert rack in the HybEZ™ Humidity Control Tray and incubate in HybEZ™ Hybridization oven for 2 h at 40 °C.
5. Remove slides from HybEZ™ Slide Rack and place in Tissue-Tek® Vertical Slide Rack submerged in Tissue-Tek® Staining Dish filled with 1× RNAscope™ Wash Buffer.
6. Wash slides twice in RNAscope™ 1× Wash Buffer for 2 min at RT by moving the slide rack up and down.
7. Remove excess liquid, place slides in HybEZ™ Slide Rack/ HybEZ™ Humidity Control Tray and add 1–2 drops (or enough to completely cover the section) of BaseScope™ *AMP 0*.
8. Incubate for 30 min at 40 °C.
9. Transfer slides to Tissue-Tek® Vertical Slide Rack and wash slides in 1× RNAscope™ Wash Buffer for 2 min at RT. Repeat with fresh 1× wash buffer.
10. Remove excess liquid, place slides in HybEZ™ Slide Rack/ HybEZ™ Humidity Control Tray and add 1–2 drops (or enough to completely cover the section) of BaseScope™ *AMP 1*.
11. Incubate for 15 min at 40 °C.

12. Transfer slides to Tissue-Tek® Vertical Slide Rack and wash slides in 1× RNAscope™ Wash Buffer for 2 min at RT. Repeat with fresh 1× wash buffer.
13. Remove excess liquid, place slides in HybEZ™ Slide Rack/HybEZ™ Humidity Control Tray and add 1–2 drops (or enough to completely cover the section) of BaseScope™ *AMP 2*.
14. Incubate for 30 min at 40 °C.
15. Transfer slides to Tissue-Tek® Vertical Slide Rack and wash slides in 1× RNAscope™ Wash Buffer for 2 min at RT. Repeat with fresh 1× wash buffer.
16. Remove excess liquid, place slides in HybEZ™ Slide Rack/HybEZ™ Humidity Control Tray and add 1–2 drops (or enough to completely cover the section) of BaseScope™ *AMP 3*.
17. Incubate for 30 min at 40 °C.
18. Transfer slides to Tissue-Tek® Vertical Slide Rack and wash slides in 1× RNAscope™ Wash Buffer for 2 min at RT. Repeat with fresh 1× wash buffer.
19. Remove excess liquid, place slides in HybEZ™ Slide Rack/HybEZ™ Humidity Control Tray and add 1–2 drops (or enough to completely cover the section) of BaseScope™ *AMP 4*.
20. Incubate for 15 min at 40 °C.
21. Transfer slides to Tissue-Tek® Vertical Slide Rack and wash slides in 1× RNAscope™ Wash Buffer for 2 min at RT. Repeat with fresh 1× wash buffer.
22. Remove excess liquid, place slides in HybEZ™ Slide Rack/HybEZ™ Humidity Control Tray and add 1–2 drops (or enough to completely cover the section) of BaseScope™ *AMP 5*.
23. Incubate for 30 min at 40 °C.
24. Transfer slides to Tissue-Tek® Vertical Slide Rack and wash slides in 1× RNAscope™ Wash Buffer for 2 min at RT. Repeat with fresh 1× wash buffer.
25. Remove excess liquid, place slides in HybEZ™ Slide Rack/HybEZ™ Humidity Control Tray and 1–2 drops (or enough to completely cover the section) of BaseScope™ *AMP 6* to cover each section.
26. Incubate for 15 min at 40 °C.
27. From this step, allow the humidity tray to equilibrate to RT.

28. Transfer slides to Tissue-Tek® Vertical Slide Rack and wash slides in 1× RNAscope™ Buffer for 2 min at RT. Repeat with fresh 1× wash buffer.
29. Briefly spin down the BaseScope™ Fast RED-B tube before opening the cap. Prepare fast RED working solution by mixing a 1:60 ratio of BaseScope™ Fast RED-B to BaseScope™ Fast RED-A (*see* **Note 8**).
30. Place slides in HybEZ™ Slide Rack/HybEZ™ Humidity Control Tray and incubate for 10 min at RT.
31. Remove excess BaseScope™ Fast RED working solution and place the slides on Tissue-Tek® Vertical Slide Rack/Tissue-Tek Staining Dish filled with tap water. Rinse twice by moving slide rack up and down.
32. To counterstain slides, place the Tissue-Tek® Vertical Slide Rack into a staining dish containing 50% hematoxylin staining solution for 2 min at RT. Slides should look purple.
33. Immediately wash slides 3 times with fresh tap water each time.
34. Immerse slides in ammonia water and move rack up and down 2–3 times. Sections should turn blue.
35. Wash slides 3–5 times in tap water.
36. After drying slides at 60 °C for at least 15 min, add 1–2 drops of VectaMount on the slide, place a coverslip over tissue section and air-dry completely for at least 5 min.
37. Evaluate samples with confocal or wide-field microscope depending on the on whether the label is fluorescence or colorimetric based. Finished slides can be stored in the dark at 4 °C for later imaging (*see* **Note 9**).

4 Notes

1. This protocol was adapted to free floating sections by drying and fixing slides. These steps allow for flat mounting of tissue sections to prevent random accumulation of BaseScope™ Fast RED signal beneath the section. Sections were cut at least 12 μm to ensure the integrity of the tissue.
2. Order at least four Tissue-Tek staining dishes to use with this protocol. One dish should be designated for hematoxylin staining, and another should be designated for ammonia water. Be sure to rinse dishes well between steps and use mild detergent to clean dishes between use.
3. Prepare the steamer ~45 min ahead to allow the temperature reach 99 °C.

4. Add 1–2 drops of RNAscope hydrogen peroxide per section is optimal for mouse brain sections, larger or smaller sections should be adjusted accordingly.
5. Ensure the barrier is not too close to the edge of the tissue for optimal staining.
6. Protease treatment was adjusted to tissue, spinal cord is a more delicate tissue than brain, so RNAscope™ Protease IV was used instead of RNAscope™ Protease III.
7. Humidity tray requires a damp paper towel in the bottom of the tray to avoid evaporation.
8. Once mixed, the Fast RED solution should be used within 5 min.
9. We have stored slides stained with BaseScope™ Fast RED with for up to 6 months; however, the intensity and resolution of the puncta slightly decrease with storage.

Acknowledgments

This work was supported by NIMH grant R00-MH099405.

References

1. Baralle FE, Giudice J (2017) Alternative splicing as a regulator of development and tissue identity. Nat Rev Mol Cell Biol 18:437
2. Rodriguez JM, Pozo F, di Domenico T, Vazquez J, Tress ML (2020) An analysis of tissue-specific alternative splicing at the protein level. PLoS Comput Biol 16:e1008287
3. Ule J, Blencowe BJ (2019) Alternative splicing regulatory networks: functions, mechanisms, and evolution. Mol Cell 76:329–345
4. Lee Y, Rio DC (2015) Mechanisms and regulation of alternative pre-mRNA splicing. Annu Rev Biochem 84:291–323
5. Lipscombe D, Lopez Soto EJ (2019) Alternative splicing of neuronal genes: new mechanisms and new therapies. Curr Opin Neurobiol 57:26–31
6. Lipscombe D, Andrade A, Allen SE (2013) Alternative splicing: functional diversity among voltage-gated calcium channels and behavioral consequences. Biochim Biophys Acta 1828:1522–1529
7. Grabowski PJ, Black DL (2001) Alternative RNA splicing in the nervous system. Prog Neurobiol 65:289–308
8. Lee CJ, Irizarry K (2003) Alternative splicing in the nervous system: an emerging source of diversity and regulation. Biol Psychiatry 54: 771–776
9. Raj B, Blencowe BJ (2015) Alternative splicing in the mammalian nervous system: recent insights into mechanisms and functional roles. Neuron 87:14–27
10. Arzalluz-Luque Á, Conesa A (2018) Single-cell RNAseq for the study of isoforms—how is that possible. Genome Biol 19:1–19
11. Bao S, Moakley DF, Zhang C (2019) The splicing code goes deep. Cell 176:414–416
12. Feng H, Moakley DF, Chen S, McKenzie MG, Menon V, Zhang C (2021) Complexity and graded regulation of neuronal cell type-specific alternative splicing revealed by single-cell RNA sequencing. bioRxiv
13. Weyn-Vanhentenryck SM, Feng H, Ustianenko D, Duffié R, Yan Q, Jacko M, Martinez JC, Goodwin M, Zhang X, Hengst U, Lomvardas S, Swanson MS, Zhang C (2018) Precise temporal regulation of alternative splicing during neural development. Nat Commun 9:2189
14. Guo X, Zhao Y, Nguyen H, Liu T, Wang Z, Lou H (2018) Quantitative analysis of alternative Pre-mRNA splicing in mouse brain sections using RNA in situ hybridization assay. J Vis Exp

15. Wang F, Flanagan J, Su N, Wang LC, Bui S, Nielson A, Wu X, Vo HT, Ma XJ, Luo Y (2012) RNAscope: a novel in situ RNA analysis platform for formalin-fixed, paraffin-embedded tissues. J Mol Diagn 14:22–29
16. Wang Z, Portier BP, Gruver AM, Bui S, Wang H, Su N, Vo H-T, Ma X-J, Luo Y, Budd GT (2013) Automated quantitative RNA in situ hybridization for resolution of equivocal and heterogeneous ERBB2 (HER2) status in invasive breast carcinoma. J Mol Diagn 15:210–219
17. Nanou E, Catterall WA (2018) Calcium channels, synaptic plasticity, and neuropsychiatric disease. Neuron 98:466–481
18. Saegusa H, Kurihara T, Zong S, Kazuno A, Matsuda Y, Nonaka T, Han W, Toriyama H, Tanabe T (2001) Suppression of inflammatory and neuropathic pain symptoms in mice lacking the N-type Ca2+ channel. EMBO J 20: 2349–2356
19. Kim C, Jeon D, Kim YH, Lee CJ, Kim H, Shin HS (2009) Deletion of N-type Ca(2+) channel Ca(v)2.2 results in hyperaggressive behaviors in mice. J Biol Chem 284:2738–2745
20. Zamponi GW (2016) Targeting voltage-gated calcium channels in neurological and psychiatric diseases. Nat Rev Drug Discov 15:19–34
21. Lipscombe D, Allen SE, Toro CP (2013) Control of neuronal voltage-gated calcium ion channels from RNA to protein. Trends Neurosci 36:598–609
22. Raingo J, Castiglioni AJ, Lipscombe D (2007) Alternative splicing controls G protein-dependent inhibition of N-type calcium channels in nociceptors. Nat Neurosci 10:285–292
23. Andrade A, Denome S, Jiang YQ, Marangoudakis S, Lipscombe D (2010) Opioid inhibition of N-type Ca2+ channels and spinal analgesia couple to alternative splicing. Nat Neurosci 13:1249–1256
24. Jiang YQ, Andrade A, Lipscombe D (2013) Spinal morphine but not ziconotide or gabapentin analgesia is affected by alternative splicing of voltage-gated calcium channel CaV2.2 pre-mRNA. Mol Pain 9:67
25. Gandini MA, Souza IA, Raval D, Xu J, Pan YX, Zamponi GW (2019) Differential regulation of Cav2.2 channel exon 37 variants by alternatively spliced μ-opioid receptors. Mol Brain 12:98
26. Bunda A, LaCarubba B, Bertolino M, Akiki M, Bath K, Lopez-Soto J, Lipscombe D, Andrade A (2019) Cacna1b alternative splicing impacts excitatory neurotransmission and is linked to behavioral responses to aversive stimuli. Mol Brain 12:81
27. Bell TJ, Thaler C, Castiglioni AJ, Helton TD, Lipscombe D (2004) Cell-specific alternative splicing increases calcium channel current density in the pain pathway. Neuron 41:127–138
28. Castiglioni AJ, Raingo J, Lipscombe D (2006) Alternative splicing in the C-terminus of CaV2.2 controls expression and gating of N-type calcium channels. J Physiol 576: 119–134
29. López Soto EJ, Lipscombe D (2020) Cell-specific exon methylation and CTCF binding in neurons regulate calcium ion channel splicing and function. eLife 9:e54879
30. Bunda A, LaCarubba B, Akiki M, Andrade A (2019) Tissue- and cell-specific expression of a splice variant in the II-III cytoplasmic loop of Cacna1b. FEBS Open Bio 9:1603–1616

Chapter 12

In Situ Imaging of mRNA Splicing Variants by SpliceRCA

Xiaojun Ren and Jinghong Li

Abstract

We present a method for multiplexed in situ imaging of individual RNA splicing variants. This method enables quantifying RNA splicing variants with single-molecule resolution, discriminating splicing isoforms with single-base precision as well as analyzing the subcellular localization of transcripts. With this technology, it is possible to study cell heterogeneity of gene expression, potentially helping to decipher rich diversity in posttranscriptional function and assist clinical monitoring.

Key words Rolling circle amplification, In situ imaging, RNA splicing variants, Fluorescence imaging

1 Introduction

Alternative splicing is a fundamental regulatory process of gene expression. This process allows producing multiple mRNA splicing variants from a single gene [1, 2]. Alternative splicing plays a critical role in transcriptome complexity, cell fate determination, and organism development [3–5]. It is known that every cell is unique, even when derived from a homogenous population, individual cells can exhibit heterogeneity arising from the stochastic nature of cellular biochemical reactions [6–8]. Cell heterogeneity has been largely described. Particularly, splicing variability between individual cells is primarily responsible for gene expression heterogeneity [9]. Such variability plays important role in affecting the drug resistance [10], metastasis of cancer cells, T-cell activation [11], the outcome of stem-cell differentiation and so on. Investigations of single-cell expression of RNA splicing isoforms offer a venue for mapping intracellular localization, classifying cell types and resolving gene regulatory circuits [12].

RNA splicing variants usually have a wide variety and high sequence homology, many alternative spliced transcripts only differ in short sequence. Moreover, some alternative spliced isoforms are lowly expressed. [13, 14] These attributes of RNA splicing variants

Peter Scheiffele and Oriane Mauger (eds.), *Alternative Splicing: Methods and Protocols*,
Methods in Molecular Biology, vol. 2537, https://doi.org/10.1007/978-1-0716-2521-7_12,

bring significant challenges for their detection and analysis. Currently, only a few methods have been developed to detect RNA splicing variants at the single-cell and single-molecule levels. Single-molecule fluorescence in situ hybridization (smFISH) is a powerful method and it had brought significant advances in the knowledge about subcellular localization of RNA [15, 16]. However, this approach is not always appropriate for alternative spliced transcripts visualization. Indeed, to ensure the labeling signals can be distinguished from the background, at least 30 different hybridization probes with the length of ~20 nucleotides (nt) are usually needed. Thus, this method usually allows for detecting RNA sequences with a minimum length of 600 nt [17, 18]. However, the human exon is with an average length of only 320 nt [19]. Therefore, large numbers of splicing isoforms with short exons can hardly be detected by smFISH. Recently, a new method named plasmonic in situ hybridization (plasmonic ISH) was proposed [20]. This method is based on gold nanoparticle labeling [21, 22]. It can differentiate splice variants with about 300 nt sequence. However, some limitations remain unsolvable, for example, the large size of gold nanoparticle can impede the delivery efficiency, and lead to particle aggregation and change the cell state. Hence, available methods to visualize alternative splice transcript isoforms remain limited. It is urgently needed for developing versatile imaging methods with high sequence resolution for specifically visualizing and distinguishing select RNA splicing isoforms.

The method described here enables multiplexed imaging RNA splicing variants at single-molecule level, termed as splice-junction anchored padlock probe mediated rolling circle amplification (SpliceRCA) [23, 24]. This method combined splice-junction anchored padlock probe with in situ rolling circle amplification (RCA) [25–27]. Given this method target splice-junction sequence, it can distinguish transcript isoforms that only differ from 30 nt. Attributed to its one-target-one-amplicon amplification, it can achieve target RNA splicing variants imaging at the single-molecule level. Moreover, the ligation reaction confers high-specificity for discrimination between highly similar sequence RNA isoforms, even with single-base resolution. Coupling the padlock probe with different fluorophores, SpliceRCA enables the simultaneous multiplex visualization of several RNA splicing variants. The method provides a single-molecule-level analysis platform of RNA splicing variants to decipher rich diversity in posttranscriptional function and assist clinical monitoring.

This chapter describes an optimized method for performing SpliceRCA on cultured cells. Instructions are provided for preparing splice junction–anchored padlock probes, for in situ amplification and imaging acquisition steps. And the overview of SpliceRCA procedure is depicted in Fig. 1. The splice junction in the target splice isoform brings close proximity between recognition modules

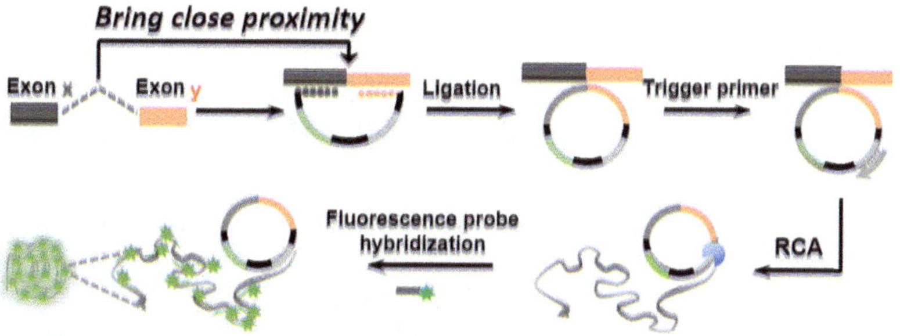

Fig. 1 Schematic illustration of in situ imaging of RNA splicing variants using SpliceRCA

targeting for exon x and exon y (Rx and Ry) in the padlock probe for circularizing. Then the padlock probe can be ligated by T4 DNA ligase. Upon the hybridization of primer with the padlock probe, in situ RCA was triggered by phi29 DNA polymerase to produce a long DNA amplicon with hundreds of repeats of the padlock probe. The amplicon naturally forms a nanoclew. Fluorophore-labeled detection probes hybridized with their complementary motifs to the RCA product. Then the amplicon from target RNA isoform could be visualized as one bright spot by the confocal microscope.

2 Materials

Purchase RNase-free tubes and tips. Prepare all solutions using RNase-free water (or treat all solutions with DEPC and then autoclave all solutions), to protect sample from RNase degradation (*see* **Note 1**). Store all enzymes at −20 °C. Store other reagents at room temperature, unless stated otherwise.

2.1 Oligonucleotide Probes

Purchase the DNA sequences and fluorophores modified DNA sequences from local companies and HPLC-purify all DNA sequences. Synthesize RNA sequences by in vitro T7 transcription reactions with ordering single-stranded DNA used as templates.

2.2 Cell Culture

- PBS (phosphate buffered saline) buffer, pH 7.2.
- Polydimethylsiloxane (PDMS) matrix: Silicone elastomer base and silicone elastomer curing agent are mixed in a ratio of 1:1. The mixture was solidified at room temperature for at least 48 h.
- DMEM medium.
- RPMI 1640 medium.
- Fetal bovine serum.
- Penicillin–streptomycin solution: 10 kU/mL penicillin and 10 mg/mL streptomycin, stored at −20 °C.

2.3 In Situ Amplification

- T4 polynucleotide kinase (10 U/μL): The enzyme is supplied in 20 mM Tris-HCl, pH 7.5, 25 mM KCl, 2 mM DTT, 0.1 mM EDTA and 50% (v/v) glycerol. Store at −20 °C. 10× Reaction Buffer: 500 mM Tris-HCl, pH 7.6 at 25 °C, 100 mM $MgCl_2$, 50 mM DTT, 1 mM spermidine.
- T4 DNA Ligase (5 U/μL): The enzyme is stored in: 10 mM Tris-HCl, pH 7.5, 50 mM KCl, 1 mM DTT, 50% (v/v) glycerol]. Store at −20 °C. 10× Reaction Buffer: 500 mM Tris-HCl, pH 7.6, 100 mM $MgCl_2$, 10 mM ATP, 10 mM DTT, 50% (w/v) polyethylene glycol 8000.
- Phi29 DNA polymerase (10 U/μL): The enzyme is supplied in 50 mM Tris-HCl, pH 7.5, 1 mM DTT, 0.1 mM EDTA, 100 mM KCl, 0.5% (v/v) Tween 20, 0.5% (v/v) Nonidet P40, and 50% (v/v) glycerol. Store at −20 °C. 10× Reaction Buffer: 330 mM Tris–acetate, pH 7.9, 660 mM K-acetate, 10 mM DTT, 100 mM Mg-acetate, 1% (v/v) Tween 20.
- ATP solution: 100 mM aqueous solution (adenosine 5-′-triphosphate), titrated to pH 7.3–7.5 with NaOH.
- dNTP mix: 10 mM each nucleotide (dATP, dCTP, dGTP, dTTP) in 0.6 mM Tris-HCl.
- Diethy pyrocarbonate (DEPC).
- Yeast tRNA solution: 10 mg/mL yeast tRNA in RNase-free water. Store at −20 °C.
- Salmon sperm DNA solution: 10 mg/mL salmon sperm DNA in RNase-free water. Store at −20 °C.
- Triton X-100 solution: 0.5% v/v Triton X-100 in 1× PBS buffer.
- 20× saline-sodium citrate (SSC) buffer: 3 M NaCl and 0.3 M sodium citrate.
- Paraformaldehyde solution: 4% (v/v) paraformaldehyde in PBS buffer.
- DEPC-PBS: 0.1% DEPC-treated buffer.
- PBS-T solution: DEPC-PBS buffer with 0.05% Tween 20.
- Gelatin solution: 0.1% gelatin in RNase-free water.
- Dithiothreitol (DTT): 100 mM in RNase-free water, freshly prepared before use.
- Fluoromount-G (containing 4′,6-diamidino-2-phenylindole [DAPI]).
- Formamide: deionized, nuclease-free.
- RiboLock RNase Inhibitor (40 U/μL): store at −20 °C.

2.4 General Equipment

- 1.5 and 0.5 mL RNase-free microfuge tubes; 15 and 50 mL RNase-free tubes.
- Pipetman.
- Cover glass: 26 × 76 mm and 22 × 22 mm.
- Water bath.
- Microcentrifuge.
- Various sizes of plastic culture dishes.
- PCR thermal cycler.
- NanoDrop 2000 UV-Vis spectrophotometer.
- Laser scanning confocal microscope.

3 Methods

3.1 Probe Design

1. The padlock probe has four modules, including the recognition region of exon junction sites (Rx, Ry), universal priming region (P), and tag motif (T) modules (Fig. 2) (*see* **Note 2**). Design the recognition region of padlock probes for spanning splicing junctions. Search the splice junction sequence of the target RNA splicing isoform from NCBI (https://www.ncbi.nlm.nih.gov/). The target recognition region (Rx + Ry) should preferably be ~30 nt long, this length works well for both GC-rich and GC-poor sequences. Design the tag motif region to be between 20 and 25 nt in length, and assure the detection sequence is not present in the genome of your studied species to minimize background fluorescence. If the padlock probe will be used in multiplexed detection, this tag motif sequence must be unique for each padlock probe. Sequences such as AACTA TACAACATACTACCTCA, TGCGTCTATTTTCTGGAGC CAT, and CTCAATTCTGCTACTGTACTAC have been used as tag motif for analysis of BRCA1 splicing variants in the MCF-7 cells (*see* **Note 3**). Design the priming region to be ~20 nt in length (e.g., GTATAGTTGTAGTACAGACAAT). Please make sure the priming region is identical for each padlock probe, to reduce sequence bias toward particular RNA splicing variants (*see* **Note 4**).
2. Analyze designed oligonucleotide sequences on BLAST (http://blast.ncbi.nlm.nih.gov/) to ensure that sequences are unique for your target RNA splicing isoform. This is significantly important for the padlock probes, and we recommend BLASTing only the target sequence without adding the linker sequence to ensure specific ligation only happens when the padlock probe hybridizes to the target of interest.

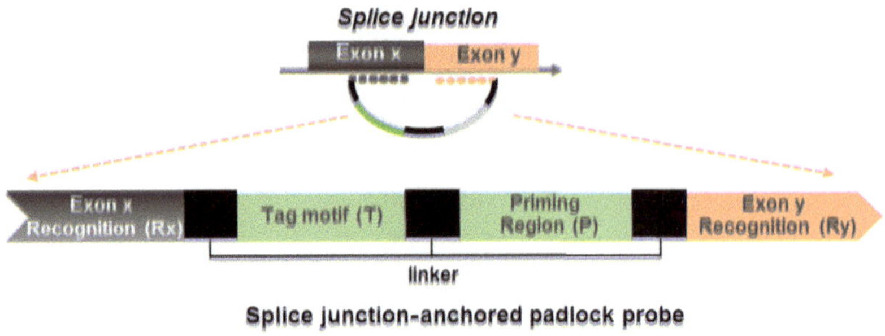

Fig. 2 Structure of padlock probe used to amplify RNA splicing variants in SpliceRCA strategy. The 5′ arm (orange) and 3′ arm (dark gray) was designed to hybridize with splice junction and have similar melting temperature. Insert a linker sequence including a tag motif (green) and priming region (light gray) to join the two arms

3. To improve the hybridization efficiency of the probe as much as possible, calculate the standard free energy ΔG and analyze the secondary structure of padlock probes by NUPACK (http://www.nupack.org/). Set salt concentration to 300 mM Na^+, 0 mM Mg^{2+} (the salt concentration of 2× SSC). The concentrations of detection probes and the monomer barcode sequence of amplicons are set to 100 nM. The temperature is set to 37 °C (the temperature at which the hybridization process is performed). The "dangles" parameter is set to "on both sides of a helix in any case." Adjust the linker sequence to ensure the highest free energy of the padlock probe (*see* **Note 5**).

3.2 Cell Culture and Padlock Probe Phosphorylation

1. Grow cells of interest in a standard cell medium (selected according to cell type) supplemented with 15% fetal bovine serum, 100 units/mL of penicillin and 100 μg/mL of streptomycin at 37 °C, 5% CO_2, and 95% air humidity. Check cell lines regularly to make sure they are alive and not infected with mycoplasma.
2. Phosphorylate the padlock probes by T4 polynucleotide kinase to serve as ligation templates. Mix the reagents in a microtube: 2 μL of 10× T4 polynucleotide kinase reaction buffer, 2 μL of Padlock probe (100 μM), 2 μL of ATP, 0.5 μL of T4 polynucleotide kinase, and 15.5 μL of RNase-free H_2O.
3. Incubate the reaction mixture at 37 °C for 1 h. Next, heat the mixture at 65 °C for 20 min to inactivate T4 polynucleotide kinase.

3.3 Cell Fixation and Permeabilization

1. Wash the 26 mm × 76 mm cover glass and PDMS with a chamber (5 mm in diameter) (Fig. 3a) twice with water and twice with ethanol (3-min sonication each).

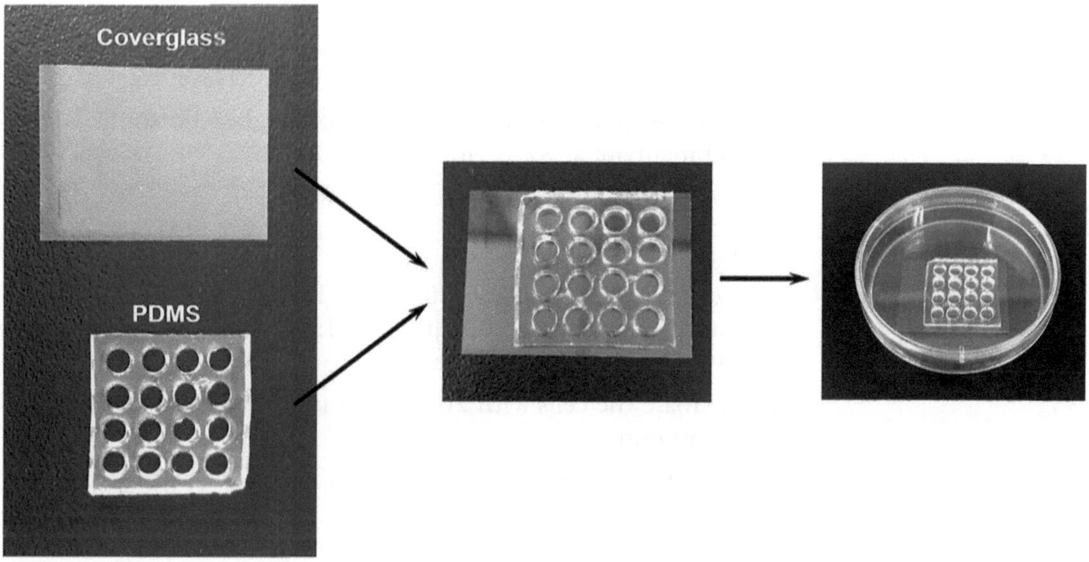

Fig. 3 Illustration of cell culture vector in this procedure. Dispose of the PDMS chamber (5 mm diameter) on the cover glass and press them tightly. The vector is placed in a sterile petri dish at 37 °C when incubating the reaction buffer with cells

2. Air-dry the cover glass and the PDMS chamber in an upright position, and expose them to UV light for 10–20 min simultaneously. Dispose of the PDMS chamber on the cover glass as shown in Fig. 3. Fit them tightly together to prevent leakage of liquid (*see* **Note 6**).
3. Immerse the chamber into 0.1% gelatin for 1 h, then wash the chamber with 1× DEPC-PBS (*see* **Note 7**).
4. Seed the cells on the gelatin-coated cover glass to allow for their adhesion. Once cells reach the desired confluency, wash them gently twice with PBS (*see* **Note 8**).
5. Fix the cells using 20 μL (for each chamber) of 4% (w/v) ice-cold paraformaldehyde in 1× PBS for 15 min at room temperature (20–25 °C), wash the cells twice with 1× DEPC-PBS (*see* **Note 9**).
6. Add 20 μL of 5% v/v Triton X-100 in 1× PBS for 5 min at room temperature. Wash the cells twice with 1× DEPC-PBS (*see* **Note 10**).

3.4 Padlock Probe Hybridization

1. Prepare the hybridization mixture: 2 μL 20× SSC buffer, 2 μL Phosphorylated padlock probe (10 μM), 1 μL DTT, 2 μL Yeast tRNA, 0.5 μL RiboLock RNase inhibitor, and 12.5 μL RNase-free H_2O.
2. Incubate the cells with the reaction mixture at 37 °C overnight.
3. Aspirate and wash the sample twice using 1× PBS-T solution (*see* **Note 11**).

3.5 T4 DNA Ligation

1. Assemble the reaction mixture in the following order: 2 μL 10× T4 ligase reaction buffer, 2 μL T4 DNA ligase, 0.5 μL RiboLock RNase inhibitor, and 15.5 μL RNase-free water.
2. Add the ligation reaction mixture to the glass-bottom chamber and incubate at 37 °C for 2 h.
3. Wash the sample one time using 1× PBS-T.

3.6 Primer Hybridization

1. Prepare the primer hybridization mixture in the following order: 2 μL 20× SSC buffer, 1 μL Primer (4 μM), 2 μL formamide, 1 μL DTT, 0.5 μL RiboLock RNase inhibitor, and 13.5 μL RNase-free water.
2. Incubate the cells with 20 μL mixture (each chamber) at 37 °C for 60 min.
3. Wash the sample 1–2 times using 1× PBS-T (*see* **Note 12**).

3.7 In Situ Rolling Circle Amplification

1. Mix the RCA reagents in a microtube: 2 μL 10× phi29 DNA polymerase reaction buffer, 1 μL Phi29 DNA polymerase, 6 μL dNTPs, 0.5 μL RiboLock RNase inhibitor, and 10.5 μL RNase-free water.
2. Add the RCA reaction mixture to the sample and incubate at 37 °C for 120 min.
3. Remove the mixture and then wash the sample using 1× PBS-T solution for 3 min at room temperature.

3.8 Fluo-Probe Hybridization

1. Assemble the reaction mixture in the following order: 2 μL 20× SSC buffer, 2 μL Formamide, 2 μL Salmon sperm DNA, 2 μL fluorophore-labeled detection probes (1 μM), and 12 μL RNase-free water.
2. Add the mixture containing 100 nM Fluo-probe to the glass-bottom of chambers and incubate at 37 °C for 30 min.
3. Wash the sample three times with 1× PBS-T solution (*see* **Notes 13** and **14**).

3.9 Mounting and Imaging

1. Immerse the sample in mounting gel solution {Fluoromount-G [containing 4′,6-diamidino-2-phenylindole (DAPI) to counterstain the cell nuclei]}.
2. Place a 22 × 22 mm cover glass on the chamber to enclose the chambers. One hour later, the sample is ready for imaging (*see* **Note 15**).
3. Image all SpliceRCA samples using a Leica TCS SP5 inverted confocal microscope (Leica, Germany) with a 63× oil-immersion objective. Use an AR+ (488 nm) laser as an excitation source for the Alexa 488–labeled probe, and use a 500–535 nm band-pass filter for fluorescence detection. Excite the Alexa 555 dye with a HeNe543 (543 nm) laser, and detect

it with a 555–620 nm band-pass filter. Excite the Cy5 dye with a HeNe633 (633 nm) laser, and detect it with a 650–750 band-pass filter. Use a 405 nm laser as an excitation source for the DAPI dye, and a 430–550 nm band-pass filter for fluorescence detection.

4. To ensure that all RCA amplicons are imaged, collect Z-stacks of images at a step size of 0.2 μm to image the entire cell.
5. Combine the Z-stacks of images to a single image by maximum intensity projection using Image J version 1.46r software (*see* **Note 16**).
6. Adjust the intensity threshold to distinguish amplicons with the superbright signal from the background.
7. To determine the copy number per cell, firstly mark out the outline of the cell from the bright field images. Next, mark the superbright spots with a pseudocolor for different RNA splicing isoforms in the fluorescence images.
8. Extract the position of RCA amplicons as well as the position and the area of cells and nucleus from the fluorescence images and bright-field images, respectively.
9. Count the bright pixels inside cells with particle analysis in ImageJ software to determine the number of RNA splicing isoforms.
10. Calculate the distance from the amplicons to the cell edge or the cell nucleus by Python 2.7.13 with the flood algorithm.

3.10 Typical/Anticipated Results

The products of SpliceRCA should appear as individual bright spots as shown in Fig. 4. The alternative splicing of BRCA1, breast cancer susceptibility gene 1 [28], is closely related with the transformation of malignant breast cancer [20, 29]. We choose three critical mRNA splicing isoforms of BRCA1, including $BRCA1_{wt}$ (all exons retained), Δ(9,10,11q) (exon 9, exon 10, and the last 333 nt of exon 11 deleted), Δ(11q) (the last 333 nt of exon 11 deleted) (Fig. 4a). The amplicons from target splice isoform could be clearly distinguished from the background autofluorescence inside cell. The images show a maximum intensity projection and therefore some signals for RNA splicing variants also appear in the nucleus. No bright spot was observed when random padlock probes were used (control group) (Fig. 4b). The average numbers of amplicons for Δ(9,10,11q), Δ(11q) and $BRCA1_{wt}$ were 4.20, 5.29, 8.02 per cell, respectively (cell number > 100).

4 Notes

1. DEPC is used for inactivating RNase in water or buffers prior to their application in experiments with easily degradable RNA. Incubate 0.1% (vol./vol.) DEPC with buffer for 1–2 h at 37 °C

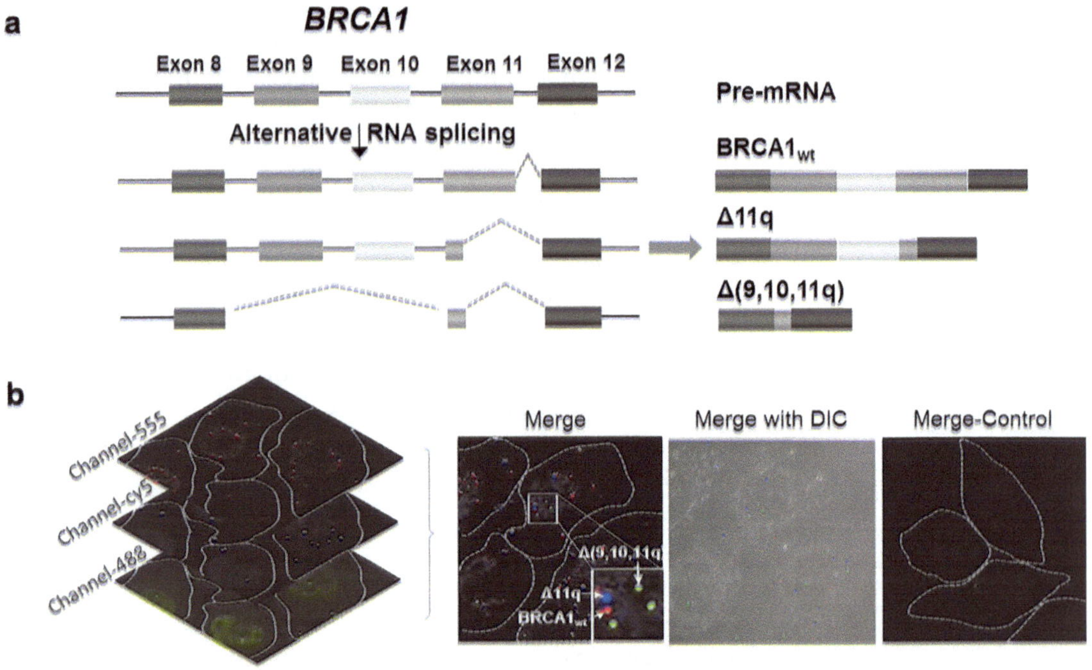

Fig. 4 Multiplexed fluorescent in situ imaging of BRCA1 splicing variants in MCF-7 cells. (**a**) Alternative splicing patterns of BRCA1; (**b**) The fluorescent image of three BRCA1 isoforms visualized by SpliceRCA. The green spots represent amplicons of Δ (9, 10, 11q), the red spots represent amplicons of BRCA1wt, and the blue spots represent amplicons of Δ (11q). The outline of MCF-7 cell is marked with a gray dot line. Scale bars: 10 μm. MCF-7 cell imaged by SpliceRCA with random padlock probes were used as control

to inhibit RNase, then autoclave for 25 min to inactivate DEPC. Since DEPC is carcinogenic, handle the DEPC with care (fume hood, wear gloves).

2. For the design of padlock probe targeting point mutations, place the mutated nucleotide position at the ultimate 3′ end of the padlock probe, which is directly adjacent to the nick that is sealed by the ligase. This can ensure the highest possible discriminatory power in the ligation reaction.
3. Avoid using fluorophores with similar excitation/emission wavelengths to minimize bleed-through of signal from one fluorescence channel to another.
4. We find that it is best to design the overall length of padlock probe at ~90 nt. Padlock probe hybridization can be hindered if the length is short. When the length is too long, amplification efficiency can be decreased. Moreover, padlock probes should have similar Tm values which could be adjusted by tuning the target complementary arm length.
5. Secondary structures of the padlock probe should be avoided as much as possible because it can influence the hybridization and amplification efficiency.

6. Completely dried the cover glass and PDMS to ensure they are attached tightly. Choose the size of chamber accordingly. We usually use chambers with a diameter of 5 mm for cell experiments.
7. Gelatin-coated dried cover glasses should be used within 1 or 2 days, or they may be stored under desiccation at room temperature for several weeks.
8. Before cell fixation, do not suck PBS solution entirely and make sure the cells are not dry, avoid the shrinking and deformation of cells.
9. The fixative solution (paraformaldehyde, 4% (vol/vol)) should be freshly prepared or used from aliquots frozen at −20 °C for one-time use. Cool PFA to 4 °C before starting the experiment.
10. From now on, “wash” means removing the liquid from the chamber, refilling it with wash buffer and then flushing two times the volume of the chamber slowly through the chamber. Add liquids to the cover glass slowly, avoiding dislodging cells from the glass. It is critical to remove the excess wash buffer but avoid drying out the sections before applying assay reagent.
11. During longer incubation times, the reaction volume can decrease below a critical value. Always use a sufficient starting reaction volume and always incubate slides in humidified reaction chambers with the openings of the Secure Seals closed.
12. The hybridization buffer should be warmed to room temperature before opening to prevent oxidation of the formamide. It is experienced that a formamide concentration of 10% works well in our lab. The concentration of formamide can be increased to enhance probe-binding stringency, especially for probes with a GC-rich content. A high formamide concentration can lead to low fluorescence signals.
13. The slides should not be dried before they are sealed.
14. From this step and onward, protect the samples from light to prevent photobleaching, especially for photolabile dyes such as Cy5.
15. Fulfill the chamber with the mounting gel solution as far as possible so that bubbles cannot appear during cover glass enclosing.
16. The distance between individual optical planes should be 0.2–0.4 μm.

Acknowledgments

This work was supported by National Natural Science Foundation of China (21904009, 21621003, 21235004, 21327806), Beijing Municipal Natural Science Foundation (2202003), and Tsinghua University Initiative Scientific Research Program.

References

1. Maniatis T, Tasic B (2002) Alternative pre-mRNA splicing and proteome expansion in metazoans. Nature 418(6894):236–243. https://doi.org/10.1038/418236a
2. Hemphill J, Liu Q, Uprety R, Samanta S, Tsang M, Juliano RL, Deiters A (2015) Conditional control of alternative splicing through light-triggered splice-switching oligonucleotides. J Am Chem Soc 137(10):3656–3662. https://doi.org/10.1021/jacs.5b00580
3. Chen M, Manley JL (2009) Mechanisms of alternative splicing regulation: insights from molecular and genomics approaches. Nat Rev Mol Cell Biol 10(11):741–754. https://doi.org/10.1038/nrm2777
4. Levesque MJ, Raj A (2013) Single-chromosome transcriptional profiling reveals chromosomal gene expression regulation. Nat Methods 10(3):246–248. https://doi.org/10.1038/nmeth.2372
5. Soreq L, Salomonis N, Israel Z, Bergman H, Soreq H (2015) Analyzing alternative splicing data of splice junction arrays from Parkinson patients' leukocytes before and after deep brain stimulation as compared with control donors. Genomics Data 5:340–343. https://doi.org/10.1016/j.gdata.2015.07.014
6. Raser JM, O'Shea EK (2004) Control of stochasticity in eukaryotic gene expression. Science 304(5678):1811–1814
7. Rosenfeld N, Young JW, Alon U, Swain PS, Elowitz MB (2005) Gene regulation at the single-cell level. Science 307(5717):1962–1965
8. Raj A, Peskin CS, Tranchina D, Vargas DY, Tyagi S (2006) Stochastic mRNA synthesis in mammalian cells. PLoS Biol 4(10):e309
9. Nilsen TW, Graveley BR (2010) Expansion of the eukaryotic proteome by alternative splicing. Nature 463(7280):457–463. https://doi.org/10.1038/nature08909
10. Spencer SL, Gaudet S, Albeck JG, Burke JM, Sorger PK (2009) Non-genetic origins of cell-to-cell variability in TRAIL-induced apoptosis. Nature 459(7245):428–432. https://doi.org/10.1038/nature08012
11. Feinerman O, Veiga J, Dorfman JR, Germain RN, Altan-Bonnet G (2008) Variability and robustness in T cell activation from regulated heterogeneity in protein levels. Science 321(5892):1081–1084. https://doi.org/10.1126/science.1158013
12. Ren X, Zhang K, Deng R, Li J (2019) RNA splicing analysis: from in vitro testing to single-cell imaging. Chem 5:2571–2592. https://doi.org/10.1016/j.chempr.2019.05.027
13. Larson DR, Singer RH, Zenklusen D (2009) A single molecule view of gene expression. Trends Cell Biol 19(11):630–637
14. Baralle FE, Giudice J (2017) Alternative splicing as a regulator of development and tissue identity. Nat Rev Mol Cell Biol 18(7):437–451. https://doi.org/10.1038/nrm.2017.27
15. Vargas DY, Shah K, Batish M, Levandoski M, Sinha S, Marras SA, Schedl P, Tyagi S (2011) Single-molecule imaging of transcriptionally coupled and uncoupled splicing. Cell 147(5):1054–1065. https://doi.org/10.1016/j.cell.2011.10.024
16. Waks Z, Klein AM, Silver PA (2011) Cell-to-cell variability of alternative RNA splicing. Mol Syst Biol 7:506. https://doi.org/10.1038/msb.2011.32
17. Raj A, Van Den Bogaard P, Rifkin SA, Van Oudenaarden A, Tyagi S (2008) Imaging individual mRNA molecules using multiple singly labeled probes. Nat Methods 5(10):877–879
18. Chen KH, Boettiger AN, Moffitt JR, Wang S, Zhuang X (2015) Spatially resolved, highly multiplexed RNA profiling in single cells. Science 348(6233):aaa6090. https://doi.org/10.1126/science.aaa6090
19. Long X, Colonell J, Wong AM, Singer RH, Lionnet T (2017) Quantitative mRNA imaging throughout the entire Drosophila brain. Nat Methods 14(7):703–706. https://doi.org/10.1038/nmeth.4309
20. Lee K, Cui Y, Lee LP, Irudayaraj J (2014) Quantitative imaging of single mRNA splice variants in living cells. Nat Nanotechnol 9(6):

474–480. https://doi.org/10.1038/nnano.2014.73

21. Reinhard BM, Siu M, Agarwal H, Alivisatos AP, Liphardt J (2005) Calibration of dynamic molecular rulers based on plasmon coupling between gold nanoparticles. Nano Lett 5(11): 2246–2252. https://doi.org/10.1021/nl051592s
22. Chen F, Lu Q, Huang L, Liu B, Liu M, Zhang Y, Liu J (2021) DNA triplex and quadruplex assembled nanosensors for correlating K (+) and pH in lysosomes. Angew Chem 60(10):5453–5458. https://doi.org/10.1002/anie.202013302
23. Ren X, Deng R, Zhang K, Sun Y, Teng X, Li J (2018) SpliceRCA: in situ single-cell analysis of mRNA splicing variants. ACS Cent Sci 4(6): 680–687. https://doi.org/10.1021/acscentsci.8b00081
24. Ren X, Deng R, Zhang K, Sun Y, Li Y, Li J (2021) Single-cell imaging of m(6) A modified RNA Using m(6) A-specific in situ hybridization mediated proximity ligation assay (m (6) AISH-PLA). Angew Chem 60(42): 22646–22651. https://doi.org/10.1002/anie.202109118
25. Deng RJ, Zhang KX, Wang LD, Ren XJ, Sun YP, Li JH (2018) DNA-sequence-encoded rolling circle amplicon for single-cell RNA imaging. Chem 4(6):1373–1386. https://doi.org/10.1016/j.chempr.2018.03.003
26. Zhang K, Deng R, Teng X, Li Y, Sun Y, Ren X, Li J (2018) Direct visualization of single-nucleotide variation in mtDNA using a CRISPR/Cas9-mediated proximity ligation assay. J Am Chem Soc 140(36): 11293–11301. https://doi.org/10.1021/jacs.8b05309
27. Ren X, Deng R, Wang L, Zhang K, Li J (2017) RNA splicing process analysis for identifying antisense oligonucleotide inhibitors with padlock probe-based isothermal amplification. Chem Sci 8(8):5692–5698. https://doi.org/10.1039/c7sc01336a
28. Orban TI, Olah E (2001) Expression profiles of BRCA1 splice variants in asynchronous and in G1/S synchronized tumor cell lines. Biochem Biophys Res Commun 280(1):32–38. https://doi.org/10.1006/bbrc.2000.4068
29. Wang H, Wang H, Duan X, Sun Y, Wang X, Li Z (2017) Highly sensitive and multiplexed quantification of mRNA splice variants by direct ligation of DNA probes at exon junction and universal PCR amplification. Chem Sci 8: 3635–3640

Chapter 13

Two-Color Fluorescent Reporters for Analysis of Alternative Splicing

John A. Calarco and Pallavi P. Pilaka-Akella

Abstract

Alternative splicing is a key layer of gene regulation that is frequently modulated in a spatiotemporal manner. As such, it is a major goal to understand the mechanisms controlling alternative splicing in specific cellular contexts. Reporters that recapitulate alternative splicing patterns of endogenous transcripts have served as excellent tools for dissecting regulatory mechanisms of splicing. In this chapter, we describe a two-color fluorescent reporter system that enables the visualization of alternative splicing patterns by microscopy at single-cell resolution in live animals. We present this reporter system in the context of the model nematode *C. elegans*.

Key words Alternative splicing, Two-color reporters, Fluorescence microscopy, Genetic screening, *C. elegans*

1 Introduction

Precursor messenger RNA (pre-mRNA) splicing is a critical processing step in the mRNA life cycle [1]. One major evolutionary advantage of the pre-mRNA splicing mechanism is that it has also enabled alternative splicing and the opportunity to generate multiple transcript variants from a single gene [2–4]. Indeed, transcriptome-wide analyses have suggested that multicellular animal genomes frequently encode transcripts with multiple isoforms, ranging from 25% of genes in invertebrates such as *C. elegans* [5] to more than 95% of genes in humans [6, 7]. This increased transcript diversity from a more limited repertoire of genes has a significant impact on proteomic diversity [8]. Moreover, alternative splicing events in pre-mRNA are often regulated in a tissue-specific manner and/or across stages of development [9, 10]. The regulation of alternative splicing in specific contexts thus provides the opportunity for cells and tissues to customize their gene expression landscape and become specialized. A striking example of the impact of

Peter Scheiffele and Oriane Mauger (eds.), *Alternative Splicing: Methods and Protocols*,
Methods in Molecular Biology, vol. 2537, https://doi.org/10.1007/978-1-0716-2521-7_13,

alternative splicing in the specialization of cell types is in the nervous system [11, 12], where several recent transcriptome profiling studies of neuronal subtypes are beginning to reveal a remarkable degree of complexity [13–17]. Given this complexity, it remains an important goal to elucidate the mechanisms that control splicing patterns in individual cell and tissue types, and how these mechanisms can go awry in diseases [18].

Reporters that faithfully recapitulate alternative splicing patterns have been instrumental in shaping our understanding of alternative splicing mechanisms. Specifically, their ease of manipulation has enabled screening of both cis elements, trans-acting factors, and small molecules, that are involved in the control of splicing patterns (*see* refs.19–22 for several recent examples). Several variations of pre-mRNA splicing reporters have been designed and employed. Typically, these reporters include a minimal portion of an endogenous gene capable of supporting splicing in a manner that is similar to the native locus [23]. Most frequently, a minimal region encompasses an alternatively spliced exon and its flanking upstream and downstream intronic sequences, but can also include neighboring constitutively spliced exons as well. This minimal sequence is then inserted into a vector capable of driving expression of pre-mRNA. Readouts of these splicing reporters can vary from measuring the mRNA isoforms directly by RT-PCR or massively parallel sequencing, or through indirect measurements of translated fusion proteins with enzymatic activity or fluorescent properties. Moreover, the choice of reporter readout can depend on the question being explored and the scale of inquiry. In most cases, if only a few reporters need to be tested, RT-PCR is generally a sufficient readout. However, large-scale screens are facilitated by either massively parallel sequencing, enzymatic, or fluorescent protein-based readouts.

One attractive reporter that has been utilized in several studies and our lab has adapted is a two-color frameshift reporter system [24, 25]. In this system, a model splicing event is cloned upstream of a sequence that encodes two fluorescent proteins that are translated in two different reading frames (Fig. 1). The model splicing event is then engineered such that the alternative exon is no longer frame preserving and its inclusion or skipping toggles between the two reading frames, leading to the translation of one fluorescent protein or the other depending on the splicing outcome (Fig. 1). In this elegant approach, the relative fluorescence of the two proteins can then be observed by fluorescence microscopy, providing an indirect readout of the spliced mRNA variants.

Our group and others have utilized two-color fluorescent reporters to study tissue-specific alternative splicing patterns in the nematode *Caenorhabditis elegans* [16, 25–29]. This model organism has an intron-rich genome with complex alternative splicing patterns and a conserved repertoire of genes encoding the

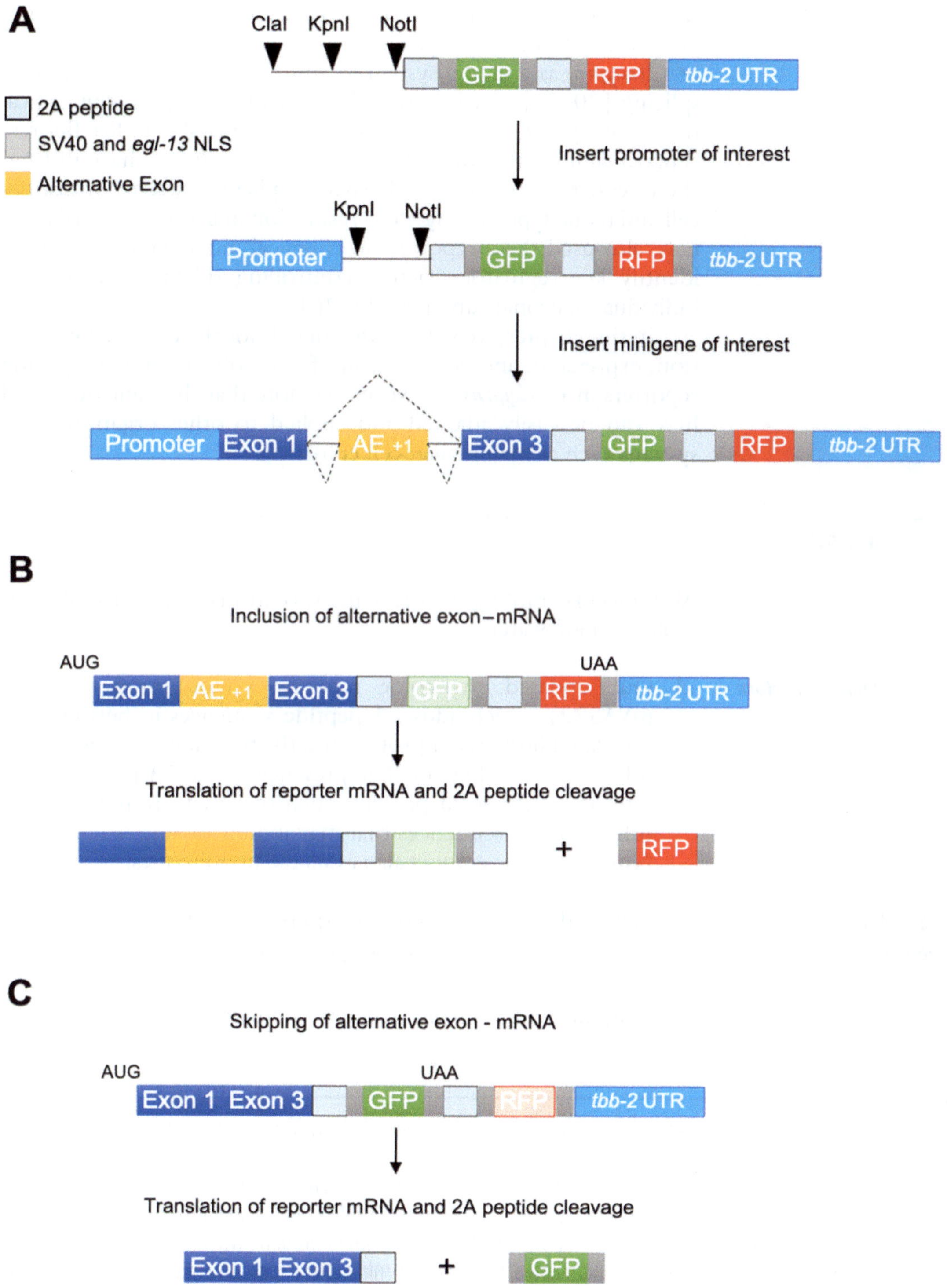

Fig. 1 Schematic of two-color frameshift reporter system (**a**) To create a new two-color reporter, a promoter of interest is cloned into available ClaI/KpnI restriction sites. Then, a minigene containing an alternatively spliced exon and neighboring introns and exons is cloned into the reporter using KpnI/ClaI sites. The alternative exon

splicing machinery and regulatory proteins such as RNA binding proteins. Indeed, several groups have now demonstrated the utility of *C. elegans* as a model system to study mechanisms of regulated splicing [30–32]. Particularly relevant to this chapter, the optical transparency of *C. elegans* renders it highly tractable to live fluorescence imaging with two-color splicing reporters. We have utilized these reporters to visualize alternative splicing patterns in distinct cell and tissue types at single cell resolution in live animals [16]. We have also used these reporters in chemical mutagenesis screens to identify key regulatory factors controlling splicing decisions in individual neuronal subtypes [25, 26].

In this chapter, we will provide details for the design, construction, expression, and visualization of two-color frameshift splicing reporters in *C. elegans*. However, we note that the logic described here can be easily adapted and applied to other organisms and systems such as mammalian cell culture.

2 Materials

All molecular biology experiments should make use of molecular biology grade water.

2.1 Available Vectors

1. Our updated two-color reporter vector (used in [16]; pWASR2), which adds 2A peptide sequences in between the translated fluorescent protein and the translated minigene peptides products (Fig. 1). The presence of the 2A peptide separates the fluorescent proteins such that they are not fused to other peptide sequences that may inhibit folding/stability (*see* **Note 1**). This vector is also available upon request.

2.2 Molecular Cloning

1. Standard molecular biology equipment (PCR machine, pipettes, tips, tubes, agarose gel setup, incubators, etc.) are required.
2. Oligonucleotides from a preferred provider.

Fig. 1 (continued) is engineered to include an extra single nucleotide in order to shift reading frame when the exon is included in the mature mRNA. Downstream of the minigene are two fluorescent protein open reading frames (encoding EGFP and mCherry) which are each flanked by two nuclear localization sequences (SV40 and *egl-13*) and 2A peptide sequences. A heterologous 3′ UTR from the *tbb-2* gene is included for proper cleavage and polyadenylation of transcripts (**b**) Depicts the mRNA and translated proteins when the alternative exon is included. The reading frame selected translates through the minigene exons and produces mCherry. (**c**) Depicts the mRNA and translated proteins when the alternative exon is skipped, resulting in the selection of the reading frame that translates EGFP. In both (**b**) and (**c**), the translation of 2A peptides upstream of each fluorescent protein enables ribosome-mediated cleavage of the fluorescent protein away from the translated upstream sequence

3. Phusion high-fidelity DNA polymerase kit (we use NEB).
4. NEBuilder HiFi DNA assembly master mix (we use NEB) for Gibson Assembly.
5. NEB-5-alpha competent *E. coli* or equivalent (we use NEB).
6. LB agar plates supplemented with 100 μg/mL ampicillin.
7. PCR cleanup and gel extraction kits.
8. Plasmid purification miniprep kit.
9. ClaI, KpnI, and NotI restriction enzymes.
10. Source of genomic DNA or synthetic DNA sequences for minigene region of reporter. We either order the DNA as synthesized double stranded DNA (we use gBlocks from IDT), or we synthesize the minigenes from purified worm genomic DNA and PCR fragments.

2.3 Microinjection of *C. elegans* for CRISPR/Cas9 Genome Editing and Screening Animals

1. High-quality plasmid DNA purification kits. Several options are available. We have used Qiagen midiprep kits or Zymo-PURE midiprep kits for purifying plasmid DNA, or we typically use the geneJET miniprep kit followed by concentration with the Zymo Clean and concentrator-5 kit.
2. Wild-type (N2) and *smg-1(r861) C. elegans* nematodes (these strains can be obtained from *Caenorhabditis* Genetics Center, CGC).
3. Nematode growth media (NGM) agar plates seeded with OP50 *E. coli* strain (CGC).
4. Microinjection apparatus. Several options are available, but we use the complete microinjection setup from Tritech Research.
5. Capillaries and needle puller for pulling needles for microinjection. Several options are available. We use a Flaming/Brown P-97 needle puller (Sutter instruments) and capillaries from World Precision Instruments to create microinjection needles.
6. 20 × 60 mm glass coverslips coated with a drop of 2% agarose (dried overnight or longer). These are injection pads used to immobilize the animals.
7. Halocarbon oil to be used for immobilization and preventing worms from desiccating while on injection pad.
8. Fluorescent stereomicroscope with excitation/emission filters to view GFP and mCherry expression.
9. Standard dissecting/stereomicroscope for routine maintenance of animals.
10. Variable temperature incubator (capable of 21–22 °C incubation) for raising animals.

2.4 Fluorescence Microscopy

1. Laser scanning confocal microscope with capacity of detecting GFP and mCherry fluorescence. Alternatively, a standard wide-field fluorescence microscope can be used for imaging also. We use a Leica SP8 confocal microscope.
2. Software for image analysis. FIJI/imageJ is a popular open-source package, but software from most major companies (Zeiss, Leica, Olympus, Nikon) will also be fine.
3. Freshly prepared 2% agarose pads on slides.
4. Coverslips.
5. M9 buffer: Combine 3 g KH_2PO_4, 6 g Na_2HPO_4, 5 g NaCl, 1 mL 1 M $MgSO_4$, and H_2O to a final volume of 1 L). Supplement with 0.01% Triton X-100 to prevent animals from sticking to plastic pipette tips.
6. 200 mM sodium azide or 100 mM levamisole for paralyzing/anesthetizing animals.
7. Fluorescent stereomicroscope (to select fluorescent animals for high-resolution microscopy).

2.5 RNA Purification and Reverse Transcription Polymerase Chain Reaction (RT-PCR)

1. RNA lysis and purification reagent, such as Tri Reagent (Sigma) or Trizol (Thermo Fisher).
2. Chloroform.
3. Nuclease-free water.
4. Anhydrous ethanol.
5. RNA purification spin column kit. Several options are available. We use the RNA Clean and Concentrator kit from Zymo Research Corporation.
6. RT-PCR kit, preferably a one-step kit. Many options are available—we use the OneStep RT-PCR kit from Qiagen.
7. Oligonucleotide primers to amplify reporter cDNA splice variants.
8. Agarose.
9. 10× TAE running buffer.
10. Ethidium bromide.
11. DNA loading dye and DNA ladder.
12. Agarose Gel Electrophoresis apparatus.
13. Microwave.
14. Camera/detector for imaging of ethidium bromide–stained gels.

3 Methods

3.1 Two-Color Reporter Design Considerations for Minigene Sequence

1. We generally clone or synthesize a smaller region of DNA sequence from an endogenous gene locus. Specifically, a good starting point is to include the alternatively spliced exon, its flanking intron sequences and neighboring exon sequences. We refer to these regions as "minigenes." In *C. elegans*, it is fairly straight forward to amplify these minigene regions by PCR because most genes contain short (less than 1 kb) introns. However, in vertebrate genomes, where intron sizes are much larger, it may be necessary to use truncated sequences. In either case, it is important to first verify that the cloned minigene recapitulates the splicing patterns of the endogenous locus. This verification process can either be performed directly with the two-color reporter vector, or by using other existing minigene splicing reporter vector systems coupled with RT-PCR as a readout (*see* Subheading 3.12 for an example RT-PCR protocol).

3.2 Synthesis of Minigene Sequence or Amplification by PCR

There are two different approaches for cloning minigene regions that will be inserted into the two-color vector. The simplest approach involves having the DNA synthesized by a company. As an alternative approach, PCR products can be amplified from genomic DNA purified from the organism of choice. Below we detail both approaches.

3.2.1 Synthesis of Minigene Sequence by Company

1. Depending on the size of the minigene sequence to be cloned, one or several synthetic double-stranded DNA molecules can be ordered. Since our protocol uses Gibson Assembly to create the final product, multiple DNA fragments can be ordered that have 25–30 nucleotide overlapping regions at their ends (and with the boundaries of the vector).
2. Once the region of interest is determined, a few other considerations are needed. If the alternative exon is frame-preserving (i.e., a multiple of three), during the design phase a single nucleotide is inserted somewhere into the region of the exon that is alternatively spliced. This single nucleotide insertion causes a shift in the reading frame such that when the alternative exon is included, the reading frame will lead to the translation of mCherry. If the alternative exon is skipped, the original reading frame will be utilized and GFP will be translated (Fig. 1; *see* **Note 2**).
3. Once the single nucleotide has been inserted into the alternative exon, care must be taken to ensure that no stop codons occur in both of the potential reading frames that are used during translation by the ribosome. Occasionally, preservation

of both the original reading frame and the second reading frame may require additional point mutations to be added to the downstream exon of the minigene.

3.2.2 Amplification of Minigene Sequence by PCR

1. If synthesis by a company is not practical or feasible, the minigene region can be amplified by PCR using genomic DNA as a template. Again, for large regions of DNA and when point mutations and nucleotide insertions are required, the PCR amplicons can be amplified as separate, smaller, overlapping fragments of DNA that can be stitched together by Gibson Assembly. As described in Subheading 3.2.1, the single nucleotide insertion required for shifting the reading frame in the alternative exon, as well as any additional point mutations required in the downstream exon to preserve reading frame, can be introduced through oligonucleotide primers during PCR.
2. An example starting point for PCR using genomic DNA as a template is listed here (*see* **Note 3**).
   (a) *Reaction*: 32.5 μL molecular biology grade water, 10 μL 5×HF Buffer, 1 μL dNTPs (10 mM), 2.5 μL forward primer (10 μM), 2.5 μL reverse primer (10 μM), 1 μL genomic DNA (50 ng/μL), 0.5 μL Phusion polymerase.
   (b) *PCR settings*.
   Step 1: 98 °C 30 s.
   Step 2: 98 °C 30 s.
   Step 3: lowest T_m + 3 °C 10–30 s (*see* **Note 4**).
   Step 4: 72 °C 1 min 45 s.
   Step 5: go to step 2, 34 times.
   Step 6: 72 °C 5 min.
   Step 7: 4 °C hold ∞.
3. Run 5 μL of PCR reaction out on a 1% agarose gel. A band of the appropriate size should be visible.
4. Purify reaction product using a DNA purification column or if necessary, by gel extraction according to manufacturer's guidelines, eluting DNA with 20 μL of water or elution buffer.

3.3 Obtaining Promoters of Interest for Expression in Specific Cell Types and Digestion of Two-Color Vector for Cloning Promoters

1. Our two-color reporter vector has unique ClaI and KpnI sites to digest the vector and insert any promoter of interest. Currently, we have several versions of our two-color reporter that drive expression in muscle cells, intestinal cells, neurons, and even neuron subtypes (all available upon request). However, any promoter of interest, including the endogenous promoter for a gene of interest with a splicing pattern to be studied, can be introduced. A new promoter DNA sequence can be

obtained through similar options as described in Subheading 3.2 for the minigene sequence. Specifically, the promoter can either be synthesized as double-stranded DNA by a company, including 25–30 nucleotides of overlapping DNA sequence with the digested vector for Gibson Assembly (*see* Subheading 3.2.1). Alternatively, the promoter can be amplified by PCR using genomic DNA as a template (*see* Subheading 3.2.2 for guidelines).

2. In parallel to obtaining the promoter DNA to be inserted, the empty two-color reporter vector is digested with ClaI and KpnI restriction enzymes in a preparative restriction digestion reaction in a PCR tubes as follows:
   (a) *Reaction*: 34 – x μL molecular biology grade water, 4 μL 10× CutSmart buffer, 1 μL ClaI, 1 μL KpnI, x μL of two-color reporter vector (total of 2 μg).
3. Incubate reactions in a thermal cycler at 37 °C for at least 4 h.
4. Clean up DNA using PCR purification column as described by manufacturer and elute in 20 μL of elution buffer or water.
5. Measure concentration of digested vector with a spectrophotometer.

3.4 Gibson Assembly Reaction to Clone Promoter into Two-Color Vector

1. Set up assembly reaction and control reaction in PCR tubes for both vectors.
   (a) *Assembly reaction*: 5 – x – y μL molecular biology grade water, x μL digested two-color vector (50 ng), y μL promoter insert (*see* **Note 5**), 5 μL 2× NEB HiFi Assembly mix.
   (b) *Control reaction*: 10 – x μL molecular biology grade water, x μL digested two-color vector (50 ng).
2. Incubate reactions at 50 °C for 1 h in a thermal cycler and transform into DH5α cells according to the guidelines of the manufacturer.
3. Plate bacteria onto LB agar plates supplemented with 100 μg/mL ampicillin. Incubate overnight at 37 °C.
4. If enrichment is observed, colonies can be screened with diagnostic restriction digest reactions. Positive clones can then be sent for Sanger sequencing to verify insertion of promoter.

3.5 Digestion of Promoter-Containing Vector to Prepare for Insertion of Minigene Fragment

1. After the promoter has been inserted, KpnI and NotI sites in the two-color reporter vector are used to digest the vector and insert the minigene sequence of interest.
   (a) *Reaction*: 34 – x μL molecular biology grade water, 4 μL 10× CutSmart buffer, 1 μL KpnI, 1 μL NotI, x μL of two-color reporter vector (total of 2 μg).

2. Incubate reactions in a thermal cycler at 37 °C for at least 4 h.
3. Clean up DNA using PCR purification column as described by manufacturer and elute in 20 μL of elution buffer or water.
4. Measure concentration of digested vector with a spectrophotometer.

3.6 Insertion of Minigene Fragment into Two-Color Reporter Vector by Gibson Assembly

1. Set up assembly reaction and control reaction in PCR tubes for both vectors.
   (a) *Assembly reaction*: 5 – *x* – *y* μL molecular biology grade water, *x* μL digested two-color vector (50 ng), *y* μL minigene insert(s) (*see* **Note 6**), 5 μL 2× NEB HiFi Assembly mix.
   (b) *Control reaction*: 10 – *x* μL molecular biology grade water, *x* μL digested two-color vector (50 ng).
2. Incubate reactions at 50 °C for 1 h in a thermal cycler and transform into DH5α cells according to the guidelines of the manufacturer.
3. Plate bacteria onto LB agar plates supplemented with 100 μg/mL ampicillin. Incubate overnight at 37 °C.
4. If enrichment is observed, colonies can be screened with diagnostic restriction digest reactions. Positive clones can then be sent for Sanger sequencing to verify the insertion of minigene.

3.7 Preparing Two-Color Reporter Injection Solutions and Microinjection

Once the fully constructed two-color splicing reporter is generated, it can be delivered into *C. elegans* by microinjection of a DNA solution. Although details of microinjection are provided here, we recommend also consulting the excellent wormbook chapter on this protocol [33].

1. Set up injection mix with high quality DNA in a microfuge tube (*see* **Note 7**).
   (a) *Injection mix (10 μL final volume)*: 10 – *x* μL molecular biology grade water, *x* μL two-color reporter vector (500 ng total, 50 ng/μL final concentration).
2. Spin for 5 min at high speed in a microcentrifuge. Collect top 6 μL of solution to a new microfuge tube.
3. Load microinjection needle with 0.5–2 μL of injection solution.
4. Place needle into pressure regulated needle holder. Break needle open using a breaking needle and clear out any air bubbles with forced air pressure. The needle should have no sign of clogging, with liquid flowing easily upon application of positive air pressure.
5. Mount adults with a single row of eggs in their uterus (*see* **Note 8**) on a 2% dry agarose pad coated with halocarbon oil. Worms

should be mounted with dorsal gonad arms in an accessible position for microinjection. Make sure to cover animals completely in oil after mounting to avoid desiccation.

6. Working quickly, proceed to microinject DNA solution into each gonad arm of the animals. After withdrawing the injection needle, apply positive pressure and flood the animal with injection solution. This will help release the animal from the agarose pad to enable recovery. With practice, one should be able to mount and inject several animals in one cycle.
7. Recover animals to a recovery plate. When executed properly, injected animals should be crawling immediately or within a few minutes after being deposited on recovery plate.
8. Repeat this procedure until roughly 10–15 animals have been injected per injection needle.
9. When finished, transfer 2–3 animals to a plate across a total of five 35 mm NGM plates. Place animals in a 21 °C incubator for 3–4 days.

3.8 Screening for Animals with Two-Color Reporter Expression and Establishing Stable Extrachromosomal Array Lines

1. After 3–4 days postinjection, plates with worms can be screened for successful expression of the two-color reporter in the F1 progeny of the injected adults. Candidate animals are screened using a fluorescent stereomicroscope capable of detecting GFP and mCherry fluorescent protein expression.
2. We recommend picking 15–30 transgenic F1 progeny and transferring three worms per plate (for a total of five-ten plates of F1 worms). This should guarantee at least two-three independent stable extrachromosomal lineages will be created.
3. Allow the F1 animals to grow for 3–4 additional days at 21 °C, which will allow the animals to develop to adulthood, self-fertilize, and F2 progeny to be born. At this point, plates can be screened for F2 progeny stably expressing the two-color reporter as an extrachromosomal array (*see* **Note 9**).

3.9 Guidelines for Imaging of Two-Color Reporters via Fluorescence Microscopy

A few considerations can be made when imaging with our two-color reporters. In many cases, we utilize these reporters to visualize differences in alternative splicing patterns across tissues. As such, we frequently coexpress the two-color reporter in multiple tissues in the same animal. Coexpression enables relative GFP/mCherry ratios to be observed in multiple tissues under the same exposure and detection settings captured in one image. If quantification is required, care must be taken to ensure exposure and detection settings are calibrated to ensure pixel saturation in both GFP and mCherry channels is minimized. Some example images of reporters expressed in multiple tissues are depicted in Fig. 2.

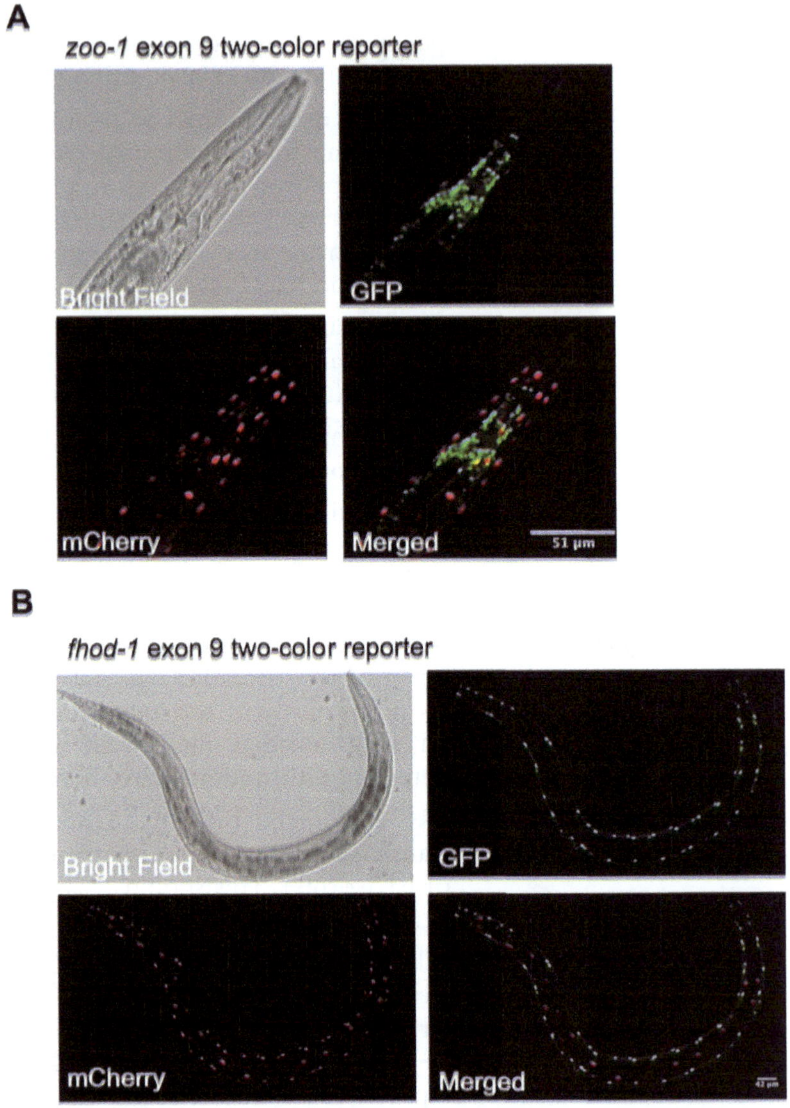

Fig. 2 Examples of two-color reporters monitoring tissue-specific alternative splicing patterns (**a**) Two-color reporter monitoring an exon skipping event (exon 9) from the *zoo-1* gene. The reporter is expressed in both neuronal cells and muscle cells. This reporter primarily undergoes exon skipping in neurons (green smaller nuclei in head region) whereas exon 9 is mostly included in muscle cells (red larger nuclei running down sides of animal). (**b**) Two-color reporter monitoring an exon skipping event (exon 9) from the *fhod-1* gene. This reporter is expressed in muscle cells and intestinal cells. In muscle cells, both included and skipped isoforms are observed (nuclei with both red and green signal running down sides of animal), while in intestinal cells the alternative exon is mainly included (red nuclei running down center of animal)

In another application, we utilize these reporters to test the effect of mutating nucleotides overlapping with putative cis elements. In this application, in order to measure quantitative differences in splicing patterns it is important to first survey all transgenic animals in the experiment. Specifically, it is critical to determine the optimal GFP and mCherry imaging settings across all reporters to ensure that pixel saturation in either channel does not occur in any of the strains being imaged. GFP/mCherry ratios can then be obtained and compared across reporters.

3.10 Preparing Animals for Imaging of Two-Color Reporters via Fluorescence Microscopy

1. Prepare molten 2% agarose dissolved in water by heating on a hotplate or by microwave. Once molten, the agarose can be kept at 80 °C on a hotplate, stirring with a stir bar.
2. Prepare two microscope slides with a few layers of laboratory tape wrapped around them. Place a third slide in the middle of the two slides with tape.
3. Add a drop of molten agarose onto the middle slide and immediately cover with a fourth slide placed perpendicular to the middle slide, with the fourth slide making contact with the two slides with the laboratory tape. This will enable the molten agarose to spread into a disc-shaped pad.
4. Once the agarose has solidified, the slides can be pulled apart and the slide with the agarose pad can be used to mount animals.
5. Add 5 μL of M9 buffer supplemented 1:1 with 200 mM of Sodium azide to the agarose pad.
6. Using a fluorescent stereomicroscope, transfer 5–10 animals expressing the two-color reporter directly into the 5 μL spot of liquid on the agarose pad.
7. Once all of the worms are transferred to the spot, cover animals gently with a coverslip.
8. Proceed to image animals on either a widefield fluorescence microscope or laser scanning confocal microscope (*see* **Note 10**).

3.11 Total RNA Extraction from Reporter Expressing Animals

This procedure works well for worms growing on two-three 35 mm plates. Volumes may need to be scaled up with increasing numbers of animals.

1. Wash worms off of plates with M9 buffer into a 1.5 mL microcentrifuge tube.
2. Spin down worms in a microcentrifuge 1500 × *g* for 2 min.
3. Aspirate off as much liquid as possible, taking care not to suck up too many worms in the pellet.
4. Add 400 μL of Tri reagent, vortex or shake tube vigorously for at least 1 min.
5. Snap freeze samples in liquid nitrogen or on dry ice. After freezing, samples can be safely stored in a −80 °C freezer for

future use for up to several months. Alternatively, if continuing with the protocol, proceed to **step 6**.

6. Thaw at 37 °C or by hand. Once sample has thawed, let stand at room temperature for 3 min.
7. Add 120 μL of chloroform to each sample. Shake sample vigorously by hand or vortex for 1 min.
8. Let samples incubate for 3–5 min at room temperature. Spin at 12,000 × *g* for 15 min in a microcentrifuge at 4 °C.
9. Take upper aqueous phase and transfer to a new microfuge tube, leaving behind and discarding interphase and bottom phase.
10. Proceed to purify total RNA using a preferred spin column kit as recommended by the manufacturer. We prefer the RNA Clean and Concentrator kit from Zymo Research Corporation.
11. During the final elution step, elute in 25 μL nuclease-free water.
12. Quantify the concentration of the RNA using a spectrophotometer.

3.12 Semiquantitative Reverse Transcription Polymerase Chain Reaction (RT-PCR)

In addition to fluorescence microscopy, reporter splicing patterns can be measured by semiquantitative RT-PCR using primers that specifically convert two-color reporter mRNA to cDNA and then amplify the cDNA by PCR. The reverse primer is designed to anneal to the GFP sequence specific to the reporter and the forward primer anneals to the upstream constitutive exon of the minigene sequence (Fig. 3a). This reaction amplifies cDNA produced from both mRNA splice variants, which can then be resolved by gel electrophoresis. We show some example RT-PCRs monitoring tissue-specific alternative splicing across four different splicing reporters in Fig. 3b. We prefer integrated single reaction RT-PCR kits over protocols that require separate cDNA synthesis and PCR steps. Below, we describe a protocol that works using the Qiagen OneStep RT-PCR kit. However, additional kits will also work.

1. Set up the RT-PCR in a PCR tube:
   (a) *Reaction mix (10 μL final volume)*: 3.9 μL nuclease-free water, 2.5 μL 5× reaction buffer, 0.4 μL 10 mM dNTP mix, 0.1 μL RiboLock ribonuclease inhibitor, 0.4 μL enzyme mix, 0.6 μL 10 μM forward primer, 0.6 μL 10 μM reverse primer, 2 μL of total RNA (at 25 ng/μL, 50 ng total).
   (b) *RT-PCR settings*:

   Step 1: 50 °C 30 min (*see* **Note 11**).

   Step 2: 95 °C 15 min.

   Step 3: 94 °C 40 s.

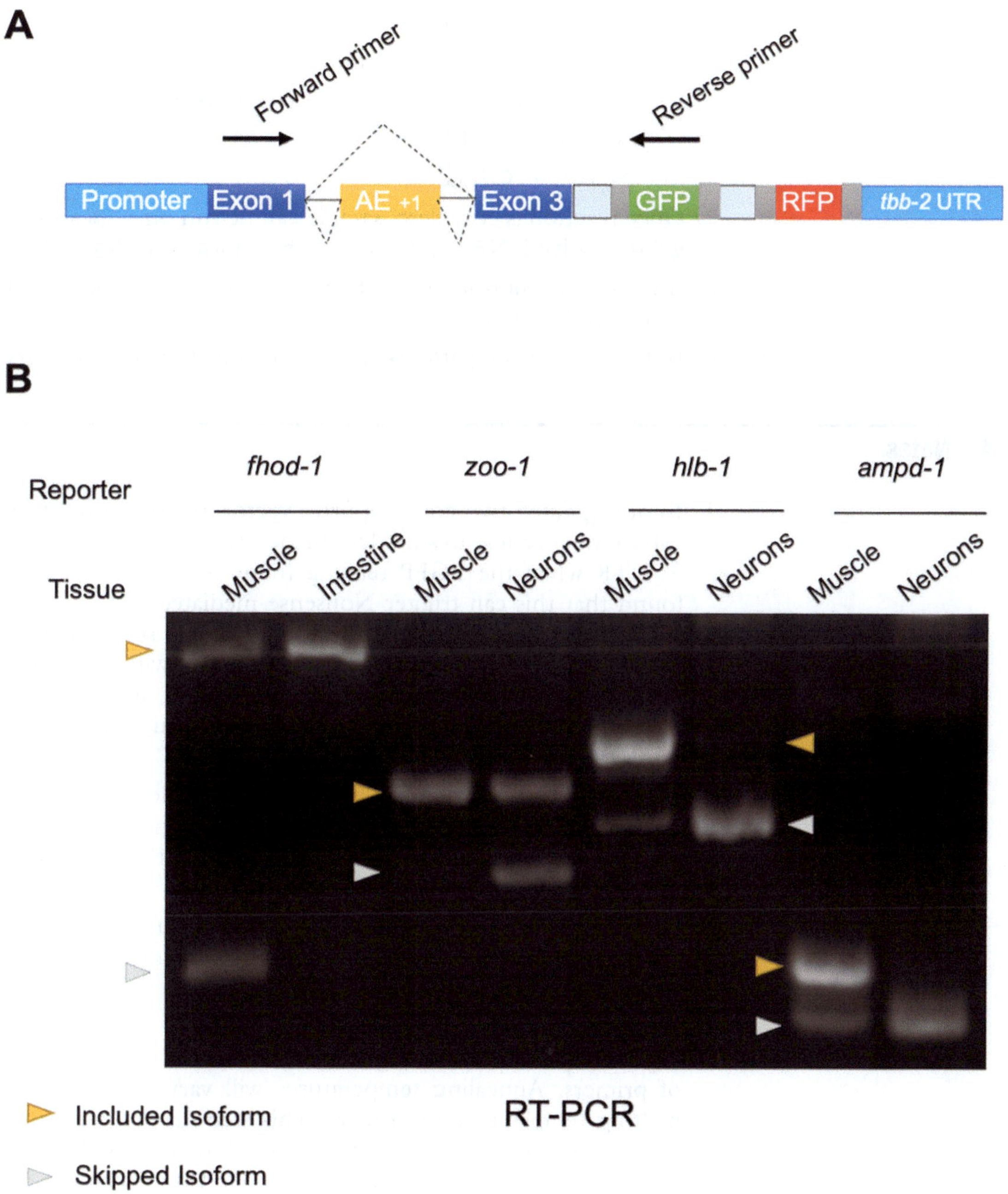

Fig. 3 Example RT-PCR results using two-color reporters to monitor tissue-specific splicing patterns (**a**) Schematic for the design of RT-PCR primers. A forward primer is designed that anneals in the constitutive exon (Exon 1 in diagram) and a reverse primer is designed to anneal in the GFP sequence. The GFP-specific reverse primer ensures that only reporter mRNA will be specifically amplified and not endogenous mRNA. (**b**) When tissue-specific promoters drive expression of the two-color reporters, tissue-specific splicing patterns can be monitored from whole animal total RNA samples. Representative RT-PCRs are displayed covering a splicing event differentially regulated between muscle cells and intestinal cells (from the *fhod-1* gene) and three examples of neuron-specific exon skipping (from the *zoo-1*, *hlb-1*, and *ampd-1* genes). Yellow arrowheads denote isoforms including the alternative exon and gray arrowheads denote isoforms skipping the alternative exon

Step 4: 58 °C 40 s (*see* **Note 12**).

Step 5: Step 4: 72 °C 1 min.

Step 6: go to step 3, 29 times (*see* **Note 13**).

Step 6: 72 °C 10 min.

Step 7: 4 °C hold ∞.

2. Once reaction is complete, sample can be run on a 2% agarose gel to resolve DNA bands from each isoform (*see* **Note 14**).
3. After electrophoresis, gel can be imaged after staining with Ethidium bromide.
4. If desired, splicing patterns can be quantified by densitometry.

4 Notes

1. In our updated two-color reporter vector, the presence of the extra sequence features has lengthened the reporter transcript 3′ UTR when the EGFP reading frame is utilized. We have found that this can trigger Nonsense-mediated mRNA decay and reduces the GFP signal. Although we still see recapitulation of splicing patterns between tissues in a wild type background, we recommend performing most imaging and screening assays in a NMD-deficient background such as in *smg-1(r861)* mutant animals.
2. When deciding on where to insert a single nucleotide into the alternative exon, we usually try to find a location that is less highly conserved and not as likely to overlap with an exonic cis element.
3. As will all PCR, some optimization may be required to amplify products including adjusting annealing temperature, magnesium, and buffer conditions.
4. For Phusion polymerase, it is recommended to set the annealing temperature to 3 °C higher than the lowest Tm of the pair of primers. Annealing temperatures will vary depending on melting temperature of primers. This step may require some troubleshooting in order to obtain homology arm amplicons in sufficient quantity. A useful site for primer Tm calculations can be found at https://tmcalculator.neb.com/#!/main
5. Generally, it is recommended to use a 2:1 insert-to-vector molar ratio for Gibson Assembly.
6. When assembling more than two fragments by Gibson assembly, we have found that increasing the amount of digested vector to 100 ng can improve success.
7. In general, we recommend using high quality DNA for preparing final vectors for microinjection. We typically use a

combination of Thermo GeneJet miniprep kits followed by a Zymo clean and concentrator kit and elute our DNA in nuclease-free water.

8. Typically, we pick late L4 stage animals the day before an injection, and incubate these animals for ~18–24 h at 21 °C. This usually yields adult animals with a single row of eggs in their uterus, which in our hands work very well for microinjection.
9. Once F2 progeny are observed, only 3–4 fluorescent animals have to be transferred each generation to maintain the lineage.
10. Optimization of imaging conditions will be required for each type of microscope and for each new reporter being imaged. It is best to develop some general guidelines and settings for two-color reporter imaging and then proceed to adjust these settings until they are optimized.
11. Manufacturer recommends keeping reaction on ice until thermal cycler block reaches 50 °C.
12. We generally design our primers to have salt-adjusted melting temperatures between 60–62 °C.
13. We have found that going beyond 30 cycles of amplification reduces the linearity of amplification and makes the readout less quantitative. We recommend keeping the amplification cycle number between 25 and 30 cycles. If dim or no signal is observed after gel electrophoresis, staining and detection, then increase the amount of total RNA added to the RT-PCR reaction.
14. We find that 2% agarose gels are sufficient to nicely resolve most alternative splicing events where the size difference between isoforms is greater than 30 nucleotides. For splice variants that differ by less than 30 nucleotides, it is recommended to consider using native polyacrylamide gel electrophoresis, which is capable of resolving differences of even a few bases.

Acknowledgments

We would like to thank Dovic King and Lisa-Marie Schneider for their help with developing our updated two-color splicing reporters. Our research is currently supported by the Natural Sciences and Engineering Research Council of Canada (NSERC), the Canadian Institutes of Health Research (CIHR), the Canada First Research Excellence Fund (CFREF), the Canada Research Chairs program (CRC), the Canada Foundation for Innovation (CFI), and the Ontario Research Foundation (ORF) and we are grateful for their generous support.

References

1. Moore MJ (2005) From birth to death: the complex lives of eukaryotic mRNAs. Science 309:1514–1518
2. Gilbert W (1978) Why genes in pieces? Nature
3. Keren H, Lev-Maor G, Ast G (2010) Alternative splicing and evolution: diversification, exon definition and function. Nat Rev Genet 11:345–355
4. Lee Y, Rio DC (2015) Mechanisms and regulation of alternative pre-mRNA splicing. Annu Rev Biochem 84:291–323. https://doi.org/10.1146/annurev-biochem-060614-034316
5. Ramani AK, Calarco JA, Pan Q et al (2011) Genome-wide analysis of alternative splicing in Caenorhabditis elegans. Genome Res 21:342–348. https://doi.org/10.1101/gr.114645.110
6. Wang ET, Sandberg R, Luo S et al (2008) Alternative isoform regulation in human tissue transcriptomes. Nature 456:470–476. https://doi.org/10.1038/nature07509
7. Pan Q, Shai O, Lee LJ et al (2008) Deep surveying of alternative splicing complexity in the human transcriptome by high-throughput sequencing. Nat Genet 40:1413–1415. https://doi.org/10.1038/ng.259
8. Nilsen TW, Graveley BR (2010) Expansion of the eukaryotic proteome by alternative splicing. Nature 463(7280):457–463
9. Kalsotra A, Cooper TA (2011) Functional consequences of developmentally regulated alternative splicing. Nat Rev Genet 12:715–729
10. Baralle FE, Giudice J (2017) Alternative splicing as a regulator of development and tissue identity. Nat Rev Mol Cell Biol 18(7):437–451
11. Raj B, Blencowe BJ (2015) Alternative splicing in the mammalian nervous system: recent insights into mechanisms and functional roles. Neuron 87:14–27
12. Vuong CK, Black DL, Zheng S (2016) The neurogenetics of alternative splicing. Nat Rev Neurosci 17:265–281
13. Furlanis E, Traunmüller L, Fucile G, Scheiffele P (2019) Landscape of ribosome-engaged transcript isoforms reveals extensive neuronal-cell-class-specific alternative splicing programs. Nat Neurosci 22:1709–1717. https://doi.org/10.1038/s41593-019-0465-5
14. Joglekar A, Prjibelski A, Mahfouz A et al (2021) A spatially resolved brain region- and cell type-specific isoform atlas of the postnatal mouse brain. Nat Commun 12(1):463. https://doi.org/10.1038/s41467-020-20343-5
15. Ha KCH, Sterne-Weiler T, Morris Q et al (2021) Differential contribution of transcriptomic regulatory layers in the definition of neuronal identity. Nat Commun 12(1):335. https://doi.org/10.1038/s41467-020-20483-8
16. Koterniak B, Pilaka PP, Gracida X et al (2020) Global regulatory features of alternative splicing across tissues and within the nervous system of C. elegans. Genome Res 30:1766–1780. https://doi.org/10.1101/gr.267328.120
17. Wang Q, Abruzzi KC, Rosbash M, Rio DC (2018) Striking circadian neuron diversity and cycling of Drosophila alternative splicing. elife 7:e35618. https://doi.org/10.7554/eLife.35618
18. Scotti MM, Swanson MS (2016) RNA mis-splicing in disease. Nat Rev Genet 17(1):19–32
19. Gonatopoulos-Pournatzis T, Wu M, Braunschweig U et al (2018) Genome-wide CRISPR-Cas9 interrogation of splicing networks reveals a mechanism for recognition of autism-misregulated neuronal microexons. Mol Cell 72:510–524.e12. https://doi.org/10.1016/j.molcel.2018.10.008
20. Stoilov P, Lin CH, Damoiseaux R et al (2008) A high-throughput screening strategy identifies cardiotonic steroids as alternative splicing modulators. Proc Natl Acad Sci U S A 105(32):11218–11223. https://doi.org/10.1073/pnas.0801661105
21. Wang Z, Rolish ME, Yeo G et al (2004) Systematic identification and analysis of exonic splicing silencers. Cell 119(6):831–845. https://doi.org/10.1016/j.cell.2004.11.010
22. Rosenberg AB, Patwardhan RP, Shendure J, Seelig G (2015) Learning the sequence determinants of alternative splicing from millions of random sequences. Cell 163(3):698–711. https://doi.org/10.1016/j.cell.2015.09.054
23. Cooper TA (2005) Use of minigene systems to dissect alternative splicing elements. Methods 37(4):331–340. https://doi.org/10.1016/j.ymeth.2005.07.015
24. Orengo JP, Bundman D, Cooper TA (2006) A bichromatic fluorescent reporter for cell-based screens of alternative splicing. Nucleic Acids Res 34(22):e148. https://doi.org/10.1093/nar/gkl967
25. Norris AD, Gao S, Norris ML et al (2014) A pair of RNA-binding proteins controls networks of splicing events contributing to specialization of neural cell types. Mol Cell 54:

946–959. https://doi.org/10.1016/j.molcel.2014.05.004

26. Thompson M, Bixby R, Dalton R et al (2019) Splicing in a single neuron is coordinately controlled by RNA binding proteins and transcription factors. elife 8:e46726. https://doi.org/10.7554/eLife.46726
27. Kuroyanagi H, Watanabe Y, Hagiwara M (2013) CELF family RNA-binding protein UNC-75 regulates two sets of mutually exclusive exons of the unc-32 gene in neuron-specific manners in Caenorhabditis elegans. PLoS Genet 9(2):e1003337. https://doi.org/10.1371/journal.pgen.1003337
28. Kuroyanagi H, Kobayashi T, Mitani S, Hagiwara M (2006) Transgenic alternative-splicing reporters reveal tissue-specific expression profiles and regulation mechanisms in vivo. Nat Methods 3(11):909–915. https://doi.org/10.1038/nmeth944
29. Amrane S, Rebora K, Zniber I et al (2014) Backbone-independent nucleic acid binding by splicing factor SUP-12 reveals key aspects of molecular recognition. Nat Commun 5: 4595. https://doi.org/10.1038/ncomms5595
30. Gracida X, Norris AD, Calarco JA (2016) Regulation of tissue-specifi c alternative splicing: *C. elegans* as a model system. In: Advances in experimental medicine and biology. Springer, pp 229–261
31. Zahler AM (2012) Pre-mRNA splicing and its regulation in Caenorhabditis elegans. WormBook 1–21
32. Wani S, Kuroyanagi H (2017) An emerging model organism Caenorhabditis elegans for alternative pre-mRNA processing in vivo. Wiley Interdiscip Rev RNA 8
33. Evans T (2006) Transformation and microinjection. WormBook. https://doi.org/10.1895/wormbook.1.108.1

Chapter 14

Quantitative Detection of Protein Splice Variants by Selected Reaction Monitoring (SRM) Mass Spectrometry

Alexander Schmidt and Dietmar Schreiner

Abstract

Molecular diversification of the cellular proteome through alternative splicing has emerged as an important biological principle. However, the lack of tools to specifically detect and quantify proteoforms (Smith et al., Nat Methods 10:186–187, 2013) is a major impediment to functional studies. Recently, biological mass spectrometry (MS) has undergone impressive advances (Mann, Nat Rev Mol Cell Biol 17:678, 2016), including the generation of a highly diverse set of biological applications (Aebersold and Mann, Nature 537:347–355, 2016), and has demonstrated to be an essential tool to address many biological questions (Savitski et al., Science 346:1255784, 2014; Rinner et al., Nat Methods 5:315–318, 2008). In particular, targeted LC-MS, with its high selectivity and specificity, is ideally suited for the precise and sensitive quantification of specific proteins and their proteoforms (Picotti and Aebersold, Nat Methods 9: 555–566, 2012). We describe in detail the application of this workflow applied to dissect the molecular diversity of the synaptic adhesion proteins and their splicing-derived proteoforms (Schreiner et al., Elife 4: e07794, 2015).

Key words Targeted mass spectrometry, Selected reaction monitoring, Proteoform quantification, Stable isotope dilution, Alternative splicing

1 Introduction

Alternative splicing of mRNA enables cells to generate an immense transcript diversity from a limited number of genes. Estimates from the sequencing analysis of the human transcriptome suggest that 92–94% of human genes undergo alternative splicing [1]. Often, the mRNA variants generated by alternative splicing differ in their untranslated regions (UTRs), which might affect mRNA stability, localization or translation, but has no direct effect on the protein function. In many other cases, alternative splice variants change the amino acid sequence of the corresponding proteins, resulting in the

Supplementary Information The online version contains supplementary material available at [https://doi.org/10.1007/978-1-0716-2521-7_14].

Peter Scheiffele and Oriane Mauger (eds.), *Alternative Splicing: Methods and Protocols*,
Methods in Molecular Biology, vol. 2537, https://doi.org/10.1007/978-1-0716-2521-7_14,

generation of distinct proteoforms, which differ in their functions (e.g., changes in enzymatic activity or binding affinity to the cognate interaction partners) and/or localization. Thus, alternative mRNA splicing vastly contributes to functional proteome diversification. High-resolution mass spectrometry (HR-LC-MS) analyses supports the notion that at least 37% of the approximately 20,000 human proteins generate multiple proteoforms [2, 3]. Alternative splicing of mRNA is an essential regulatory mechanism during development [4]. Finally, alterations in alternative splicing, caused by mutations in RNA cis-regulatory elements or components of spliceosome machinery, are known to cause a range of human diseases [5].

Alternative splicing programs can generate families of tens, hundreds, or even thousands of closely related proteoforms frequently distinguished by only a single peptide [6, 7]. The detection of individual differentially spliced mRNA isoforms at the RNA level can be assessed (qualitatively and quantitatively) by PCR-based approaches, second generation or long-read sequencing methods. However, such a detection of alternatively splicing events at mRNA level does not provide information about corresponding proteoforms. The challenge with exploring proteoform-specific functions is often the lack of suitable tools to detect or quantify endogenous protein variants.

One possibility to circumvent these problems is the application of targeted LC-MS approaches [8–11], such as selected reaction monitoring (SRM; also referred to as MRM for "multiple reaction monitoring") [12, 13] if carried out on low resolution triple quadrupole LC-MS platforms or parallel reaction monitoring (PRM) if performed on high-resolution Q-TOF or Orbitrap based MS systems [14, 15]. Since these three closely related targeted approaches follow the same analytical principles, the methods described in this chapter are applicable to all of them and we just refer to the term "SRM" for all approaches in the following paragraphs. We will discuss the slight differences at the end of this chapter. In contrast to conventional shotgun proteomic LC-MS that sample a random portion of the proteome, SRM assays use optimized and specific separation and detection parameters for a subset of preselected peptides (termed proteotypic peptides or PTPs). In combination with stable isotope labeled (SIL) peptides, targeted MS considerably improves consistent and precise quantification over standard shotgun LC-MS [16–18] and provides the sensitivity and selectivity needed to detect and precisely quantify low-abundance proteins and proteoforms within the complex proteome of eukaryotes [19].

A concert of different software tools/data repositories have been developed [20–22] assisting in each step of this LC-MS approach, including the selection of suitable peptides [23, 24], the assay generation [25, 26], and the data analysis [27, 28]. Despite these recent improvements and the overall higher sensitivity, setup of targeted MS analyses still remains more time demanding compared to standard shotgun LC-MS approaches. Therefore, it is still only used by a few expert proteomics laboratories and for specific projects. Thus, there is strong need to further facilitate the single steps and application to make this promising and powerful analytical technique more popular among the larger proteomics community.

Figure 1 shows the general outline of this targeted MS analysis workflow for detection and quantification of neurexin splice variants from mouse brain tissue. Neurexins represent a highly diverse protein family of synaptic adhesion proteins generated by cell type specific alternative splicing program [29]. Most of the neurexin proteoforms differ typically in small, but functionally important, peptide insertions. Detection of different neurexin proteoforms using classical antibody-based techniques is therefore challenging, making detection by SRM the method of choice.

In the following chapter we provide a detailed protocol for SRM based assay development and application for detection of neurexin splice variants in mouse brain [7]. The methodology and workflow described in this protocol can be easily adapted for other proteins of interest or tissues.

2 Materials

All materials used are of the highest quality. HPLC-grade solvents (water, acetonitrile) were from Fisher scientific. Proteomics grade trypsin is obtained from Promega (Dübendorf, Switzerland), Lys-c from Wako Pure Chemical Corporation. All solid chemicals are from Sigma-Aldrich and of the highest purity available. Stable isotope labeled peptides are ordered from JPT (www.JPT.com, SpikeTides-L). Trifluoroacetic acid (TFA), sequencing grade, is ordered from Thermo Fisher Scientific.

2.1 Selection of Peptide Sequences for SRM Assays

1. Targeted mass spectrometry software Skyline (current version 20.2, available here: https://skyline.ms/project/home/software/Skyline/begin.view).

2.2 SRM Assay Generation

1. Stable isotope labeled (SIL) peptides (JPT).
2. *LC Buffer A*: 99.9% water (HPLC-grade)–0.1% formic acid.
3. Reverse phase HPLC column (75 μm × 30 cm) packed in house with C18 resin (Reprosil-AQ Pur, 1.9 μm, Dr. Maisch).
4. EASY nano-LC system (Proxeon Biosystems, Thermo Scientific).
5. LTQ-Orbitrap Elite mass spectrometer equipped with a nano electrospray ion source (Thermo Scientific).

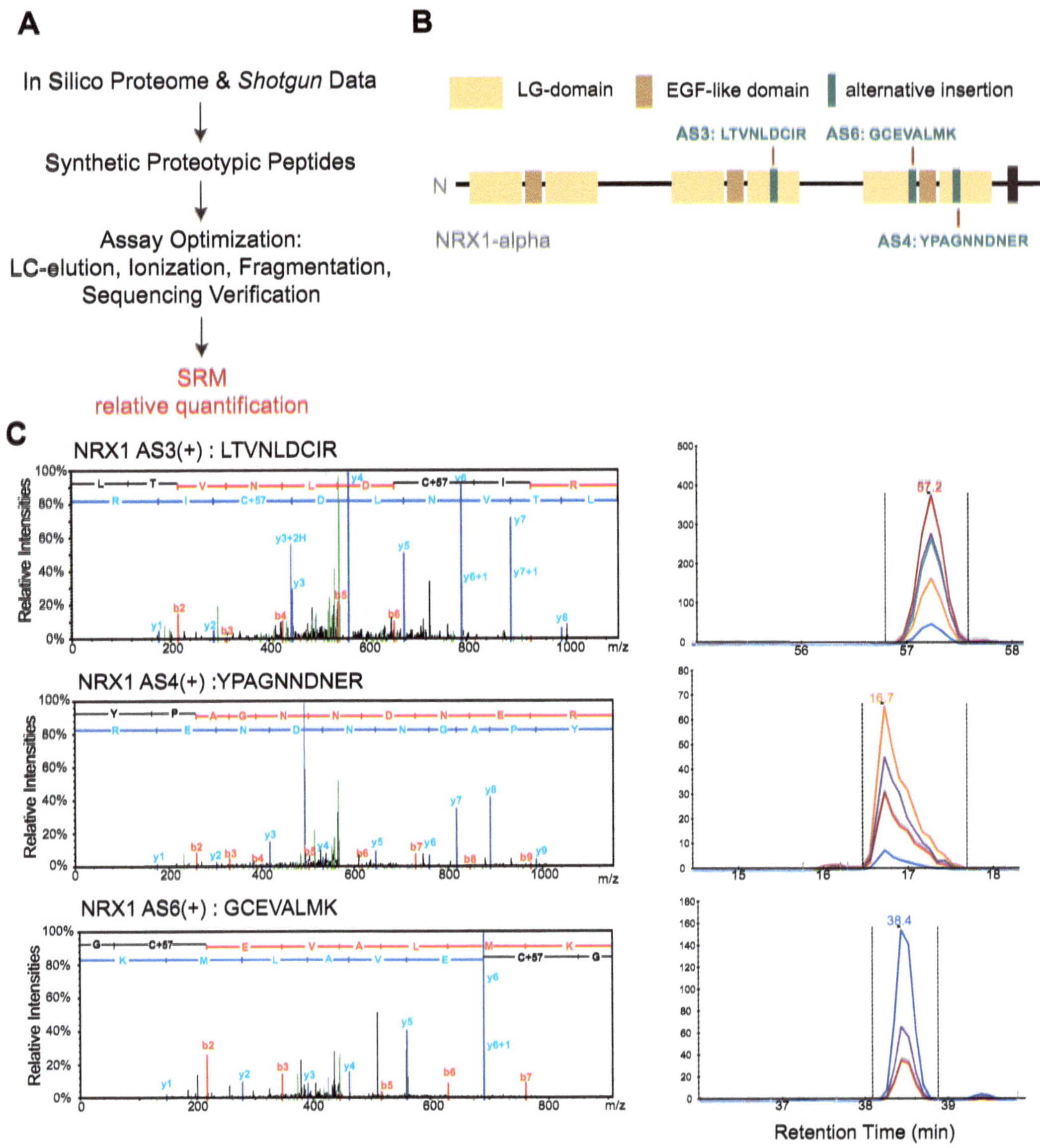

Fig. 1 (**a**) Workflow for the quantitative SRM-based protein analysis. (**b**) Proteotypic peptides for relative quantification of neurexin-1-alpha (NRXN1-alpha) proteoforms containing insertions at alternatively spliced segments (AS) 3, 4, and 6 (for the complete list of proteotypic peptides for detection of different neurexins and their splice variants please refer to [7]). The position of peptides within the overall domain structure of NRX1-alpha protein is indicated (laminin-G and EGF-domains are marked, transmembrane domain (TMD) shown as box, sites modified by alternative splicing in dark green). (**c**) *Left panel:* Example MS/MS spectra of three endogenous peptides used for detection and quantification of NRX1 proteoforms containing insertions at alternatively spliced segments 3, 4, and 6. Letters above the peaks indicate the amino acid sequence of the corresponding peptides (blue = y-ions; red = b-ions). *Right panel:* Retention times and transition patterns of corresponding synthetic heavy peptides. (Figure modified from [7]/original figure was published under Creative Commons Attribution license)

2.3 Preparation of Synaptic Proteins Enriched Triton-Resistant Membranes (TRM) from Mouse Brain

1. Tabletop cooled centrifuge (e.g., Eppendorf 5415 R).
2. Tabletop ultracentrifuge (e.g., Beckman Coulter/Optima MAX-XP).
3. TLA 50 or TLA120.2 rotor for ultracentrifuge.
4. Glass–Teflon homogenizer.
5. Protease and phosphatase inhibitors.
6. *Homogenization buffer*: 0.32 M sucrose, 1 mM $MgCl_2$, 0.5 mM $CaCl_2$, 1 mM $NaHCO_3$ complemented with protease inhibitors.
7. *Triton resistant membrane (TRM) buffer*: 12 mM Tris-HCl, pH 8.1, 1% Triton X-100 complemented with protease inhibitors.
8. *Solubilization buffer*: 50 mM ammonium bicarbonate, 1% sodium deoxycholate.
9. Pierce™ BCA Protein Assay Kit—Reducing Agent Compatible (ThermoFisher Scientific).

2.4 Protein Digestion and Preparation for SID-SRM Analysis

1. Approximately 10 μg of protein extract in *Solubilisation buffer*.
2. *Reduction buffer*: 0.2 M Tris-(2-carboxyethyl)-phosphine hydrochloride (TCEP) and 0.1 M HEPES, pH 8.5 ± 0.1*.
3. *Alkylation buffer*: 0.4 M Iodoacetamide and 0.1 M HEPES, pH 8.5 ± 0.1*.
4. *Alkylation stop buffer*: 0.5 M N-acetyl-cysteine and 0.1 M HEPES, pH 8.5 ± 0.1*.
5. Trypsin, recombinant, proteomics grade.
6. 5% TFA in water (HPLC-grade).
7. BioPureSPN Mini™ SPE cartridges (HUM S18V, The Nest Group, Inc.™, MA, USA).
8. *C18 Equilibration buffer*: 0.1% TFA in water (HPLC-grade).
9. *C18 Washing buffer*: 0.1% TFA, 5% acetonitrile (HPLC-grade) in water (HPLC-grade).
10. *C18 Elution buffer*: 0.1% TFA, 50% acetonitrile (HPLC-grade) in water (HPLC-grade).

*—Use 0.1 M NaOH or 0.1 M HCl for pH adjustment.

2.5 HPLC Buffers for SID-SRM Analysis

1. *LC Buffer A*: 99.9% water (HPLC-grade), 0.1% formic acid.
2. *LC Buffer B*: 80% acetonitrile (HPLC-grade), 19.9% water (HPLC-grade), 0.1% formic acid.

3 Methods

Below we describe a basic protocol for the extraction of synaptic proteins enriched Triton-resistant membranes from mouse brain and their targeted SRM-MS analysis to dissect and quantify splice variants of the neurexin protein family. The protocol provides detailed general description of the SRM pipeline starting from the selection of proteotypic peptides, development of SRM based assay and experimental procedures and can be easily adopted for any protein of interest. Additionally, we briefly discuss recent developments in the targeted LC-MS for protein and proteoform monitoring [30].

3.1 Selection of Peptide Sequences for SRM Assays

A key step in SRM-based quantification of proteins is the initial selection of the peptides that will be used for quantification. Several public resources provide large numbers of existing LC-MS datasets (SRM Atlas [23], Peptide Atlas SRM Experiment Library (PASSEL) [31], CPTAC Assay Portal [32], Panorama Public [33], MRMAssayDB [34, 35], ProteomicsDB [36], Massive, PRIDE, MaxQB [21]) as well as prediction algorithms [37] that facilitate the selection of the most suited peptides for SRM analysis. However, it is often not possible to predict whether a peptide will be suitable for quantification in a particular biological matrix or instrumental setup. On top, the distinct quantification of specific proteoforms relies on the detection of small sequence stretches, sometimes with only one amino acid difference, greatly limiting peptide choice.

1. Peptides containing the specific amino acid sequences of splice variants of interest are selected. We use SwissProtein/TrEMBL (https://www.uniprot.org/) for obtaining standard protein sequences. Splice variants containing protein sequences are directly translated from mRNA using standard DNA editing software (e.g., ApE, SnapGene) or using online tools (e.g., https://web.expasy.org/translate/).
2. Protein sequences are imported into Skyline (*see* **Note 1**). The software automatically performs an in silico tryptic digest and generates a list of all peptide sequences matching predefined filtering criteria (*see* **Note 2**).
3. All peptide sequences are checked for uniqueness using a BLAST algorithm against the predicted proteome of the relevant organism. We usually use the algorithms provided by UniProt (https://www.uniprot.org/blast/) or NCBI (https://blast.ncbi.nlm.nih.gov/Blast.cgi?PAGE=Proteins). Only peptides having a unique sequence are termed proteotypic peptides (PTPs) and are used for SRM assay generation.

4. If several PTPs are available for a specific protein or proteoform, not all have to be analyzed. Priority is given to PTPs with high MS response. This information can be obtained from public repositories [21, 36, 38] using the peptide-spectrum matches (PSM) provided (*see* **Note 3**). Peptides with high PSM numbers are often detected and are preferred. While this might not directly reflect the actual MS response of the peptide ion, it is still a straightforward strategy to find suitable PTPs. A better option, if available, is to select for peptide ions with high MS1 peak intensities (*see* **Note 4**).
5. Evaluate selected PTPs regarding possible modifications that might affect their correct quantification. While these cannot be completely avoided, following the simple rules below will at least minimize these risks. As a rule of thumb, peptides containing known post-translational modifications (e.g., listed in UniProt), methionine (oxidation), n-terminal glutamine (formation of pyroglutamic acid) or ragged-ends (KK, RR, KR, RK, prone to missed cleavages) should be avoided, if possible. For more detailed information about criteria for choice of proteotypic peptides *see* [14].
6. Some PTPs from reference proteins, like actin or tubulin, should be included in the final SRM assay panel to control for variations in analyzed sample amount during SRM analysis.

3.2 SRM Assay Generation

1. Obtain, if possible, three highly pure stable isotope labeled (SIL) peptides [16] corresponding to the proteins you wish to measure (*see* **Note 5**).
2. Prepare a working solution containing all SIL peptides at a concentration of 50 fmol/peptide/μL in *LC buffer A* (*see* **Note 6**).
3. Analyze 100 fmol (2 μL) of SIL working solution by standard shotgun LC-MS analysis as described [7]. Here, peptides are separated by an EASY nano-LC system equipped with a reverse phase HPLC column (75 μm × 30 cm) packed in house with C18 resin using the gradient indicated in Table 1.
4. LC-MS/MS analysis of separated peptides is done using an LTQ-Orbitrap Elite mass spectrometer equipped with a nano electrospray ion source.
5. The mass spectrometer is operated with the following settings.
   (a) Distal spray voltage: 2.5 kV.
   (b) Top 10 method using HCD and a resolution of 15,000 FWHM (at 400 *m/z*).
   (c) Cycle time is approximately 2 s.
   (d) Normalized collision energy is set to 35%.

Table 1
Setup of the nanoflow pump gradient

Time (min)	LC buffer A (%)	LC buffer B (%)	Flow rate (μL/min)
0	95	5	0.2
75	72	28	0.2
90	60	40	0.2
92	95	5	0.2
110	95	5	0.2

6. MS data is searched using MaxQuant [39], against publicly available protein sequence databases containing the sequences of all selected PTPs (*see* **Note 7**). We use the *Mus musculus* Swiss-Prot (reviewed) database (www.uniprot.org) and add all splice variant and proteoform sequences.
7. The search results are imported into Skyline by generating a spectral library from the search results (msms.txt file) (*see* **Note 8**). For more details, please refer to Skyline instruction https://skyline.ms/project/home/software/Skyline/begin.view.
8. Once the spectral library is activated, the software automatically selects all precursor ions and most intense fragment masses for each identified peptide.
9. These are exported as a transition list and imported into the acquisition software operating the SRM-LC-MS triple quadrupole platform.
10. Before analyzing the samples, each SRM assay is optimized regarding collision energies, parent ion masses and fragment ion selection in pilot experiments using the SIL peptide mix and the Skyline software [28].
11. The final transition list employed for targeted LC-MS analysis of selected neurexin splice variant in synaptic enriched mouse brains samples that can directly be important into the Vantage QQQ MS is provided in Electronic Supplementary Table 1.

3.3 Preparation of Synaptic Proteins Enriched Triton-Resistant Membranes (TRM) from Mouse Brain

The detection of many proteins and/or proteoforms by SRM can be performed directly on crude tissue or cell extracts (*see* **Note 9**). For robust and reliable detection of low-abundance proteins an enrichment strategy is recommended. For synaptic proteins and proteoforms, we have been applying isolation of Triton X-100 resistant membrane fractions (TRM) from a crude synaptosomal fraction. This rapid enrichment strategy allows for high sample throughput and reliable SRM-based quantification of synaptic proteins [7].

1. Dissect brain region of interest (at this point tissue can be snap frozen and stored at −80 °C).
2. Homogenize tissue in a glass–Teflon homogenizer in 1:20 (w/v) of homogenization buffer complemented with protease inhibitors and (if desired) phosphatase inhibitors).
3. Centrifuge homogenate at 1400 × *g* at 4 °C for 10 min.
4. Transfer supernatant into new tubes and centrifuge at 14,000 × *g* at 4 °C for 10 min.
5. Discard supernatant and resuspend pellet by pipetting up and down in 1:5 (w/v) (weight of the starting material) in Triton X-100 containing buffer (12 mM Tris–HCl, pH 8.1, 1% Triton X-100 complemented with protease (if necessary) phosphatase inhibitors).
6. Incubate samples at 4 °C for 10 min with rotation.
7. Transfer samples into ultracentrifuge suitable tubes and centrifuge at 100,000 × *g* at 4 °C for 1 h.
8. Discard supernatant and resuspend pellet representing TRM in 1:1 (w/v) of solubilisation buffer. Incubate samples in a heat-block at 65 °C with shaking at 1200 rpm for ca. 10 min.
9. Determine protein concentration with BCA assay.

3.4 Tryptic Digest and Column Purification of Peptide

1. Prepare TRM extract aliquots containing around 10 μg total protein.
2. Add solubilisation buffer to a final sample volume of 40 μL.
3. Reduce protein samples by addition of 1 μL of *Reduction buffer* and incubation at 95 °C for 10 min.
4. Alkylate reduced proteins by addition of 1 μL of *Alkylation buffer* at 25 °C for 30 min in the dark.
5. Destroy remaining iodoacetamide by addition of 1 μL of *Alkylation stop buffer* and incubation at RT for 10 min.
6. Add trypsin at a final enzyme–substrate concentration of 1:50 (w/w), and digest overnight at 37 °C (*see* **Note 10**).
7. The resulting peptide mixture is acidified with 10 μL 5% TFA and the precipitated sodium deoxycholate removed by centrifugation at 20,000 × *g* for 5 min.
8. The supernatant containing the peptides is solid phase extracted using micro spin columns, dried under vacuum and stored at −80 °C until further use.

3.5 SRM LC-MS Analysis

1. Dissolve dried peptide samples in LC buffer A at a concentration of 0.5 μg/μL.
2. In a first experiment, the linear dynamic range of each SRM assay should be determined by a dilution experiment spiking different amounts of SIL peptides into the sample matrix.
3. Therefore, prepare a pool of sample aliquots, split the pool into six aliquots and add 100, 10, 1, 0.1 and 0.01 fmol of SIL peptides/μg of total peptides.
4. Analyse all samples by SIL-SRM-MS as described in [7] using the same elution parameters as described above (Table 1) and the optimized transitions provided in Electronic Supplementary Table 1 starting with the pure pool aliquot to the highest SIL concentration. Each sample should be analysed in technical duplicates.
5. The mass spectrometer is operated with the following settings.
   (a) Distal spray voltage: 2.5 kV.
   (b) Q1 and Q3 peak width is set to 0.7 *m/z*.
   (c) Cycle time is set to 3 s.
   (d) Collision gas pressure is set to 1.2 mTorr.
6. All SRM LC-MS acquired raw files are imported into the Skyline software for quantitative data analysis.
7. All selected transitions are manually checked and the ratios of each endogenous/SIL peptide pair calculated.
8. Plot the ratios according to the SIL concentration and determine assay linearity, lower limits of quantification (LLOQ) and lower limits of detection (LLOD) as described in [40].
9. For the real experiment, add 10 fmol of SIL peptides/ug of total peptides to each sample and carry out the SIL-SRM-MS analysis as described above.
10. Always start the sample queue with an SRM analysis of 50 fmol/SIL pure peptide mix. This allows to check the entire workflow and facilitates the detection of the SIL peptides in the samples by comparing elution times and transition order.
11. Check the pure SIL peptide mix for possible "light" contaminations. If present, subtract these ratios from the sample ratios before further analysis.
12. Use the included reference protein ratios to control for sample amount variations.
13. Check that the ratios are within the linear range of the corresponding SRM assay and are above the limit of quantification.
14. After these calculations, the peptide ratios can be used for differential analysis between samples. If quantified SIL peptides are used, absolute endogenous peptide concentrations can be determined using the calculated ratios.

4 Notes

Targeted MS in combination with stable isotope dilution (SID) is a well-established technique for precise quantification of small molecules. It has recently gained momentum in the proteomics field for precise quantification of selected protein subsets and proteoforms that are challenging to monitor over large sample numbers. While targeted MS approaches currently used in proteomics, namely, selected- or multiple-reaction monitoring SRM/MRM and parallel-reaction monitoring (PRM), follow the same analytical workflows, there are a few differences in performance due to the different MS instruments used that should be considered when performing targeted LC-MS analyses. SRM/MRM is carried out on relatively cheap and very robust, low resolution triple quadrupole instruments. These LC-MS platforms provide excellent sensitivity, selectivity and duty cycles and are ideally suited to monitor large number of targets. In contrast, PRM is carried out on high-resolution LC-MS platforms, like Q-TOF and Orbitrap analysers. Due to the acquisition of full tandem mass spectra, the duty cycle of PRM is lower compared to SRM and therefore, a lower number of targets can be monitored per LC-MS analysis. However, the high-resolution of the instrument increases the overall selectivity of the approach, leading to a higher dynamic range within complex peptide samples, like digested cell/tissue lysates obtained from higher eukaryotes [41, 42]. Thus, depending on the application, either approach should be considered. In general, targeted SIL-MS has matured to a robust and sensitive quantification tool in proteomics. Due to its high precision, sensitivity and straightforward assay generation (including modifications, SNPs, etc.), targeted MS is also an attractive alternative to standard immunoassays. The application of this state-of-the-art technology to diverse biological settings provide a unique opportunity for a more complete biological understanding.

1. To get deeper insights into Skyline, the MacCoss lab provides very informative online tutorials, webinars, and courses detailing Skyline functionality (https://skyline.ms/project/home/software/Skyline/begin.view).
2. In case Skyline fails to generate the desired peptide sequences, these can also be generated by other tools (e.g., PeptideCutter (https://web.expasy.org/peptide_cutter/) or the Protein digestion simulator (https://omics.pnl.gov/software/protein-digestion-simulator) and easily imported into Skyline.
3. Protein sequences can be obtained from publicly available databases (like www.uniprot.org) or from own sequencing experiments. If specific sequence stretches need to be analyzed, sequence fragments including sequence part(s) of interest

should be extended by constitutive sequences to the respective endoproteinase cleavage sites should be used.

4. Some public repositories exist that directly provide peptide rankings based on acquired LC-MS data or predictions [23, 24, 35]. This is very useful and can speed up the PTP selection process if the relevant proteins are present in the database. Since many splice variants are not yet presented in public protein sequence databases, PTPs for these proteoforms are often not available by these tools. In this case, peptide suitability has to be predicted for the selection process [38] and five instead of only three SIL peptides should be synthesized and used for SRM assay generation to increase the chances of developing at least two well performing assays.

5. SIL peptides must contain isotopically heavy lysine or arginine. This protocol employs peptides containing Lys-[13C6, 15 N2] or Arg-[13C6, 15 N4]. Heavy labeling should be >97%. Peptides do not have to be purified. For relative quantification (described in this protocol), SIL peptides do not have to be accurately quantified. For absolute protein/proteoform quantification, SIL peptides need to be quantified before use. This can be achieved by amino acid analysis or fluorescent tags [43]. When performing targeted SIL-MS analyses, the content of peptide containing no heavy amino acids should be as low as possible (ideally <0.1%) to avoid false positive identifications and a loss of assay sensitivity. With the high dynamic range of targeted MS, even trace contaminations become detectable, in particular when using PRM at higher resolution settings. Therefore, the content of light/unlabeled peptide needs to be determined for each SIL peptide assay by analyzing a pure SIL peptide mix of sufficient concentration. 50–100 fmol/SIL peptide on column has shown to provide strong enough MS signals to even detect trace amounts of light contaminations. This light content has to be subtracted from the SIL peptide ratios determined from the real samples. Along these lines, and depending on the light contamination content of the SIL peptides, the amount of spiked in SIL peptides should be kept at a minimum to avoid interference with the endogenous peptide ion signal. If the light contamination is found to be high (>1%), the concentration of each SIL peptidePeptides has to adjusted to the levels of its endogenous counterpart to avoid false positive identifications.

6. Always use low-binding tubes and tips when pipetting SIL peptides do minimize peptide absorbance to the hydrophobic surfaces and resulting peptide losses. This is particularly critical when using quantified SIL peptides for absolute protein abundance measurements. In addition to using low-binding materials, try to keep samples volumes as small and the peptide

concentrations as high as possible. This will also reduce the risk of peptide losses and false quantifications. Aliquot SIL peptide mixes to avoid multiple freeze-thaw cycles. Stocks should be stored at −80 °C.

7. To increase the number of identified SIL peptides, we set the FDR thresholds to 1 on peptide and protein level in the MaxQuant software. We also set arg10 and lys8 as standard modifications and not as label for quantification (default).

8. Skyline also accepts other database search engines for spectral library building. If MaxQuant fails to detect some SIL, other search algorithms can be tried instead. Even when lowering thresholds and using different search engines, some peptides might escape detection by conventional database searching. This does not necessarily mean they are not suited for targeted MS assays. In this case, all single charged y-ions of the double and triple charged precursor ion, respectively, should be selected for analysis in Skyline before running the optimization. This usually includes the best transitions for these peptides and the overall performance can then be evaluated during the assay optimization steps.

9. In our experimental setup we have been routinely using Triton-insolubility based strategy for synaptic proteins enrichment. In general, any type of biochemical or affinity based enrichment (e.g., density gradient based isolation of cellular compartments and organelles or immunoprecipitations) could be applied, which might be useful if detecting low-abundance protein and/or their proteoforms. However, in many cases detection of desired protein could be performed on crude protein extracts. In this case cells or tissue could be directly homogenized in Solubilization buffer.

10. While Trypsin is the preferred protease to generate sensitive targeted LC-MS assays, sometimes it fails to produce peptides of amenable size for targeted LC-MS analysis. Then, other proteases, like endoproteinase Lys-c, Glu-c or Chymotrypsin, can be used instead or also in combination with trypsin to generate peptides of suited length. For absolute protein quantification using quantified SIL peptides, a combinatorial digestion using Lys-c (enzyme–substrate: 1:100 [w/w], 4 h at 37 °C) followed by trypsin (enzyme–substrate: 1:50 [w/w], 37 °C overnight) should be performed to efficiently cleave peptides with ragged ends and to ensure accurate peptide quantities [44].

References

1. Wang ET, Sandberg R, Luo S, Khrebtukova I, Zhang L, Mayr C, Kingsmore SF, Schroth GP, Burge CB (2008) Alternative isoform regulation in human tissue transcriptomes. Nature 456(7221):470–476. https://doi.org/10.1038/nature07509
2. Kim MS, Pinto SM, Getnet D, Nirujogi RS, Manda SS, Chaerkady R, Madugundu AK, Kelkar DS, Isserlin R, Jain S, Thomas JK, Muthusamy B, Leal-Rojas P, Kumar P, Sahasrabuddhe NA, Balakrishnan L, Advani J, George B, Renuse S, Selvan LD, Patil AH, Nanjappa V, Radhakrishnan A, Prasad S, Subbannayya T, Raju R, Kumar M, Sreenivasamurthy SK, Marimuthu A, Sathe GJ, Chavan S, Datta KK, Subbannayya Y, Sahu A, Yelamanchi SD, Jayaram S, Rajagopalan P, Sharma J, Murthy KR, Syed N, Goel R, Khan AA, Ahmad S, Dey G, Mudgal K, Chatterjee A, Huang TC, Zhong J, Wu X, Shaw PG, Freed D, Zahari MS, Mukherjee KK, Shankar S, Mahadevan A, Lam H, Mitchell CJ, Shankar SK, Satishchandra P, Schroeder JT, Sirdeshmukh R, Maitra A, Leach SD, Drake CG, Halushka MK, Prasad TS, Hruban RH, Kerr CL, Bader GD, Iacobuzio-Donahue CA, Gowda H, Pandey A (2014) A draft map of the human proteome. Nature 509(7502): 575–581. https://doi.org/10.1038/nature13302
3. Smith LM, Kelleher NL, Consortium for Top Down P (2013) Proteoform: a single term describing protein complexity. Nat Methods 10(3):186–187. https://doi.org/10.1038/nmeth.2369
4. Baralle FE, Giudice J (2017) Alternative splicing as a regulator of development and tissue identity. Nat Rev Mol Cell Biol 18(7): 437–451. https://doi.org/10.1038/nrm.2017.27
5. Scotti MM, Swanson MS (2016) RNA mis-splicing in disease. Nat Rev Genet 17(1): 19–32. https://doi.org/10.1038/nrg.2015.3
6. Chen BE, Kondo M, Garnier A, Watson FL, Puettmann-Holgado R, Lamar DR, Schmucker D (2006) The molecular diversity of Dscam is functionally required for neuronal wiring specificity in Drosophila. Cell 125(3): 607–620. https://doi.org/10.1016/j.cell.2006.03.034
7. Schreiner D, Nguyen TM, Russo G, Heber S, Patrignani A, Ahrne E, Scheiffele P (2014) Targeted combinatorial alternative splicing generates brain region-specific repertoires of neurexins. Neuron 84(2):386–398. https://doi.org/10.1016/j.neuron.2014.09.011
8. Mann M (2016) Origins of mass spectrometry-based proteomics. Nat Rev Mol Cell Biol 17(11):678. https://doi.org/10.1038/nrm.2016.135
9. Aebersold R, Mann M (2016) Mass-spectrometric exploration of proteome structure and function. Nature 537(7620): 347–355. https://doi.org/10.1038/nature19949
10. Savitski MM, Reinhard FB, Franken H, Werner T, Savitski MF, Eberhard D, Martinez Molina D, Jafari R, Dovega RB, Klaeger S, Kuster B, Nordlund P, Bantscheff M, Drewes G (2014) Tracking cancer drugs in living cells by thermal profiling of the proteome. Science 346(6205):1255784. https://doi.org/10.1126/science.1255784
11. Rinner O, Seebacher J, Walzthoeni T, Mueller LN, Beck M, Schmidt A, Mueller M, Aebersold R (2008) Identification of cross-linked peptides from large sequence databases. Nat Methods 5(4):315–318. https://doi.org/10.1038/nmeth.1192
12. Picotti P, Aebersold R (2012) Selected reaction monitoring-based proteomics: workflows, potential, pitfalls and future directions. Nat Methods 9(6):555–566. https://doi.org/10.1038/nmeth.2015
13. Schreiner D, Simicevic J, Ahrne E, Schmidt A, Scheiffele P (2015) Quantitative isoform-profiling of highly diversified recognition molecules. elife 4:e07794. https://doi.org/10.7554/eLife.07794
14. Lange V, Picotti P, Domon B, Aebersold R (2008) Selected reaction monitoring for quantitative proteomics: a tutorial. Mol Syst Biol 4: 222. https://doi.org/10.1038/msb.2008.61
15. Maiolica A, Junger MA, Ezkurdia I, Aebersold R (2012) Targeted proteome investigation via selected reaction monitoring mass spectrometry. J Proteome 75(12):3495–3513. https://doi.org/10.1016/j.jprot.2012.04.048
16. Gerber SA, Rush J, Stemman O, Kirschner MW, Gygi SP (2003) Absolute quantification of proteins and phosphoproteins from cell lysates by tandem MS. Proc Natl Acad Sci U S A 100(12):6940–6945. https://doi.org/10.1073/pnas.0832254100
17. Gygi SP, Rist B, Gerber SA, Turecek F, Gelb MH, Aebersold R (1999) Quantitative analysis of complex protein mixtures using isotope-coded affinity tags. Nat Biotechnol 17(10): 994–999. https://doi.org/10.1038/13690
18. Ong SE, Blagoev B, Kratchmarova I, Kristensen DB, Steen H, Pandey A, Mann M (2002)

Stable isotope labeling by amino acids in cell culture, SILAC, as a simple and accurate approach to expression proteomics. Mol Cell Proteomics 1(5):376–386. https://doi.org/10.1074/mcp.m200025-mcp200

19. Picotti P, Bodenmiller B, Mueller LN, Domon B, Aebersold R (2009) Full dynamic range proteome analysis of S. cerevisiae by targeted proteomics. Cell 138(4):795–806. https://doi.org/10.1016/j.cell.2009.05.051
20. Wilhelm M, Schlegl J, Hahne H, Gholami AM, Lieberenz M, Savitski MM, Ziegler E, Butzmann L, Gessulat S, Marx H, Mathieson T, Lemeer S, Schnatbaum K, Reimer U, Wenschuh H, Mollenhauer M, Slotta-Huspenina J, Boese JH, Bantscheff M, Gerstmair A, Faerber F, Kuster B (2014) Mass-spectrometry-based draft of the human proteome. Nature 509(7502):582–587. https://doi.org/10.1038/nature13319
21. Schaab C, Geiger T, Stoehr G, Cox J, Mann M (2012) Analysis of high accuracy, quantitative proteomics data in the MaxQB database. Mol Cell Proteomics 11(3):M111.014068. https://doi.org/10.1074/mcp.M111.014068
22. Perez-Riverol Y, Csordas A, Bai J, Bernal-Llinares M, Hewapathirana S, Kundu DJ, Inuganti A, Griss J, Mayer G, Eisenacher M, Perez E, Uszkoreit J, Pfeuffer J, Sachsenberg T, Yilmaz S, Tiwary S, Cox J, Audain E, Walzer M, Jarnuczak AF, Ternent T, Brazma A, Vizcaino JA (2019) The PRIDE database and related tools and resources in 2019: improving support for quantification data. Nucleic Acids Res 47(D1):D442–D450. https://doi.org/10.1093/nar/gky1106
23. Kusebauch U, Campbell DS, Deutsch EW, Chu CS, Spicer DA, Brusniak MY, Slagel J, Sun Z, Stevens J, Grimes B, Shteynberg D, Hoopmann MR, Blattmann P, Ratushny AV, Rinner O, Picotti P, Carapito C, Huang CY, Kapousouz M, Lam H, Tran T, Demir E, Aitchison JD, Sander C, Hood L, Aebersold R, Moritz RL (2016) Human SRMAtlas: a resource of targeted assays to quantify the complete human proteome. Cell 166(3):766–778. https://doi.org/10.1016/j.cell.2016.06.041
24. Zauber H, Kirchner M, Selbach M (2018) Picky: a simple online PRM and SRM method designer for targeted proteomics. Nat Methods 15(3):156–157. https://doi.org/10.1038/nmeth.4607
25. Brusniak MY, Kwok ST, Christiansen M, Campbell D, Reiter L, Picotti P, Kusebauch U, Ramos H, Deutsch EW, Chen J, Moritz RL, Aebersold R (2011) ATAQS: a computational software tool for high throughput transition optimization and validation for selected reaction monitoring mass spectrometry. BMC Bioinformatics 12:78. https://doi.org/10.1186/1471-2105-12-78
26. Nahnsen S, Kohlbacher O (2012) In silico design of targeted SRM-based experiments. BMC Bioinformatics 13(Suppl 16):S8. https://doi.org/10.1186/1471-2105-13-S16-S8
27. Reiter L, Rinner O, Picotti P, Huttenhain R, Beck M, Brusniak MY, Hengartner MO, Aebersold R (2011) mProphet: automated data processing and statistical validation for large-scale SRM experiments. Nat Methods 8(5):430–435. https://doi.org/10.1038/nmeth.1584
28. MacLean B, Tomazela DM, Shulman N, Chambers M, Finney GL, Frewen B, Kern R, Tabb DL, Liebler DC, MacCoss MJ (2010) Skyline: an open source document editor for creating and analyzing targeted proteomics experiments. Bioinformatics 26(7):966–968. https://doi.org/10.1093/bioinformatics/btq054
29. Gomez AM, Traunmuller L, Scheiffele P (2021) Neurexins: molecular codes for shaping neuronal synapses. Nat Rev Neurosci 22(3):137–151. https://doi.org/10.1038/s41583-020-00415-7
30. Shi T, Song E, Nie S, Rodland KD, Liu T, Qian WJ, Smith RD (2016) Advances in targeted proteomics and applications to biomedical research. Proteomics 16(15–16):2160–2182. https://doi.org/10.1002/pmic.201500449
31. Kusebauch U, Deutsch EW, Campbell DS, Sun Z, Farrah T, Moritz RL (2014) Using PeptideAtlas, SRMAtlas, and PASSEL: comprehensive resources for discovery and targeted proteomics. Curr Protoc Bioinformatics 46:13.25.11–13.25.28. https://doi.org/10.1002/0471250953.bi1325s46
32. Whiteaker JR, Halusa GN, Hoofnagle AN, Sharma V, MacLean B, Yan P, Wrobel JA, Kennedy J, Mani DR, Zimmerman LJ, Meyer MR, Mesri M, Rodriguez H, Clinical Proteomic Tumor Analysis C, Paulovich AG (2014) CPTAC assay portal: a repository of targeted proteomic assays. Nat Methods 11 (7):703–704. https://doi.org/10.1038/nmeth.3002
33. Sharma V, Eckels J, Schilling B, Ludwig C, Jaffe JD, MacCoss MJ, MacLean B (2018) Panorama public: a public repository for quantitative data sets processed in skyline. Mol Cell Proteomics 17(6):1239–1244. https://doi.org/10.1074/mcp.RA117.000543

34. Bhowmick P, Mohammed Y, Borchers CH (2018) MRMAssayDB: an integrated resource for validated targeted proteomics assays. Bioinformatics 34(20):3566–3571. https://doi.org/10.1093/bioinformatics/bty385
35. Bhowmick P, Roome S, Borchers CH, Goodlett DR, Mohammed Y (2021) An update on MRMAssayDB: a comprehensive resource for targeted proteomics assays in the community. J Proteome Res 20(4):2105–2115. https://doi.org/10.1021/acs.jproteome.0c00961
36. Schmidt T, Samaras P, Frejno M, Gessulat S, Barnert M, Kienegger H, Krcmar H, Schlegl J, Ehrlich HC, Aiche S, Kuster B, Wilhelm M (2018) ProteomicsDB. Nucleic Acids Res 46 (D1):D1271–D1281. https://doi.org/10.1093/nar/gkx1029
37. Gessulat S, Schmidt T, Zolg DP, Samaras P, Schnatbaum K, Zerweck J, Knaute T, Rechenberger J, Delanghe B, Huhmer A, Reimer U, Ehrlich HC, Aiche S, Kuster B, Wilhelm M (2019) Prosit: proteome-wide prediction of peptide tandem mass spectra by deep learning. Nat Methods 16(6):509–518. https://doi.org/10.1038/s41592-019-0426-7
38. Desiere F, Deutsch EW, King NL, Nesvizhskii AI, Mallick P, Eng J, Chen S, Eddes J, Loevenich SN, Aebersold R (2006) The PeptideAtlas project. Nucleic Acids Res 34(Database issue): D655–D658. https://doi.org/10.1093/nar/gkj040
39. Cox J, Mann M (2008) MaxQuant enables high peptide identification rates, individualized p.p.b.-range mass accuracies and proteome-wide protein quantification. Nat Biotechnol 26(12):1367–1372. https://doi.org/10.1038/nbt.1511
40. Addona TA, Abbatiello SE, Schilling B, Skates SJ, Mani DR, Bunk DM, Spiegelman CH, Zimmerman LJ, Ham AJ, Keshishian H, Hall SC, Allen S, Blackman RK, Borchers CH, Buck C, Cardasis HL, Cusack MP, Dodder NG, Gibson BW, Held JM, Hiltke T, Jackson A, Johansen EB, Kinsinger CR, Li J, Mesri M, Neubert TA, Niles RK, Pulsipher TC, Ransohoff D, Rodriguez H, Rudnick PA, Smith D, Tabb DL, Tegeler TJ, Variyath AM, Vega-Montoto LJ, Wahlander A, Waldemarson S, Wang M, Whiteaker JR, Zhao L, Anderson NL, Fisher SJ, Liebler DC, Paulovich AG, Regnier FE, Tempst P, Carr SA (2009) Multi-site assessment of the precision and reproducibility of multiple reaction monitoring-based measurements of proteins in plasma. Nat Biotechnol 27(7):633–641. https://doi.org/10.1038/nbt.1546
41. Peterson AC, Russell JD, Bailey DJ, Westphall MS, Coon JJ (2012) Parallel reaction monitoring for high resolution and high mass accuracy quantitative, targeted proteomics. Mol Cell Proteomics 11(11):1475–1488. https://doi.org/10.1074/mcp.O112.020131
42. Gallien S, Duriez E, Crone C, Kellmann M, Moehring T, Domon B (2012) Targeted proteomic quantification on quadrupole-orbitrap mass spectrometer. Mol Cell Proteomics 11(12):1709–1723. https://doi.org/10.1074/mcp.O112.019802
43. Schnatbaum K, Zerweck J, Nehmer J, Wenschuh H, Schutkowski M, Reimer U (2011) SpikeTides™—proteotypic peptides for large-scale MS-based proteomics. Nat Methods 8(3):i–ii. https://doi.org/10.1038/nmeth.f.337
44. Glatter T, Ludwig C, Ahrne E, Aebersold R, Heck AJ, Schmidt A (2012) Large-scale quantitative assessment of different in-solution protein digestion protocols reveals superior cleavage efficiency of tandem Lys-C/trypsin proteolysis over trypsin digestion. J Proteome Res 11(11):5145–5156. https://doi.org/10.1021/pr300273g

Chapter 15

Probing the Interactions of Splicing Regulatory Small Molecules and Proteins with U1 snRNP Using NMR Spectroscopy

Sébastien Campagne, Tebbe de Vries, and Frédéric H. -T. Allain

Abstract

Alternative RNA splicing is an essential part of gene expression that not only increases the protein diversity of metazoan but also provides an additional layer of gene expression regulation. The U1 small ribonucleoparticle (U1 snRNP) plays an essential role in seeding spliceosome assembly and its binding on weak 5′-splice sites is regulated by transient interactions with splicing factors. Recent progress in allele specific splicing correction has shown the therapeutic potential offered by small molecule splicing modifiers that specifically promotes the recruitment of U1 snRNP to modulate alternative splicing and gene expression. Here, we described a method to reconstitute U1 snRNP in vitro and to study labile interactions with protein or synthetic splicing factors using solution state NMR spectroscopy. This approach allowed us to validate direct interactions between splicing regulators and U1 snRNP and could also be useful for the screening of small molecules acting on splicing regulation.

Key words Splicing regulation, U1 snRNP, Splicing factors, Small molecule splicing modifier, NMR spectroscopy

1 Introduction

Regulation of mRNA splicing mainly occurs at the initial stages of spliceosome assembly, when the U1 small nuclear ribonucleoprotein (U1 snRNP) recognizes the 5′-splice site (5′-SS) or when the U2 small nuclear ribonucleoprotein (U2 snRNP) and its associated factors (U2AFs) bind to the branch point and the 3′-splice site, respectively [1]. The 5′-SS is one of the key *cis* RNA elements of the intron since it directly base pairs with the 5′-end of the U1 snRNA (the RNA component of U1 snRNP) to seed spliceosome assembly [2]. The optimal sequence of the 5′-SS forms a perfect RNA duplex with the 5′-end of U1 snRNA (strong 5′-SS). However, when the 5′-SS sequence diverges from this consensus or when its accessibility is reduced by the formation of secondary structures, splicing

Peter Scheiffele and Oriane Mauger (eds.), *Alternative Splicing: Methods and Protocols*, Methods in Molecular Biology, vol. 2537, https://doi.org/10.1007/978-1-0716-2521-7_15,

regulation is prone to occur [3]. In these cases, splicing factors bind specifically to the surroundings of the splice site and modulate the recruitment of U1 snRNP. The communication between the splicing factors and the splicing machinery is based on labile protein-protein and protein-RNA interactions that are difficult to study at atomic resolution [4–9]. Other and we have recently shown that the recruitment of U1 snRNP on weak 5'-SS can be modulated by small molecule splicing modifiers that either destabilize inhibitory secondary structures or that bind to specific tertiary structures formed by the U1 snRNA and the 5′-SS intermolecular RNA helix [10–15]. The small molecules can be viewed as minimal splicing factors since they provide an additional link between the U1 snRNA and the pre-mRNA. These seminal findings opened a quest for the identification of novel splicing modifiers acting on different weak 5′-SS, which could have therapeutic applications.

U1 snRNP is composed of 10 proteins and the 164-nucleotides long U1 snRNA which adopts a trefoil fold containing three stem loops [16–18]. Among the ten proteins, the seven Sm proteins (B/B′, D_1, D_2, D_3, E, F, and G) are common to all UsnRNPs and form an heptameric ring (Sm core) around a short single stranded RNA region of the U1 snRNA called the Sm site (Fig. 1a, b). In the cell, Sm core assembly is chaperoned by the SMN complex and several associated factors in the cytoplasm, however, it was previously shown that the Sm core can be assembled spontaneously in vitro [19, 20]. Furthermore, U1 snRNP contains three U1-specific proteins (U1-70K, U1-A, and U1-C) that assemble on the preparticles (U1 snRNA—Sm core) when translocated back to the nucleus. The RNA recognition motif (RRM) domain of U1-70K interacts specifically with the U1 snRNA stem loop 1 and its N-terminal extension wraps around the Sm site to contact U1-C with its N-terminal extremity. U1-A interacts with U1 snRNA stem loop 2 via its first RRM domain. Finally, the U1-C protein is anchored to the Sm core through its interaction with Sm D_3 and plays a crucial role during 5′-SS recognition since its N-terminal zinc finger binds the minor groove of the intermolecular RNA helix formed between the 5′-SS and the 5′-end of U1 snRNA (Fig. 1c). In order to get insights into the molecular mechanisms of alternative splicing, we reconstitute U1 snRNP in vitro in large amounts to probe interactions with synthetic or natural splicing factors by NMR spectroscopy. We build our approach upon previous methods [19, 20] to reconstitute a particle that contains mostly full length proteins, except for U1-70K and Sm B/B′ that lack the RS domain and the last 50 amino acids, respectively (Fig. 1d).

In this protocol, the protein components of U1 snRNP are produced in bacteria and purified by various chromatographic approaches. The U1 snRNA is transcribed in vitro using T7 RNA polymerase. The particle is assembled essentially following the same

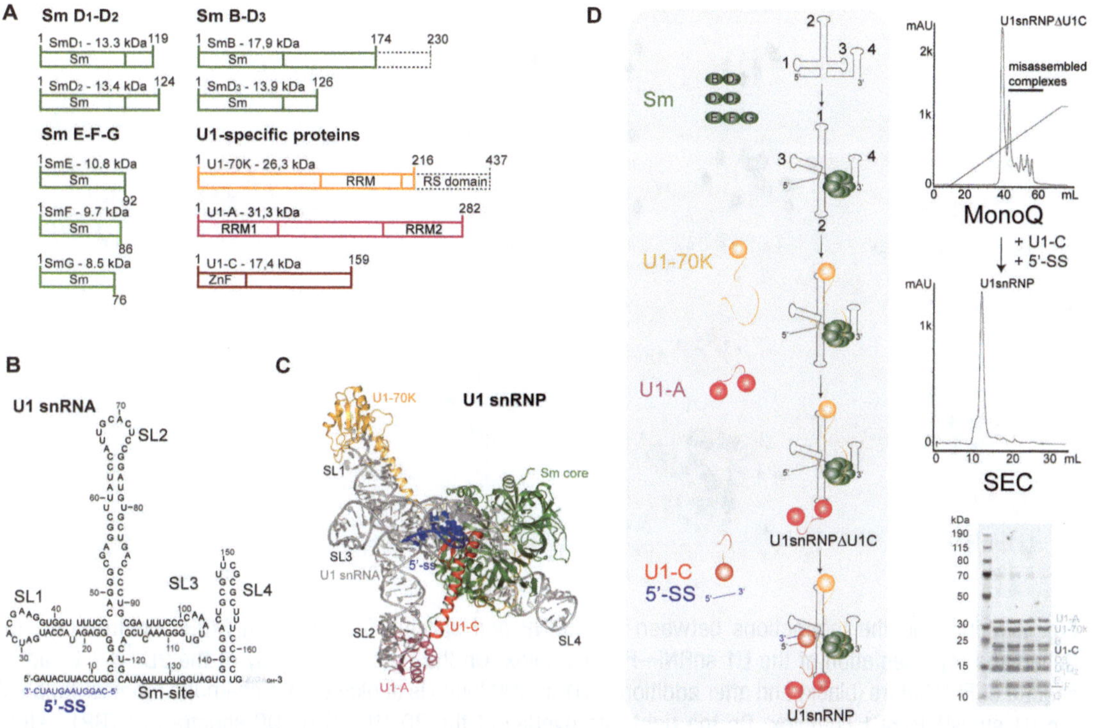

Fig. 1 In vitro reconstitution of U1 snRNP. (**a**) Schematic representation of the U1 snRNP protein components. (**b**) Schematic representation of the U1 snRNA and the 5′-splice site. (**c**) Structure of U1 snRNP [16–18]. (**d**) On the left, the assembly of U1 snRNP is schematically illustrated. On the right, the chromatograms obtained for the anion exchange and size exclusion purification steps are shown. At the bottom right, a SDS-PAGE gel shows the composition of the in vitro reconstituted U1 snRNP from several fractions of the size exclusion chromatography

order as in the cell, however, in the absence of chaperones [21, 22]. Firstly, a U1 snRNP lacking U1-C is assembled and purified using anion exchange chromatography. The purified pre-particles are then combined with U1-C and the 5′-SS and further purified using size exclusion chromatography (Fig. 1d). The reconstitution approach has been validated by gel electrophoresis, crosslinking of segmentally isotope-labeled RNA and MS/MS or CLIR-MS [23], Small Angle Neutron Scattering and cryogenic electron microscopy [24].

The in vitro reconstituted U1 snRNP has a molecular weight of 220 kDa and tumbles slowly in solution. In order to monitor the labile interactions with splicing regulators, we developed an NMR spectroscopy based approach in which the NMR resonances of the ligand (splicing regulators) are followed upon addition of the particle. By comparing the NMR fingerprints of the ligand before and after addition of the reconstituted U1 snRNP, the presence of chemical shift changes informs about a direct interaction. In the

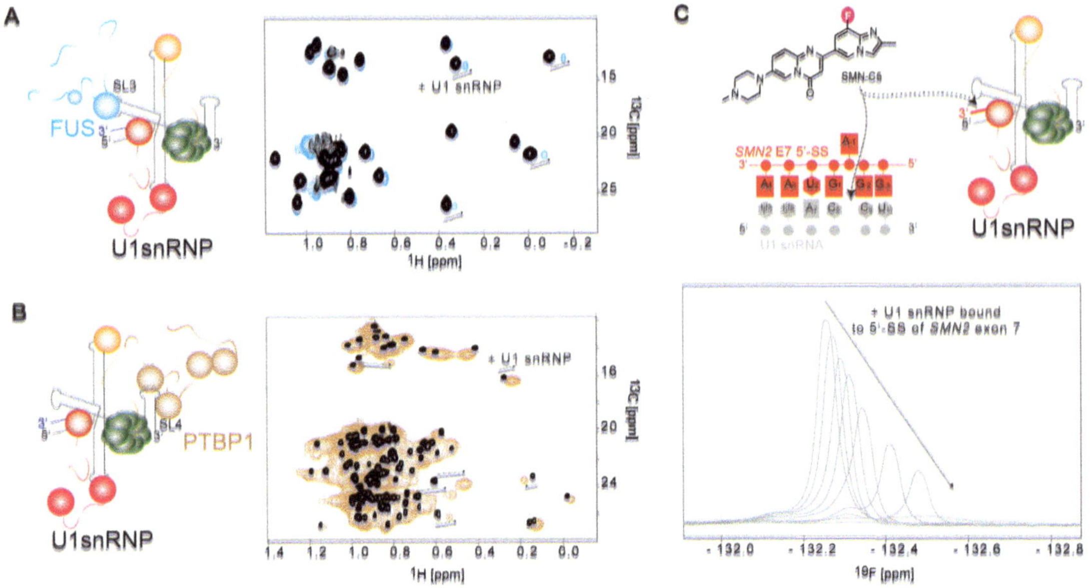

Fig. 2 Monitoring the interactions between U1 snRNP and splicing factors using NMR spectroscopy. (**a**) Schematic representation of the U1 snRNP–FUS complex. On the right, the overlay of the 2D ^{1}H-^{13}C HMQC spectra of FUS before (black) and after addition of U1 snRNP (blue) is depicted. (**b**) Schematic representation of the U1 snRNP–PTBP1 complex. On the right, the overlay of the 2D ^{1}H-^{13}C HMQC spectra of PTBP1 before (black) and after addition of U1 snRNP (marron) is shown. (**c**) The structure of the SMN2 splicing modifier SMN-C5 and the binding pocket for SMN-C5 are schematically illustrated. At the bottom, an overlay of the 1D ^{19}F spectra of SMN-C5 observed upon addition of U1 snRNP bound to the 5′-SS of *SMN2* exon 7 is shown

case of the *SMN2* splicing modifiers [12], such a binding reaction was monitored by ^{19}F NMR spectroscopy, since most studied splicing modifiers contained at least one fluorine nucleus (Fig. 2). Using this approach, we decipher the specificity of the small molecules for the U1 snRNP bound to a specific 5′-SS. This approach is very powerful as the method is highly sensitive and the NMR spectra are straightforward to analyze. In the case of protein splicing factors, the observation of the ligand requires specific ^{13}C-labeling on the methyl groups of side chains in a deuterated background (i.e., Isoleucine, Leucine, and Valine [ILV]). The titration is monitored by following the chemical shifts of the methyl groups in 2D ^{1}H-^{13}C HMQC spectra (Fig. 1). Using this approach, we successfully detected the direct interactions between U1 snRNP and protein splicing factors (Fig. 2a, b), including FUS and PTBP1 [5, 24]. Altogether, our approach allows the identification of direct interactions between U1 snRNP and synthetic or natural splicing factors that regulate splicing.

2 Materials

All solutions are prepared using reagents of the highest available purity and with ultrapure de-ionized water. Particular care should always be taken to avoid RNase contamination during preparation of reagents. Plasmids for the expression of protein and RNA constructs can be obtained by the authors upon request.

2.1 Expression and Purification of the U1 snRNP Protein Components

1. *Escherichia coli* BL21 DE3 competent cells.
2. pET26bII polycistronic genes SmB-D_3 (6xHis on Sm D_3), Sm D_1–D_2, and Sm E-F-G (6xHis on Sm G).
3. pET26bII U1-A (1–282), U1-A RRM1 (1–127) (6xHis, Thrombin).
4. pET28a U1-C (1–159) and U1-70K (1–216) (C-terminal 6xHis).
5. Lysogeny broth (LB) medium and common antibiotics (Kanamycin and Carbenicillin).
6. Isopropyl-thio-β-D-galactopyranoside (IPTG).
7. Shaking incubator, sterile 1 L and 5 L Erlenmeyer flasks.
8. Centrifuge with temperature control that allows cooling to 4 °C, 1 L centrifugation bottles to pellet bacterial culture (3500 × *g*) and 50 mL bottles to clarify cell lysates (30,000 × *g*).
9. Microfluidizer (e.g., Microfluidics) or French press.
10. 5 mL HisTrap column (GE Healthcare).
11. 5 mL HiTrap SP column (GE Healthcare).
12. Centrifugal filters (5 or 10 kDa cutoff, e.g., from Amicon).
13. Size exclusion S75 (GE Healthcare) or equivalent.
14. Äkta prime at 4 °C and Äkta pure or purifier at room temperature.
15. Dialysis membrane (3.5 kDa cutoff).
16. 80 units of thrombin per digestion (Sigma).
17. TEV protease [25].
18. DNAse I and lysozyme.
19. EDTA-free protease inhibitors (Roche).
20. Buffer A: 20 mM Hepes, pH 7.8, 1 M NaCl, 2 M urea, 2.8 mM β-mercaptoethanol.
21. Buffer B: 20 mM Hepes, pH 7.8, 1 M NaCl, 500 mM imidazole, 2.8 mM β-mercaptoethanol.
22. Buffer C: 10 mM sodium phosphate, pH 6, 5 mM DTT.

23. Buffer D: 10 mM sodium phosphate, pH 6, 2 M NaCl, 5 mM DTT.
24. Buffer E: 10 mM sodium phosphate, pH 6, 5 mM DTT.
25. Buffer F: 20 mM Hepes, pH 7.4, 0.5 M NaCl, 0.5 M urea,, 10 mM DTT, 1 mM EDTA.
26. Buffer G: 20 mM Hepes, pH 7.4, 2 M NaCl, 10 mM DTT, 1 mM EDTA.
27. Buffer H: 10 mM sodium phosphate buffer, pH 6.8, 50 mM NaCl, 5 mM DTT.
28. Buffer I: 20 mM Hepes, pH 7.5, 0.5 M NaCl, 0.5 M urea, 10 mM imidazole, 2.8 mM β-mercaptoethanol.
29. Buffer J: 20 mM Hepes, pH 7.5, 0.5 M NaCl, 500 mM imidazole, 2.8 mM β-mercaptoethanol.
30. Buffer K: 20 mM Hepes, pH 7.5, 0.5 M NaCl, 5 mM DTT.
31. Buffer L: 20 mM Hepes, pH 7.5, 1 M NaCl, 0.5 M urea, 10 mM imidazole, 2.8 mM β-mercaptoethanol.
32. Buffer M: 20 mM Hepes, pH 7.5, 0.5 M NaCl, 300 mM imidazole, 2.8 mM β-mercaptoethanol.
33. Buffer N: 20 mM Hepes, pH 7.5, 0.5 M NaCl, 0.5 M urea.
34. Buffer O: 20 mM Hepes, pH 7.5, 0.5 M NaCl, 300 mM imidazole.
35. Buffer P: 20 mM Hepes, pH 7.5, 0.1 M NaCl.
36. Buffer Q: 20 mM Hepes, pH 7.5, 2 M NaCl.
37. Buffer R: 10 mM sodium phosphate buffer, pH 6.8, 50 mM NaCl, 0.5 mM EDTA, 5 mM DTT.
38. Buffer S: 10 mM Hepes, pH 7.8, 0.1 M NaCl, 0.5 M urea, 2.8 mM β-mercaptoethanol.
39. Buffer T: 10 mM Hepes, pH 7.8, 0.1 M NaCl, 0.5 M urea, 250 mM imidazole, 2.8 mM β-mercaptoethanol.
40. Buffer U: 10 mM sodium phosphate buffer, pH 6.8, 50 mM NaCl, 5 mM DTT.
41. Buffer Y: 20 mM Hepes, pH 7.5.

2.2 Production of the U1 snRNA and Variants Using In Vitro Transcription

1. *Escherichia coli* TOP10 or DH5α competent cells.
2. Luria–Bertani medium and carbenicillin.
3. pUC19-U1 snRNA (contains a T7 promoter followed by the hammerhead ribozyme in fusion with the sequence coding for U1 snRNA and a *SalI* cleavage site).
4. Plasmid preparation kit.
5. *SalI* HF restriction enzyme.
6. In house produced T7 RNA polymerase. We are producing the enzyme according to the previously described protocol [26].

7. 20 mM NTP solutions and 80 mM GMP solution.
8. 20× Transcription buffer: 800 mM Tris–HCl, pH 8.1, 20 mM spermidine, 0.2% Triton X-100, 100 mM DTT.
9. Pyrophosphatase (inorganic phosphatase dissolved in 20 mM Tris–HCl, pH 8, 100 mM KCl, 50% glycerol, 0.1 mM EDTA, 1 mM DTT at a concentration of 100 U/mL).
10. 0.5 M EDTA previously autoclaved and sterile water.
11. Water bath set at 37 °C.
12. 0.22 μm filters for syringe.
13. High-Performance Liquid Chromatography (HPLC) with a preparative anion-exchange chromatography column Dionex DNAPac PA-100 column (22 × 250 mm). The system should be coupled to water bath in order to perform chromatography runs at 85 °C.
14. Eluent A: 12.5 mM Tris–HCl, pH 8.0, 6 M urea.
15. Eluent B: 12.5 mM Tris–HCl, pH 8.0, 0.5 M $NaClO_4$, 6 M urea.
16. *n*-Butanol.
17. Small centrifuge for falcon tubes, chemical hood, heat block, liquid nitrogen and lyophilizer.

2.3 Assembly and Purification of U1 snRNP

1. Solution of purified Sm B-D_3: 100 μM stock in buffer E.
2. Solution of purified Sm D_1-D_2: 100 μM stock in buffer H.
3. Solution of purified Sm E-F-G: 100 μM stock in buffer K.
4. Solution of purified U1-70K: 10 μM stock in buffer M.
5. Solution of purified U1-A or U1-A RRM1: 100 μM stock in buffer R.
6. Solution of purified U1-C: 10 μM stock in buffer U.
7. Solution of purified U1 snRNA: 100 μM stock in buffer R.
8. Recombinant RNAse inhibitors (Invitrogen).
9. Deprotected 5′-splice site oligonucleotide (Dharmacon) dissolved in sterile H_2O (100 μM stock).
10. Sterile buffer V: 10 mM Hepes, pH 7.5, 250 mM KCl, 5 mM DTT.
11. 1 mL MonoQ column (Pharmacia).
12. Sterile buffer W: 10 mM Hepes, pH 7.5, 2 M KCl, 5 mM DTT.
13. Sterile buffer X: 10 mM Sodium phosphate, pH 6.8, 100 mM NaCl, 5 mM DTT.
14. Preparative size exclusion column Superdex 200 (GE Healthcare).

2.4 Methyl Labelling of Splicing Factors

1. Shaking incubator, sterile 500 mL and 5 L Erlenmeyer flasks.
2. *Escherichia coli* BL21 DE3 previously transformed with the plasmid allowing for the expression of the splicing factor.
3. Lysogeny broth agar plates complemented with kanamycin and heat block.
4. Trace elements 100×: 5 g/L EDTA, 6 g/L $CaCl_2 \cdot 2H_2O$, 440 mg/L $CuSO_4 \cdot 5H_2O$, 1.2 g/L $MnCl_2 \cdot 4H_2O$, 20 mg/L H_3BO_3, 700 mg/L $ZnSO_4 \cdot 7H_2O$, 6 g/L $FeSO_4 \cdot 7H_2O$, 200 mg/L L-ascorbic acid.
5. 100× MEM Vitamin Solution (Gibco).
6. M9 medium prepared with H_2O supplemented with 2 mM $MgSO_4$, 100 μM $CaCl_2$, 0.5 g/L EDTA, 2 g/L α-D-glucose, 1 g/L NH_4Cl, trace elements 1×, MEM vitamin mix 1×, and the required antibiotic.
7. M9 medium prepared with D_2O (99.9% Sigma) supplemented with 2 mM $MgSO_4$, 100 μM $CaCl_2$, 0.5 g/L EDTA, 2 g/L α-D-glucose, 1 g/L NH_4Cl, trace elements 1×, MEM vitamin mix 1× and the required antibiotic.
8. α-Ketobutyric acid (methyl-^{13}C, 99%; 3,3-D2, 98%, Cambridge Isotope Laboratory) and α-ketoisovaleric acid (^{13}C5, 98%; 3-D1, 98%, Cambridge Isotope Laboratory) for ^{13}C labelling of ILV methyl groups.

2.5 NMR Spectroscopy Analysis of U1 snRNP Ligand Binding

1. High field NMR spectrometer (≥700 MHz) equipped with a cryo-probe.
2. 5 mm NMR tubes and ultrapure D_2O.
3. 500 μL solution of the ligand (protein or small molecule) at 20 μM containing 10% D_2O in buffer X.
4. 50–100 μL of reconstituted U1 snRNP at 200 μM stored in buffer X.

2.6 Time Considerations

1. The preparation of the U1 snRNP protein components is relatively tedious and requires 2–3 weeks of work, including 13 chromatography steps for the preparation of the proteins.
2. The preparation of the U1 snRNA requires 3–4 days of wet-lab work.
3. The assembly and purification of the particles needs 3 additional days of work.
4. The NMR titration requires 3–5 days according to the quality of the NMR spectra. The formed complexes have a molecular weight ranging from 220–250 kDa and tumble slowly, in respect to the NMR timescale. Accordingly, high quality NMR spectra with good signal-to-noise ratio often require many scans.

5. In total, probing the interaction between a splicing factor and U1 snRNP using the method presented here takes about one month of work.

3 Methods

3.1 Expression and Purification of the Sm Proteins

Expressions of Sm B-D_3, Sm D_1-D_2 and Sm E-F-G are performed in *E. coli* BL21 DE3 at 20 °C, 25 °C and 37 °C during 12 h, 12 h and 24 h, respectively (*see* **Note 1**). Before induction by 1 mM IPTG, all cells are grown at 37 °C until an OD of 0.6–0.8 is reached. All the proteins are expressed in LB medium. Protein expression is stopped by pelleting the cells at 3,500 × *g* for 10 min. Cell pellets are solubilized in their respective buffers supplemented with 10 μg/mL DNase I, 10 μg/mL lysozyme, and EDTA-free protease inhibitors. Cell lysis is performed using a microfluidizer with 3–5 consecutive cycles at 15,000 psi. The cell lysates are clarified by centrifugation (30,000 × *g* at 4 °C for 45 min). At the end of the purifications, all Sm proteins and U1-A are concentrated to 100 μM, flash-frozen and stored at −80 °C. U1-C and U1-70K are stored at a concentration of 10 μM. Each purification is checked on a gel electrophoresis (*see* **Note 2**) and each protein solution is tested for the presence of RNAse activity (*see* **Note 3**).

3.1.1 Sm B-D_3

1. Lyse the cells of the Sm B-D_3 expression culture in 50 mL buffer A.
2. Load the supernatant on a 5 mL HisTrap column with a flow rate of 0.5 mL/min at 4 °C. Wash the column with 100 mL buffer A and 50 mL of 10% B (50 mL) and step-elute with 20 mL of 100% buffer B using the same flow rate.
3. Cleave the heterodimer with 80 units of thrombin and dialyze into 2 L of buffer C at 4 °C.
4. Perform a cation exchange by eluting with a gradient of buffer D.
5. Dialyze the protein into 2 L buffer E and perform a size exclusion chromatography in buffer E.
6. Adjust the concentration to 100 μM ($\varepsilon_{280nm} = 7700\ M^{-1}\ cm^{-1}$) with buffer E before storage.

3.1.2 Sm D_1-D_2

1. Lyse the cells of the Sm D_1-D_2 expression culture in 50 mL buffer F.
2. Load the supernatant on a 5 mL HiTrap SP column with a flow rate of 2 mL/min at 4 °C. Wash the column with 100 mL of buffer F and elute with 20 mL of buffer G.

3. Perform a size exclusion chromatography in buffer H.
4. Adjust the concentration to 100 μM with buffer H ($\varepsilon_{280\text{nm}} = 8500\ \text{M}^{-1}\ \text{cm}^{-1}$) before storage.

3.1.3 Sm E-F-G

1. Lyse the cells of the Sm E-F-G expression culture in 50 mL of buffer I.
2. Load the supernatant on a 5 mL HisTrap column with a flow rate of 2 mL/min at room temperature. Wash the column with 100 mL of buffer I and elute with 20 mL of buffer J.
3. Cleave the heterotrimer Sm E-F-G with thrombin and dialyze in 2 L of buffer K.
4. Perform a size exclusion chromatography in buffer K.
5. Adjust the concentration to 100 μM with buffer K ($\varepsilon_{280\text{nm}} = 24{,}410\ \text{M}^{-1}\ \text{cm}^{-1}$) before storage (*see* **Note 4**).

3.2 Expression and Purification of U1 Specific Proteins

Expression of U1-A, U1-C and U1-70K are performed in *E. coli* BL21 DE3 in LB medium over-night (14–16 h) at 37 °C, 25 °C and 20 °C, respectively. Before induction by 1 mM IPTG, all cells are grown at 37 °C until an OD of 0.6–0.8 is reached. 1 mM $ZnSO_4$ is added to the U1-C expression culture at the same time than IPTG.

3.2.1 U1-70K

1. Lyse the cells of the U1-70K expression culture in 50 mL of buffer L.
2. Load the supernatant on a 5 mL HisTrap column with a flow rate of 2 mL/min at room temperature. Wash the column with buffer 150 mL of buffer L and elute with 20 mL of buffer M.
3. Collect the fractions containing the protein, estimate the protein concentration ($\varepsilon_{280\text{nm}} = 24{,}410\ \text{M}^{-1}\ \text{cm}^{-1}$) and directly store them at −80 °C (concentration is about ~10 μM).

3.2.2 U1-A

1. Lyse the cells of the U1-A expression culture in 50 mL of buffer N.
2. Load the supernatant on a 5 mL HisTrap column with a flow rate of 2 mL/min at room temperature. Wash the column with 100 mL of buffer N and elute with 20 mL of buffer O.
3. Cleave-off the 6xHis-tag with TEV protease at room temperature and dilute five times with buffer Y to reduce the NaCl concentration to 100 mM.
4. Load the protein on a 5 mL HiTrap SP column previously equilibrated with buffer P with a flow rate of 2 mL/min at room temperature, wash with 50 mL of buffer P and elute with 30 mL of buffer Q.

5. Perform a size exclusion chromatography in buffer R.
6. Adjust the concentration to 100 μM with buffer R ($\varepsilon_{280nm} = 7450\ M^{-1}\ cm^{-1}$) before storage (*see* **Note 5**).

3.2.3 U1-C

1. Lyse the cells of the U1-C expression culture in 50 mL of buffer S.
2. Load the supernatant on a 5 mL HisTrap column with a flow rate of 2 mL/min at 4 °C. Wash the column with 100 mL of buffer S, 30 mL of 10% buffer T and elute with 20 mL of buffer T.
3. Perform a size exclusion chromatography in buffer U.
4. Estimate the concentration ($\varepsilon_{280nm} = 12{,}950\ M^{-1}\ cm^{-1}$) before storage. The protein was generally stored at 10 μM (*see* **Note 6**).

3.3 Preparation of the U1 snRNA

1. Perform a large-scale plasmid preparation *of E. coli* TOP10 cells containing the pUC19-U1 plasmid. 700 ng of plasmid are required to perform a 10 mL transcription reaction.
2. Linearize the plasmid with *SalI* overnight at 37 °C.
3. In vitro–transcribe the U1 snRNA using T7 RNA polymerase. A typical reaction contains: 30 mM $MgCl_2$, 6 mM of each NTP, 4 mM GMP, 35 ng/μL linearized plasmid, and 1.7 μM T7 RNA Polymerase in a transcription buffer containing 40 mM Tris–HCl, pH 8.0, 1 mM spermidine, 0.01% Triton X-100 and 5 mM DTT. 1 U/mL pyrophosphatase is added to reduce magnesium phosphate accumulation. After 4 h, the reaction is stopped by adding 100 mM EDTA pH 8.0. The mixture is then centrifuged (4000 × *g*, 10 min) and filtered with 0.22 μm filter.
4. Purify the transcription mixture with a high-performance liquid chromatography system that allows RNA separation on an anion exchange column at 85 °C and in the presence of 6 M urea. Typically, we are using a gradient of 30–75% eluent B within 18 min or 0–65% B gradient within 26 min, 260 nm.
5. Perform a butanol extraction [27] of fractions containing U1 snRNA to remove salts and urea.
6. Verify the purity of the U1 snRNA on a 12% acrylamide–urea gel stained using Toluidine.
7. Denature and refold the U1 snRNA, lyophilize and store at −20 °C (*see* **Note 7**).

3.4 Reconstitution and Purification of U1 snRNP for NMR Spectroscopy

1. The U1 snRNP in vitro reconstitution is usually performed in 20 nmoles scale (*see* **Note 8**). In a falcon tube, add 2 mL of buffer V, 30 nmoles of each Sm protein oligomers (300 μL of Sm B-D_3, 300 μL of Sm D_1-D_2, and 300 μL of Sm E-F-G, all stocks being at 100 μM), 5 μL of RNase inhibitors, and

20 nmoles of the U1 snRNA (200 μL at 100 μM). Incubate the mixture at 30 °C for 30 min and 15 min at 37 °C. After that, incubate the mixture on ice for 5 min and add 20 nmoles of U1-70K (2 mL at 10 μM). After 15 min on ice, add 20 nmoles of U1-A (200 μL at 100 μM), dilute the mixture five-fold with buffer V and leave the mixture on ice overnight.

2. The next day, purify U1 snRNPΔU1-C using anion exchange chromatography at 4 °C or using ice cold buffers if the purification system is not in a cold room. Equilibrate the MonoQ column with buffer V and load the mixture on the column (*see* **Note 9**). Elute the particle from the column using a linear gradient of buffer W. The particle elutes at 24% of buffer W while the misassembled complexes and the free U1 snRNA will elute at 28–35% and 37% of buffer W.
3. Dialyze U1 snRNPΔU1-C against buffer X (*see* **Note 10**), concentrate the preparticle and estimate its concentration based on RNA absorbance at 260 nm (ε = 2,200,000 M^{-1} cm^{-1}). According to the determined concentration of U1 snRNPΔU1-C, add 1.5 molar equivalent of U1-C, 1.2 molar equivalent of the 5′-splice site oligonucleotide and 5 μL of RNAse inhibitors.
4. Load the sample on the preparative size exclusion column (Superdex S200) previously equilibrated with buffer X. Perform an isocratic elution. Pool the fractions containing the particle of interest and concentrate to 50 μL using centrifugal filters at 12 °C. Measure the concentration using OD_{260nm} and check the particle composition on a SDS-PAGE stained first with Coomassie Blue for the observation of the proteins and then with Toluidine to observe the RNA (*see* **Note 11**).
5. The reconstitution/purification yields range from 20 to 60% with respect to the starting material. Accordingly, in order to obtain 20 nmoles of in vitro reconstituted U1 snRNP, the protocol has to be performed at least two times.

3.5 Preparation of the U1 snRNP Ligand

3.5.1 Preparation of ILV Labelled Splicing Factors

Here, we present a generalized protocol for the preparation of protein splicing factors with ^{13}C methyl labelling on ILV. The protein is produced in a deuterated background. Since the protocol for the purification of the splicing factor is specific to each splicing factor, it is not described in this protocol.

1. On day 1, transform the plasmid allowing the expression of the splicing factor under the control of a T7 promoter in *E. coli* BL21 DE3 and select for transformants using LB-Agar plates complemented with the appropriate antibiotic (*see* **Note 12**).
2. On day 2, start a 20 mL preculture in M9-H_2O medium in the morning. In the afternoon, transfer 1 mL of the M9-H_2O in 20 mL of fresh M9-D_2O medium and leave the cells grow until day 3 (*see* **Note 13**).

3. On day 3, centrifuge gently the preculture (1000 × *g*, 10 min) and transfer the cell in fresh M9-D_2O medium (0.5–2 L culture). Leave the cell grow until OD_{600nm} is around 0.4 and add the precursors for methyl labelling: 100 mg/L of alpha-ketobutyric acid and 60 mg/L of alpha-ketoisovaleric acid. Adjust the temperature of the shaker for optimal expression of the protein of interest (*see* **Note 14**).
4. After 1 h, induce the expression of the splicing factor by adding the appropriate amount of IPTG to the culture. At the end of the overexpression, pellet your cells by centrifugation (3500 × *g*, 12 min).
5. Purify the splicing factor under RNase-free conditions, prepare a 500 μL NMR sample of the protein at 20 μM in buffer X 10% D_2O and place the sample into a 5 mm NMR tube.

3.5.2 Preparation of Small Molecule Ligands

1. Resuspend the small molecule in Buffer X at a concentration of 20 μM. Prepare a 500 μL NMR sample containing 10% D_2O. The sample should appear clear and no traces of precipitation should visible. If the molecule aggregates, addition of up to 2% ^2H DMSO are tolerated by U1 snRNP.
2. Record a 1D ^1H and 1D ^{19}F NMR spectrum to confirm the narrow linewidths of the small molecule.

3.6 NMR Titrations

1. Record the reference NMR spectra of the free splicing factor (1D ^1H and 2D ^{13}C-^1H HMQC) or small molecule (1D ^1H and 1D ^{19}F) at 37 °C (concentration 20 μM in buffer X 10% D_2O).
2. Add in a stepwise manner the U1 snRNP solution to the splicing factor or small molecule sample and record another set of the same NMR experiments after each addition.
3. The observation of significant changes in the positions of the NMR signals of the ligand reveal a direct interaction between U1 snRNP and the splicing factor or the small molecule.
4. To quantify the binding affinity using NMR spectroscopy, plot the chemical shift perturbations against the concentration of U1 snRNP and model the curves with previously described equations [28].

4 Notes

1. The typical yield for the protein purification of Sm B-D_3, Sm D_1-D_2, and Sm E-F-G are 8, 10, and 18 mg/L of culture. To produce large stocks of Sm proteins, we recommend to use 4 L of culture for Sm B-D_3 and Sm D_1-D_2 while 2 L of culture for Sm E-F-G should be sufficient. In the case of U1-specific

proteins, the expected yield are 2, 3, and 20 mg/L of culture for U1-70K, U1-C, and U1-A, respectively. To produce large stocks of U1-specific proteins, we recommend to use 4 L of culture for U1-70K and U1-C, while 2 L of culture is enough for the production of U1-A.

2. We recommend taking aliquots before and after induction and at various points during the protein purification to check for the presence and purity of the protein by SDS PAGE (12–15% acrylamide). Similarly, denaturing urea PAGE (12% acrylamide) is recommended to analyze RNA fractions of anion-exchange chromatography to select the fractions that contain pure U1 snRNA.
3. Each purified protein is checked for the presence of RNase activity using the RNaseAlert Lab Test kit (Ambion) that detects RNase activity conveniently and fast.
4. Sm E-F-G are three proteins with very similar molecular weight. In order to resolve the three bands on a gel electrophoresis, we recommend to use the precasted NuPAGE 4–12% Bis Tris gels from Invitrogen and to run those with the MES-based running buffer. Similar gels are used to check the protein composition of the reconstituted U1 snRNP.
5. Minimal constructs of U1-A containing only the RRM1 can be used for the reconstitution of U1 snRNP. The protein is purified in the same manner than the full-length protein.
6. U1-C is a difficult protein to purify and is prone to aggregation and degradation. We observed that performing both purification steps consecutively (the same day) increase the yield and reduce protein degradation.
7. All the U1 snRNAs produced in this study lack the pseudo uridines at the 5′-end and the modified cap. The modifications could be added to U1 snRNA by segmental labelling [29]. We also produced a U1 snRNA derivative that contains an optimal 5′-splice site fused to the U1 snRNA sequence. This duplex is stabilized by an apical GNRA type stem loop.
8. The reconstitution was also performed at higher scale but in this case, we recommend to split the sample at the stage of the MonoQ column to avoid overloading. The same applies for the final step of size exclusion chromatography.
9. For the initial step of purification by anion exchange, a large volume has to be loaded on the column. For this purpose, the use of automatic sample pump loading system is strongly recommended.
10. To skip the dialysis, it is also possible to reduce the salt concentration of the sample by successive cycles of concentration/dilution using appropriate centrifugal filters.

11. The observation of all the U1 snRNP components is challenging due to the overlap of the different proteins. To cope with this, we recommend to run precasted gradient (4–12% acrylamide) SDS-PAGE gels in the appropriate running buffer to achieve maximal resolution for small proteins.
12. We recommend to start at the transformation step for every deuterated cultures. Indeed, metabolism adaptation to high concentrations of D_2O is a critical step for the expression of deuterated proteins.
13. We recommend to keep the M9-H_2O preculture during day 2 in case bacteria do not grow in the M9-D_2O preculture.
14. In order to obtain specific labeling in a deuterated background and to be able to record methyl-TROSY experiments without carbon decoupling, the ^1H from position 3 of the precursors can be readily and quantitatively exchanged to ^2H by incubation at high pH in D_2O before use (at pH 12.5, ~2–3 h for α-ketoisovalerate, and at pH 10.5, ~12–14 h for α-ketobutyrate) [30]. Alternatively, in case the precursors are still protonated on position 3, enhanced sensitivity 2D ^1H-^{13}C HMQC experiments can be recorded using the recently described XL-ALSOFAST pulse program [31].

References

1. Black DL (2003) Mechanisms of alternative pre-messenger RNA splicing. Annu Rev Biochem 72:291–336
2. Mount SM, Pettersson I, Hinterberger M et al (1983) The U1 small nuclear RNA-protein complex selectively binds a 5′ splice site in vitro. Cell 33:509–518
3. Roca X, Krainer AR, Eperon IC (2013) Pick one, but be quick: 5′ splice sites and the problems of too many choices. Genes Dev 27: 129–144
4. Cho S, Hoang A, Sinha R et al (2011) Interaction between the RNA binding domains of Ser-Arg splicing factor 1 and U1-70K snRNP protein determines early spliceosome assembly. Proc Natl Acad Sci U S A 108:8233–8238
5. Jutzi D, Campagne S, Schmidt R et al (2020) Aberrant interaction of FUS with the U1 snRNA provides a molecular mechanism of FUS induced amyotrophic lateral sclerosis. Nat Commun 11:6341
6. Mourao A, Bonnal S, Soni K et al (2016) Structural basis for the recognition of spliceosomal SmN/B/B' proteins by the RBM5 OCRE domain in splicing regulation. Elife 5:e14707
7. Sharma S, Maris C, Allain FH et al (2011) U1 snRNA directly interacts with polypyrimidine tract-binding protein during splicing repression. Mol Cell 41:579–588
8. Staknis D, Reed R (1994) SR proteins promote the first specific recognition of pre-mRNA and are present together with the U1 small nuclear ribonucleoprotein particle in a general splicing enhancer complex. Mol Cell Biol 14:7670–7682
9. Subramania S, Gagné L, Campagne S et al (2019) SAM68 interaction with U1A modulates U1 snRNP recruitment and regulates mTor pre-mRNA splicing. Nucleic Acids Res 47:4181–4197
10. Naryshkin NA, Weetall M, Dakka A et al (2014) *SMN2* splicing modifiers improve motor function and longevity in mice with spinal muscular atrophy. Science 345:688–693
11. Sivaramakrishnan M, McCarthy KD, Campagne S et al (2017) Binding to *SMN2* pre-mRNA-protein complex elicits specificity for small molecule splicing modifiers. Nat Commun 8:1476

12. Campagne S, Boigner S, Rudisser S et al (2020) Structural basis of a small molecule targeting RNA for a specific splicing correction. Nat Chem Biol 15:1191–1198
13. Palacino J, Swalley SE, Song C et al (2015) SMN2 splice modulators enhance U1-pre--mRNA association and rescue SMA mice. Nat Chem Biol 11:511–517
14. Garcia-Lopez A, Tessaro F, Jonker HRA et al (2018) Targeting RNA structure in SMN2 reverses spinal muscular atrophy molecular phenotypes. Nat Commun 9:2032
15. Chen JL, Zhang P, Abe M et al (2020) Design, optimization, and study of small molecules that target tau pre-mRNA and affect splicing. J Am Chem Soc 142:8706–8727
16. Stark H, Dube P, Luhrmann R et al (2001) Arrangement of RNA and proteins in the spliceosomal U1 small nuclear ribonucleoprotein particle. Nature 409:539–542
17. Pomeranz Krummel DA, Oubridge C, Leung AK et al (2009) Crystal structure of human spliceosomal U1 snRNP at 5.5 a resolution. Nature 458:475–480
18. Weber G, Trowitzsch S, Kastner B et al (2010) Functional organization of the Sm core in the crystal structure of human U1 snRNP. EMBO J 29:4172–4184
19. Will CL, Rumpler S, Klein Gunnewiek J et al (1996) In vitro reconstitution of mammalian U1 snRNPs active in splicing: the U1-C protein enhances the formation of early (E) spliceosomal complexes. Nucleic Acids Res 24:4614–4123
20. Muto Y, Krummel DA, Kambach C et al (2001) Reconstitution of the spliceosomal U1 snRNP from all recombinant subunits and its characterisation by ionspray Q-tof mass-spectrometry. Nucleic Acids Res Suppl 1:275–276
21. Raker VA, Plessel G, Luhrmann R (1996) The snRNP core assembly pathway: identification of stable core protein heteromeric complexes and an snRNP subcore particle in vitro. EMBO J 15:2256–2269
22. Will CL, Luhrmann R (2001) Spliceosomal UsnRNP biogenesis, structure and function. Curr Opin Cell Biol 13:290–301
23. Dorn G, Leitner A, Boudet J et al (2017) Structural modeling of protein-RNA complexes using crosslinking of segmentally isotope-labeled RNA and MS/MS. Nat Methods 14:487–490
24. Campagne S, de Vries T, Malard F et al (2021) An in-vitro reconstituted U1 snRNP allows the study of the disordered regions of the particle and the interactions with proteins and ligands. Nucleic Acids Res 49:gkab135. https://doi.org/10.1093/nar/gkab135
25. Tropea JE, Cherry S, Waugh DS (2009) Expression and purification of soluble His(6)-tagged TEV protease. Methods Mol Biol 498: 297–307
26. Price SR, Ito N, Oubridge C et al (1995) Crystallization of RNA-protein complexes. I. Methods for the large-scale preparation of RNA suitable for crystallographic studies. J Mol Biol 249:398–408
27. Cathala G, Brunel C (1990) Use of n-butanol for efficient recovery of minute amounts of small RNA fragments and branched nucleotides from dilute solutions. Nucleic Acids Res 18:201
28. Williamson MP (2013) Using chemical shift perturbation to characterise ligand binding. Prog Nucl Magn Reson Spectrosc 73:1–16
29. Duss O, Maris C, von Schroetter C et al (2010) A fast, efficient and sequence-independent method for flexible multiple segmental isotope labeling of RNA using ribozyme and RNase H cleavage. Nucleic Acids Res 38:e188
30. Tugarinov V, Kanelis V, Kay LE (2006) Isotope labeling strategies for the study of high-molecular-weight proteins by solution NMR spectroscopy. Nat Protoc 1:749–754
31. Rossler P, Mathieu D, Gossert AD (2020) Enabling NMR studies of high molecular weight systems without the need for Deuteration: the XL-ALSOFAST experiment with delayed decoupling. Angew Chem Int Ed Engl 59:19329–19337

Chapter 16

Genome-Wide CRISPR Screening to Identify Mammalian Factors that Regulate Intron Retention

Anna M. Scarborough, Ashwin Govindan, and Nicholas K. Conrad

Abstract

Intron retention (IR) regulates gene expression to control fundamental biological processes like metabolism, differentiation, and cell cycle. Despite a wide variety of genes controlled by IR, few techniques are available to identify regulators of IR in an unbiased manner. Here, we describe a CRISPR knockout screening method that can be applied to uncover regulators of IR. This method uses GFP reporter constructs containing a retained intron from a gene of interest such that GFP signal is regulated by IR in the same fashion as the endogenous gene. The GFP levels are then used as a readout for genome-wide CRISPR screening. We have successfully used this approach to identify novel regulator of IR of the MAT2A transcript and propose that similar screens will be broadly applicable for the identification of novel factors that control IR of specific transcripts.

Key words CRISPR screens, Intron retention, Splicing, Intron detention, Posttranscriptional regulation, MAT2A, OGT

1 Introduction

Intron retention (IR) regulates as much as 80% of coding genes across various tissue types and controls fundamental biological processes like metabolism, differentiation, and cell cycle [1–8]. There are a surprisingly wide variety of fates for transcripts with retained introns. They can be exported to the cytoplasm where they code for distinct protein isoforms (e.g., [9]) or undergo nonsense mediated decay (NMD) due to the presence of intronic premature termination codons (e.g., [10]). In mammals, transcripts containing retained introns are often retained in the nucleus (e.g., [11, 12]), and this subset of IR events has been dubbed "intron detention" [13]. Transcripts with detained introns can be dead-end RNAs subject to nuclear RNA decay machinery (e.g., [14]) or they can serve as reservoirs that are later spliced upon cellular stimulus (e.g., [15, 16]). In any case, the direct regulation

Peter Scheiffele and Oriane Mauger (eds.), *Alternative Splicing: Methods and Protocols*,
Methods in Molecular Biology, vol. 2537, https://doi.org/10.1007/978-1-0716-2521-7_16,

of specific genes or families of related genes by IR requires factors that repress and/or promote splicing of a specific intron(s) in different cellular environments. However, the mechanisms by which cells regulate IR remain the least explored of alternative splicing events.

Large-scale genetic screens allow researchers to "ask the cell" what factors are involved in a specific process. These unbiased assays can connect previously unlinked pathways and reveal new gene functions. Historically restricted to model organisms, genetic screens can now be applied to mammalian cells using CRISPR technology. In principle, an array of genome-wide CRISPR schemes can be applied to uncover novel mechanisms of IR. Here, we describe one strategy that we have successfully employed to uncover unexpected roles for the 3′-end formation factor CFI_m in regulating IR of the MAT2A gene [17]. We discuss critical aspects of reporter design, integration, and validation, as well as provide a detailed protocol for the screen.

1.1 General Strategy

Our scheme uses a CRISPR knockout library to inactivate genes, and our readout is GFP production from a reporter gene that contains the retained intron of interest. The screens can be performed under activating conditions in which the reporter is mostly spliced, or repressing conditions in which the intron is largely retained. If the screen is performed under activating conditions, GFP levels will be high (Fig. 1a, top). If splicing is promoted by a splicing inducer, sgRNAs that knockout this factor will reduce GFP (Fig. 1a, bottom). Conversely, a knockout screen can be performed in repressing conditions where IR is favored (Fig. 1b, top). In this case, knockout of any negative factors that promote IR will lead to enhanced GFP signal (Fig. 1b, bottom). While we focus on knockout screens here, the approach can be extended using gain-of-function overexpression screens (e.g., CRISPRa) [18–21]. It is important to keep in mind that in addition to direct splicing regulators, other upstream and downstream components of the pathway under investigation may be identified. This can be tremendously powerful in revealing many facets of a regulatory pathway. However, it also underscores that rigorous validation and mechanistic follow-up needs to be performed to understand a candidate gene's function in the pathway of interest. In addition, factors that affect reporter expression but are not related to IR (e.g., CMV promoter) will need to be excluded by follow-up experiments.

1.2 Reporter Design

The information obtained from any screen is only as good as the reporter used to query the genome. For an ideal reporter, the regulation of its retained intron mirrors the response of the retained intron in the endogenous transcript. Practically, the reporter must be easily monitored and selectable from the population. We use GFP because it requires no manipulation of cells, and differentially

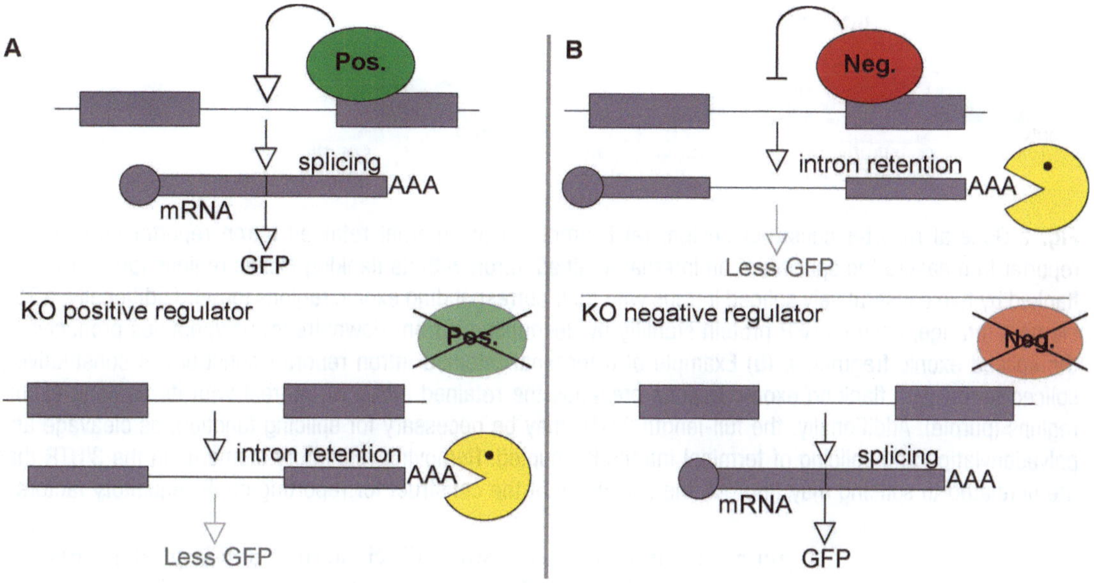

Fig. 1 CRISPR screening for positive and negative regulators of IR. (**a**) Under conditions in which splicing of a retained intron is favored, a positive regulator may promote splicing of a specific retained intron in a transcript thereby increasing GFP levels (top). Knockout of this positive regulator results in increased intron retention and a reduction of GFP intensity (bottom). (**b**) Under conditions in which intron retention is favored, a negative regulator may reduce splicing efficiency of a specific retained intron in a transcript resulting in nuclear retention and decay of the RNA (top). Knockout of that negative regulator results in increased splicing efficiency and increased GFP intensity (bottom)

expressing cells can be sorted by fluorescent activated cell sorting (FACS). Besides having an easily assayed reporter product like GFP, additional factors must be considered in reporter design. As described below, reporters with efficiently spliced introns, 2A "self-cleaving" peptides, and minimal responsive regions have been reliable in our hands. The general schemes we have used for genes with terminal or internal retained introns are depicted in Fig. 2.

Cells couple splicing with nearly all other phases of gene expression, so inclusion of an efficiently spliced intron is important for heterologous gene expression [22–24]. In our reporters, we insert a constitutively spliced intron(s) to support transcript expression even under conditions when the splicing of the retained intron is inefficient. Moreover, retained introns are normally components of genes with constitutively spliced introns, so placing the retained intron into a similar arrangement in the reporters better reflects the natural context. For a gene with an internal retained intron, we placed a constitutive intron upstream and downstream of the retained intron (Fig. 2a). For a gene with a terminal retained intron, we placed a constitutively spliced intron upstream of the retained intron (Fig. 2b). We use the efficiently spliced second intron of β-globin in our reporters, but any constitutively spliced intron should work. We also include flanking exonic regions of the

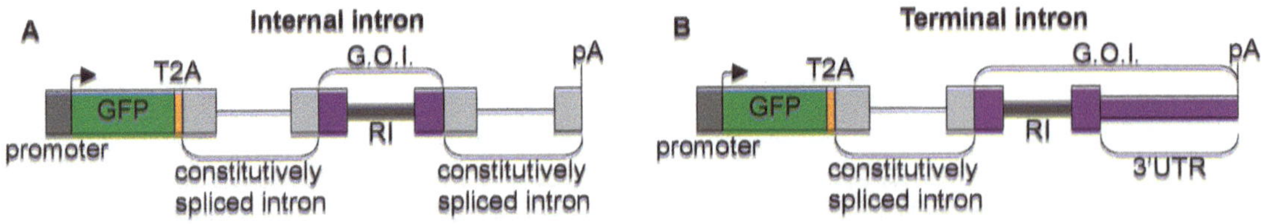

Fig. 2 General reporter construct design. (**a**) Example of an internal retained intron reporter construct. A reporter to measure the splicing of an internal retained intron with its flanking exonic regions (purple) will be flanked by two constitutively spliced introns with their corresponding exonic regions (gray). Additionally, a T2A element (orange) ensures GFP protein stability by decoupling it from downstream polypeptides produced by the spliced exonic fragments. (**b**) Example of a terminal retained intron reporter construct. A constitutively spliced intron with flanking exonic regions precedes the retained intron of interest with its flanking exonic regions (purple). Additionally, the full-length 3′UTR may be necessary for splicing function, as cleavage and polyadenylation and splicing of terminal introns is coupled. Removing regulatory elements in the 3′UTR that are unrelated to splicing may improve the specificity of the construct for reporting on IR regulatory factors

gene of interest to be sure all cis-acting splicing elements are present (see below). Importantly, these exons are in frame with the rest of the reporter to prevent NMD.

Insertion of the T2A "self-cleaving" peptide between the GFP and spliced portions of a reporter ensures GFP output from the reporter construct (Fig. 2, orange). T2A is one of a family of virus-derived small peptide sequences that do not actually self-cleave as the name implies, but instead induce ribosomal skipping to generate two separate proteins from one mRNA [25]. Inclusion of this peptide produces a GFP with a small T2A tag and a separate downstream polypeptide. An early version of one of our reporters excluded this element, and the fusion of GFP to β-globin and additional sequences from the C-terminus of the MAT2A protein produced an unstable GFP product [17].

Since the goal of this screen is to identify factors specifically related to IR, we attempt to minimize the contributions of other cis-acting sites that affect expression of the gene of interest independently of IR. To do so, use the smallest endogenous sequence that faithfully recapitulates the regulated IR response. Prior to a screen, we determine the minimal responsive sequence using transient transfections. Of course, this strategy requires an assay that can be used to activate (or repress) splicing of the retained intron in the gene of interest. For example, intracellular SAM levels regulate intron retention of the SAM-synthetase MAT2A, and we can indirectly deplete SAM by removing methionine from cell media. Using this response, we were able to test regulation of the MAT2A retained intron reporters by examining their responses to methionine depletion [26]. Similarly, we defined elements in the O-GlcNAc transferase (OGT) retained intron using drugs that inhibit OGT [27]. If no such assays are available, a reasonable starting point would be to include the entire retained intron and flanking exons.

1.3 Cell Type Selection, Reporter Integration, and Validation

The cell type chosen for CRISPR screens should be selected based on genomic stability, ease of manipulation, and its relevance for your gene of interest. We use HCT116 colorectal cancer cells in our screens because they have a stable, near-diploid genome [28]. The rationale is that the CRISPR knockout may be more efficient since only two alleles need to be targeted instead of the multiple alleles found in aneuploid lines. Furthermore, HCT116 cells can be transfected, used in siRNA experiments, and grow well in culture. Some IR regulation may only occur in particular cell types, so cell line choice ultimately depends on the specific gene and regulatory response being tested.

To integrate the reporter into the genome, we used TALEN-mediated integration into the AAVS1 safe harbor locus in the PPP1R12C gene [29–31]. Alternatively, reporters can be integrated using lentiviral constructs, or by any other means of stable integration. Interestingly, when screening for loss of GFP expression, sgRNAs that target near the integration site (or even on the same chromosome) are often recovered ([17]; Drs. Joshua Mendell and Hema Manjunath, personal communication). This is presumably due to CRISPR-induced DNA breaks leading to recombination events between chromosomes that remove the reporter. Knowing the integration site is advantageous in that it allows the researcher to disregard these genes as candidates. On the other hand, it is disadvantageous in that candidates from an entire chromosome may need to be discarded [17], so performing biological replicates using clonal lines with different integration sites allows a more complete query of the genome.

As validation, it is convenient to have conditions that induce and/or repress IR. For example, depletion of methionine, and therefore SAM, induces splicing of the MAT2A retained intron. For our studies of MAT2A, we selected clonal reporter cell lines that had a high differential GFP expression dependent on methionine concentrations in the media. In some cases, such treatments will be unavailable, but it is reasonable to presume that cells will respond to inactivation of a gene of interest by increasing its expression. For genes regulated by IR, this includes increasing the efficiency of splicing of the retained intron. Therefore, upon knockdown of the mRNA from endogenous gene of interest, the reporter should be efficiently spliced leading to more GFP. In Fig. 3, we show two examples of this strategy. One reporter line has an internal retained intron (OGT) and another with a terminal retained intron (MAT2A). In both cases, siRNA-mediated knockdown of the endogenous mRNA led to enhanced GFP expression demonstrating that this gene-specific assay can replace pathway-specific treatments.

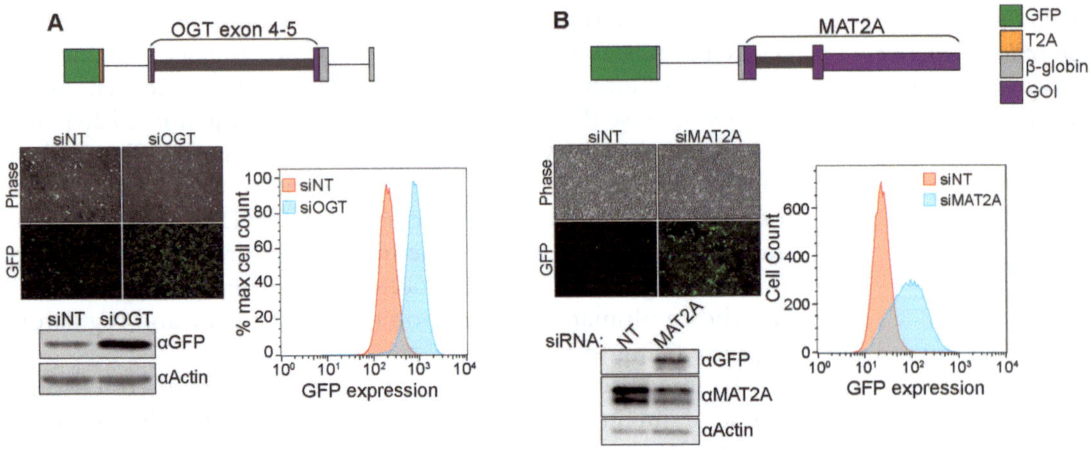

Fig. 3 Example validations of clonal reporter cell lines for CRISPR screens. (**a**) An OGT reporter construct consisting of GFP, a T2A peptide, and the retained OGT intron (intron 4) flanked by β-globin exon and intron sequences was inserted into HCT116 cells at the AAVS1 site. A clonal reporter cell line was validated using an siRNA against the endogenous OGT (siOGT) compared to a nontargeting siRNA (siNT) control. GFP production was analyzed by fluorescent microscopy, western blotting, and flow cytometry. (**b**) A MAT2A reporter construct consisting of MAT2A, a β-globin intron with flanking exonic regions, the MAT2A detained intron and flanking exonic regions, and the full length 3′UTR was inserted into HCT116 cells at the AAVS1 site then used to produce a clonal cell line. This reporter did not contain a T2A element, resulting in unstable GFP-transgene protein product for many clonal cell lines [17]. However, the clonal cell line chosen for CRISPR screening responded to stimuli in a manner identical to the endogenous MAT2A. To validate the reporter cell line, endogenous MAT2A was knocked down using siRNA (siMAT2A). As expected, cells responded to decreased MAT2A levels by increasing MAT2A splicing efficiency, which was reflected in the GFP reporter. We validated GFP up-regulation by fluorescent microscopy, western blotting, and flow cytometry

Finally, cryptic splicing is common with exogeneous gene reporters [32], so we recommend verifying that the reporter splices properly. Isoforms can be analyzed by northern blot, RT-PCR, and sequencing to be sure that proper splicing patterns are observed under the corresponding inducing or repressing conditions.

1.4 Lentiviral Library Selection

We used the Brunello library to perform a genome-wide knockout screen [33]. The library has four optimized sgRNAs targeting each human protein coding gene and multiple non-targeting controls. Cas9 and the sgRNAs are expressed from the same lentivirus, so only a single transduction and selection step is required. Conveniently, the library is available from Addgene and a detailed protocol for preparing lentivirus is provided. To reduce variance between replicate screens, it is a good idea to make enough lentivirus so that all replicates can be performed using the same lentiviral preparation. As mentioned above, CRISPRa or other types of screens may also be informative [18–21].

1.5 CRISPR Screen Timing

The timeline for CRISPR screens may vary with to the gene of interest being investigated and with the need for identification of essential genes. Specifically, the length of time between lentiviral transduction and cell selection may need to be optimized. Guide RNAs that target essential genes will be lost from the total cell population over time since cells that have deleted these genes will die. Therefore, opting for a shorter screen length (~7–12 days post-transduction) will allow for better recovery of sgRNAs that target essential genes. That said, we note that essential genes can be detected after as much as 14 days [34]. A shorter length of time will reduce bias against essential genes, but the trade-off is that sufficient time must been given for candidate genes to be knocked out and the protein of interest to be lost. The timing CRISPR screens can be based on the GFP response after knockout of known factors involved in your regulatory pathway. For example, when screening for MAT2A splicing factors, we performed a time course in which we assayed GFP production after transduction with the Brunello lentivirus targeting the essential METTL16 gene, a known regulator of MAT2A IR. If there are no known factors regulating splicing of your pathway, a time course of induction can be performed in which the endogenous gene of interest is knocked out to induce splicing of the reporter (Fig. 3). It is important to keep in mind that the timing of GFP response observed from these controls does not necessarily reflect all proteins in the pathway, so no single time course will be ideal to assess functional knockout of all genes.

2 Materials

The materials necessary for screening depend on screen coverage, length, and cell type. Coverage is the number of times each sgRNA in the library is queried. Therefore, for a 500× coverage screen, each sgRNA would be transduced into a single cell 500 times. Coverage must be maintained in subsequent flow cytometry selection and sequencing steps as well. The protocol here is for a nine-day, 500× coverage screen of an HCT116 cell-derived reporter line. However, the materials necessary and timing can be adjusted dependent on the particular screen. The amount of materials listed is required for three replicates of viral titering and a single replicate of a CRISPR screen.

2.1 Determination of CRISPR Lentiviral Library Titer

1. 9 mL Brunello lentiviral library. This can be purchased from Addgene as viral stocks or as a DNA library then prepared according to Doench et al. [33]. If preparing the virus in house, psPAX2 and pMD2.G (Addgene) will be necessary. Library should be aliquoted and stored at −80 °C before titering.

2. Reporter cell line (HCT116 derivative).
3. 150 mL growth media: Dulbecco's Modified Eagle Medium (DMEM), 2 mM L-glutamine, penicillin–streptomycin, and 10% fetal bovine serum (FBS).
4. 15 mL infection media: DMEM, 2 mM L-glutamine, penicillin–streptomycin, 20 mM pH 7.2 HEPES, 3% FBS, 4 μg/mL polybrene.
5. 60 mL phosphate buffered saline (PBS).
6. 24 mL 1× trypsin (1× trypsin solution contains 0.5 g/L porcine trypsin, 0.2 g/L EDTA in PBS).
7. 9 × 6-well plates.
8. 3× opaque-walled 96-well plates, compatible with luminometer.
9. 4 mL CellTiter-Glo® Luminescent Cell Viability Assay.
10. Shaker for mixing multiwell plates.
11. Luminometer.
12. Tissue culture incubator at 37 °C and with 5% CO_2.

2.2 CRISPR Screen

1. Reporter cell line.
2. 114 × 15 cm plates.
3. 9 L growth media: DMEM media, 2 mM L-glutamine, penicillin–streptomycin, and 10% FBS.
4. 400 mL infection media: DMEM media, 2 mM L-glutamine, penicillin–streptomycin, 20 mM HEPES, pH 7.2, 3% FBS, 4 μg/mL polybrene.
5. 800 mL PBS with calcium chloride and magnesium chloride (PBS+/+).
6. 2 L PBS.
7. 750 mL 1× trypsin.
8. 1 mg/mL puromycin stock.
9. Titered Brunello lentivirus preparation. The required volume of media containing lentivirus is calculated in Subheading 3.1, **step 28**.
10. 6 × 15 mL conical tubes.
11. 16 × 50 mL conical tubes.
12. 2 × sterile, disposable 500 mL flasks or equivalent for pooling cells for short periods of time.
13. 100 mL FACS buffer: 2% FBS, 0.5 mM EDTA, 50 mM HEPES, pH 7.2 in PBS.
14. 15 mL collection buffer: 50% FBS, 50 mM HEPES, pH 7.2 in PBS.

15. 1 × 100 μM disposable nylon cell strainer.
16. Moflo® MLS flow cytometer, or equivalent.
17. 1 mL tissue lysis buffer: 10 mM Tris–HCl, pH 8.0, 100 mM NaCl, 0.4% SDS, 11 mM EDTA, 0.2 mg/mL Proteinase K.
18. Centrifuge capable of cooling to 4 °C and holding 15 and 50 mL conical tubes.
19. Tissue culture incubator at 37 °C and with 5% CO_2.
20. 2× microcentrifuge tubes.
21. Tabletop centrifuge capable of chilling to 4 °C.

2.3 DNA Preparation

1. Microcentrifuge tubes.
2. Heat block with shaker function.
3. 10 μL 2 mg/mL RNase A solution.
4. 6 × phase lock tubes, such as MaXtract High Density 1.5 mL tubes.
5. 4 mL phenol–chloroform–isoamyl alcohol (25:24:1, v/v; PCA).
6. 1 mL chloroform.
7. 4 mL 75% ethanol.
8. 3 μL GlycoBlue coprecipitate.
9. Nanodrop spectrophotometer.
10. Tabletop centrifuge.
11. Ultrapure water, such as Milli-Q.

2.4 Library Amplification

1. P5-PCR#1 primer 5′-tatcttgtggaaaggacga-3′.
2. P7-PCR#1 primer 5′-gagccaattcccactcctttc-3′.
3. P5 stagger primer mix, consisting of eight P5 primers combined in equal molar ratios [33].
4. One unique barcoded P7 sequencing primer per sample, including inputs [33].
5. Ex Taq DNA polymerase kit. Up to 70 μL of enzyme is necessary per replicate of a screen.
6. Thermocycler capable of holding 96 × 0.1 mL PCR tubes.
7. 5× DNA loading dye.
8. 0.1 mL PCR tubes.
9. GeneRuler 1 kbp Plus DNA ladder.
10. Agarose gel electrophoresis apparatus.
11. Agarose.
12. Ethidium bromide.
13. Ultrapure water.

2.5 PCR Cleanup and Sequencing

1. 4 mL AMPure XP.
2. Microcentrifuge tubes.
3. Magnetic strip.
4. 70% ethanol.
5. Ultrapure water.
6. Next-generation sequencing kit, such as Illumina Next Seq 500.

3 Methods

3.1 Determination of CRISPR Lentiviral Library Titer

1. If you purchased pretitered virus, skip to **step 28**. If not, proceed with the protocol (*see* **Notes 1–6** for lentiviral handling and storage).
2. Plate the reporter cell line at 4×10^5 cells per well of a 6-well plate.
3. Incubate the cells in the tissue culture incubator for 24 h.
4. Thaw the lentivirus at 37 °C in a secondary container in the water bath. The lentivirus is ready immediately after it thaws. Do not let it incubate at 37 °C longer than necessary.
5. Make four, two-fold serial dilutions of lentivirus using infection media; each diluted sample should be at least 1.2 mL.
6. Replace the media on the cells with 1 mL per well of undiluted lentivirus, diluted lentivirus from **step 5**, or infection media.
7. Incubate at 37 °C with 5% CO_2 overnight.
8. Add 1 mL of growth media per well.
9. Incubate at 37 °C with 5% CO_2 overnight.
10. Split and dilute the cells ten-fold into two wells of a 6-well plate, one with and one without 1 μg/mL puromycin.
11. Incubate at 37 °C with 5% CO_2 overnight.
12. Replace the media to fresh growth media +/− 1 μg/mL puromycin.
13. Incubate at 37 °C with 5% CO_2 overnight.
14. To determine the number of live cells, begin by equilibrating CellTiter-Glo reagent to room temperature (RT) then gently vortex to mix.
15. Wash cells with PBS then trypsinize with 400 μL trypsin per well.
16. Quench the trypsin using 600 μL room-temperature growth media for a total volume of 1 mL per well.
17. Plate 100 μL cell mix per well of a 96-well plate.
18. Plate 100 μL media without cells to serve as a blank.

19. Add 100 μL CellTiter-Glo per well with cells or media.
20. Mix the 96-well plate for 10 min on an orbital shaker at RT.
21. Measure the luminescent signal using the luminometer.
22. Divide the intensity of the samples from cells selected with puromycin to that of the no antibiotic selector for a given lentiviral infection to determine the fraction of cells infected for each condition (*see* **Note 7**).
23. Repeat **steps 1–22** twice more, for a total of three replicates.
24. Plot the three replicates on a scatter plot, with the fraction of cells infected on the *y*-axis and volume of lentivirus used on the *x*-axis. The titering experiments should show a linear trend, with little variance between replicates.
25. Average the percent of cells infected for each lentiviral condition.
26. Multiply the number of cells plated in **step 1** by the calculated fraction of cells infected (**step 22**). Only apply this calculation to volumes of lentivirus that had 5–30% of cells infected and fit the linear trend (**step 24**).
27. Divide the number of cells infected (**step 26**) by the volume of lentivirus used for the given dilution to calculate viral titer.

$$\frac{(Number\ cells\ plated) \times (Fraction\ cells\ infected)}{volume\ of\ lentivirus}$$
$$= infectious\ viral\ units\ per\ mL$$

28. Determine the volume of lentivirus necessary for the CRISPR screen. This will depend on the number of sgRNA in the library selected and desired library coverage. For a 500× coverage screen of the Brunello library, if the titer is 500,000 viral units per mL, ~77 mL lentivirus will be necessary per replicate of the CRISPR screen.

$$\frac{(sgRNA/library) \times (library\ coverage)}{infectious\ viral\ units\ per\ mL}$$
$$= volume\ lentivirus\ per\ screen$$

3.2 CRISPR Screen

1. Use the equation below to calculate the number of cells necessary to plate to reach the desired coverage for the infection scheme. Coverage is the coverage of the screen; sgRNA/library is the number of sgRNA in the selected lentiviral library (ex. Brunello library is 76,441 sgRNA/library). The percent of cells to be infected should be ~30% of cells.

$$\frac{Coverage \times \frac{sgRNA}{library}}{\%cells\ to\ be\ infected} = numer\ of\ cells\ necessary\ to\ plate$$

Using the Brunello library, a minimum of 1.27×10^8 cells will be needed for 500× coverage (*see* **Notes 8** and **9**).

2. Seed twenty-six 15 cm plates with 5×10^6 reporter cells per plate. This density should ensure that the cells will be approximately 100% confluent in 3 days.
3. Incubate the cells overnight at 37 °C, 5% CO_2.
4. (Day 0) Thaw the necessary volume of lentivirus, calculated in **step 28** of Subheading 3.1, at 37 °C in a secondary container. Use the lentivirus immediately after thawing.
5. Dilute the lentivirus with infection media. The volume of infection media will depend on your lentiviral titer, but should equal the total volume necessary for each 15 cm plate to be covered with 15 mL of media:

$$(15\ mL \times 26\ plates) - (mL\ lentivirus) = volume\ infection\ media\ to\ add$$

6. Replace the reporter cell media with 15 mL infection media containing the diluted lentivirus.
7. Incubate the cells overnight at 37 °C, 5% CO_2.
8. (Day 1) Add 15 mL growth media to the cells to bring the volume to 30 mL per plate.
9. Incubate the cells overnight at 37 °C, 5% CO_2.
10. (Day 2) The cells should be ~100% confluent at this point. Wash the cells with PBS, then trypsinize using 5 mL trypsin per plate.
11. Quench the trypsin with 5 mL growth media per plate.
12. Plate 10 mL cell mix per plate among 26 × 15 cm plates, each with 30 mL growth media with 1 μg/mL puromycin total. This step does not dilute cell number per plate. However, it promotes faster selection because puromycin is somewhat less effective on 100% confluent cells. Furthermore, this maintains reasonable cell numbers on plates after most cells (~70%) die.
13. Incubate the cells overnight at 37 °C, 5% CO_2.
14. (Day 3) Wash the cells once with 15 mL PBS+/+. This wash step and the wash step below (**step 17**) remove dying cells and debris from puromycin selection.
15. Replace the media with 30 mL growth media with 1 μg/mL puromycin.
16. Incubate the cells overnight at 37 °C, 5% CO_2.
17. (Day 4) Wash the cells once with 15 mL PBS+/+.
18. Replace the media with 30 mL growth media with 1 μg/mL puromycin.

19. Incubate the cells overnight at 37 °C, 5% CO_2.
20. (Day 5) At this point, cells should be ~100% confluent. Wash the cells once with 15 mL PBS per plate.
21. Trypsinize the cells with 5 mL trypsin per plate.
22. Quench the trypsin with 5 mL growth media per plate.
23. Pool the quenched cells from all 26 plates in a 500 mL flask.
24. Plate 40 × 15 cm plates with 2 mL cell mix per plate into 30 mL growth media with 1 μg/mL puromycin (*see* **Note 10**).
25. Incubate the cells overnight at 37 °C, 5% CO_2.
26. (Day 6) Replace the media with fresh 30 mL growth media with 1 μg/mL puromycin.
27. Incubate the cells overnight at 37 °C, 5% CO_2.
28. (Day 7) At this point, cells will be ~100% confluent. Wash the cells once with 15 mL PBS per plate.
29. Trypsinize the cells with 5 mL trypsin per plate.
30. Quench the trypsin with 5 mL growth media per plate.
31. Pool the trypsinized cells in a 500 mL flask.
32. Plate 48 × 15 cm plates with 2 mL cell mix per plate into 30 mL growth media with 1 μg/mL puromycin.
33. Incubate the cells overnight at 37 °C, 5% CO_2.
34. (Day 8) Replace the media with fresh 30 mL growth media with 1 μg/mL puromycin.
35. Incubate the cells overnight at 37 °C, 5% CO_2.
36. (Day 9) Approximately 1 h before sorting for the CRISPR screen, warm trypsin and chill the centrifuge.
37. Chill FACS buffer and collection buffer on ice.
38. Prepare 6 × 15 mL conical tubes with 2.5 mL collection buffer per tube. Store on ice.
39. Trypsinize the cells using 5 mL of trypsin per plate.
40. Dilute the trypsinized cells with 10 mL ice-cold PBS per plate.
41. Pool the cells into 50 mL conical tubes, three plates of cells per tube.
42. Pellet the cells at 1000 × *g*, 3 min, 4 °C.
43. Resuspend and pool the cells in ice-cold FACS buffer. The buffer volume will depend on the number of cells to be sorted and the instrument used for sorting. See the FACS operator manual for cell concentrations for optimal sorting. We typically use 1 mL of FACS buffer per 15 cm dish.
44. Gently pipet the cells up and down to ensure cells are thoroughly suspended.

45. Strain the cells by passing through a 100 μm Nylon filter.
46. Save a population of cells as an input in a microcentrifuge tube. The number of cells will depend on the screen coverage. We saved a minimum 4×10^7 cells for a 500× screen of the Brunello library.

$$(sgRNA/library) \times (coverage) = minimum\ number\ of\ cells\ for\ input$$

47. Pellet the input at 1000 × *g*, 4 °C, 3 min.
48. Resuspend the input in 500 μL tissue lysis buffer. Store at −20 °C.
49. Sort the remaining cells, collecting the highest or lowest 0.1–1.5% GFP expressing cells dependent on your screen design; we generally use 1% as a cutoff. Keep the cells on ice or chilled at 4 °C for the duration of the screen. A sufficient number of single cells must be sorted to maintain the desired coverage which is determined by multiplying the coverage by the total number of sgRNAs in the library. For a 500× screen using the Brunello library, 3.8×10^7 cells must be sorted and 3.8×10^5 cells will be collected for a 1% GFP cutoff. A portion of the sample is lost in DNA and library preparation, so it is preferred to sort more than the minimum number of cells. Collect the cells in the 15 mL conical tubes containing 2.5 mL FACS collection buffer. Using the cell growth scheme provided above, you will most likely not need to sort all of your sample to reach 500× coverage. However, it is much better to have an excess of cells, as some cells will be lost during straining, sorting, and preparative steps.
50. Pellet the cells at 1000 × *g* for 3 min at 4 °C in the 15 mL conical tubes.
51. Aspirate the supernatant.
52. Resuspend and pool the pellet in 1 mL ice-cold PBS+/+.
53. Transfer to a microcentrifuge tube.
54. Pellet the cells at 1000 × *g* for 3 min at 4 °C.
55. Resuspend the cell pellet in 500 μL tissue lysis buffer.
56. Store at −20 °C until ready to begin DNA preparation.

3.3 DNA Preparation

DNA preparation methods have been adapted from others [34–36]. If conducting multiple replicates of CRISPR screens, we recommend waiting to prepare the DNA until all replicates are ready to diminish batch effects in sample processing. Samples are stable for several months in tissue lysis buffer at −20 °C.

1. Incubate samples resuspended in tissue lysis buffer at 55 °C with ~1 × *g* mixing for 3 h to overnight.
2. Cool the tubes for ~3 min at RT.

3. Add 5 μL 2 mg/mL RNase A to each tissue lysis buffer sample.
4. Vortex for ~5 s to mix.
5. Incubate at 37 °C, with ~1 × *g* mixing for 1 h.
6. Precentrifuge three MaXtract phase lock tubes at 16,000 × *g* for 1 min per 500 μL sample. Do not add samples to the tubes yet.
7. Add 500 μL PCA to each sample in tissue lysis buffer.
8. Vortex the sample for ~30 s.
9. Quickly transfer the homogeneous PCA–tissue lysis buffer mix equally between two precentrifuged phase lock tubes per sample.
10. Centrifuge at 16,000 × *g*, 5 min, RT.
11. Add an additional 500 μL PCA to each phase lock tube.
12. Invert the tubes vigorously for 30 s to mix.
13. Centrifuge at 16,000 × *g*, 5 min, RT.
14. Combine the aqueous layers of the two precentrifuged phase lock tubes per sample to a fresh precentrifuged phase lock tube.
15. Add 500 μL chloroform per tube.
16. Invert the tubes vigorously for 30 s to mix.
17. Centrifuge at 16,000 × *g*, 5 min, RT.
18. Transfer the aqueous layer to a fresh microcentrifuge tube.
19. Add 1.3 μL GlycoBlue then invert to mix.
20. Add 1 mL 100% ethanol per tube then invert to mix.
21. Incubate at −80 °C, 15 min to 1 h, or −20 °C overnight.
22. Pellet the DNA at 16,000 × *g*, 10 min, 4 °C.
23. Aspirate the supernatant.
24. Wash the pellet with 1 mL 75% EtOH.
25. Centrifuge at 16,000 × *g*, 5 min, 4 °C.
26. Aspirate the supernatant.
27. Centrifuge the tubes at maximum speed ~5 s.
28. Remove the residual ethanol by gently pipetting with a gel-loading tip.
29. Allow the pellet to air-dry briefly (~1–2 min). If the pellet is completely dry it will be difficult to resuspend.
30. Add 100 μL sterile water to the sorted sample and 500 μL to the input.
31. Incubate at RT for 30 min or 4 °C overnight.
32. Pipet gently ~20 times to mix.

33. Check the DNA concentration by nanodrop. Typically, our concentrations range from 150–250 ng/μL for selected cells at ~500× coverage, with a total of 15–25 μg total DNA, and approximately 1000–2000 ng/μL for the input. Dilute the input to approximately the concentration of the sorted sample using sterile water.

3.4 Library Amplification

For library amplification, we use a two-step PCR protocol adapted from Shalem et al. and Golden et al. [34, 37]. If multiple replicates of a screen are to be sequenced as biological replicates, we recommend amplifying all replicates at the same time in the same thermocycler to diminish batch effects (*see* **Notes 11–14**).

1. Set up twenty PCR#1 reactions for the input and as many PCR#1 reactions for the selected sample as recovered DNA permits, using all the selected sample DNA collected. If you have less than 13.2 μg DNA for the sorted sample, divide the DNA equally between two 100 μL reactions. Create a single reaction omitting the DNA as a water control. PCR#1 setup: 10 μL 10× ExTaq buffer, 8 μL dNTP mix (ExTaq kit), 5 μL 10 μM Forward primer/P5, 5 μL 10 μM Reverse primer/P7, 6.6 μg DNA, 1.5 μL ExTaq polymerase, H_2O (milliQ) to 100 μL.
2. Amplify using the following thermocycler settings: 95 °C 1 min, [95 °C × 30 s, 53 °C × 30 s, 72 °C × 30 s] × 18 cycles, 72 °C × 10 min, hold 4 °C.
3. Pool the reactions for each sample.
4. Prior to generating the final libraries with PCR#2 (**steps 10–12**), perform an analytical PCR to determine the appropriate cycle number for each sample. Create a 200 μL PCR#2 master mix for each sample (including the water control) by doubling the standard 100 μL PCR#2 reaction: 10 μL 10X ExTaq buffer, 8 μL dNTP mix (ExTaq kit), 5 μL 10 μM P5 stagger primer mix, 5 μL 10 μM P7 barcode primer, 5 μL PCR#1 product, 1.5 uL ExTaq polymerase, H2O (milliQ) to 100 μL.
5. Divide the PCR#2 master mixes into seven 25 μL reactions per sample to be used as a cycle test.
6. Repeat the thermocycler settings in **step 2**, removing tubes immediately after cycles 8, 9, 10, 11, 12, 14, and 18 complete.
7. Mix 5 μL DNA loading buffer per 25 μL reaction.
8. Load 25 μL per reaction onto a 2% DNA agarose gel with 0.3 μg/mL ethidium bromide using 5 μL GeneRuler 1 kbp Plus ladder as a molecular weight marker.
9. Compare the band intensities of the expected ~300 bp PCR#2 product for each sample to the 300 bp marker of the GeneRuler 1 kbp Plus ladder. The cycle number that generates a

product with roughly the same intensity as this marker should be used in the subsequent PCR#2 reactions for library preparation. Importantly, the 18-cycle water control sample should not have a band.

10. Set up four PCR#2 reactions per sample using the conditions found in **step 4**. Each sample should have a unique P7 barcode primer.
11. As a negative control, create a single PCR#2 reaction per sample, replacing the template with the PCR#1 water control and using the same sample P7 barcode primers as in **step 10**.
12. Use the following thermocycler settings and the sample-specific cycle number as determined in **steps 4–9**: 95 °C 1 min, [95 °C × 30 s, 53 °C × 30 s, 72 °C × 30 s] × variable cycles, hold 4 °C. Remove tubes after the desired cycle has completed when the thermocycler has begun to ramp to 95 °C for the subsequent cycle.
13. Pool the four 100 μL PCR#2 reactions in a microcentrifuge tube for each reaction.
14. Remove 25 μL of pooled PCR#2 and mix with 5 μL DNA loading buffer in a fresh tube.
15. Analyze 25 μL per reaction onto a 2% DNA agarose gel with 0.3 μg/mL ethidium bromide using 5 μL GeneRuler 1 kbp Plus ladder as a molecular weight marker. The input and sorted sample PCRs should have produced a ~300 bp fragment, while the water control should be blank.
16. Store samples at −20 °C or continue to the PCR cleanup and sequencing step using AMPureXP following the Broad Institute protocol [38].

3.5 PCR Cleanup and Sequencing

1. Resuspend AMPure XP beads by shaking. AMPure XP beads separate differing DNA sizes based on the concentration of AMPure beads in the sample, so it is essential to maintain ratios of AMPure XP beads to the sample added. Contaminating genomic DNA remains on the beads in **steps 3–7** below, while the desired 300 bp PCR product will bind the beads in **steps 9–16**.
2. Add 187.5 μL AMPure XP beads (0.5× volumes) per 375 μL sample.
3. Mix the samples well by pipetting up and down until homogeneous.
4. Incubate for 5 min, at RT.
5. Separate the beads on a magnet for 2 min.
6. Keeping the tubes on the magnetic strip, collect the supernatant containing the amplified DNA product.

7. Resuspend AMPure XP beads by shaking.
8. Add 675 μL AMPure XP beads (1.8× volumes) per sample.
9. Mix well by pipetting up and down until homogeneous.
10. Separate the beads on a magnet for 3 min.
11. Discard the supernatant; the 300 bp DNA product remains on the beads.
12. Wash the beads with 1 mL 70% ethanol for 1 min while still on the magnetic strip.
13. Aspirate the wash solution.
14. Repeat **steps 12** and **13** to wash the beads a second time.
15. Use a gel-loading pipette tip to remove residual ethanol.
16. Allow the beads to air-dry for 2–5 min.
17. Resuspend the beads in 375 μL of water and mix by flicking the tube.
18. Incubate for 2 min, at RT.
19. Separate the beads on a magnetic strip for 2 min.
20. Collect the supernatant containing the DNA product.
21. Repeat **steps 1–16** once more.
22. Resuspend the beads in 50 μL water by flicking.
23. Incubate for 2 min, at RT.
24. Separate the beads on a magnetic strip for 2 min.
25. Collect the supernatant, being extra careful to not collect any beads. This contains the purified product.
26. Measure the samples for quality control using nanodrop, qBit, TapeStation, and qPCR, or as recommended by your next-generation sequencing protocols.
27. Sequence using Illumina next-generation sequencing [38] (*see* **Note 15**).
28. Identify reads overrepresented in the selected pool relative to the unselected input pool using MAGeCK analysis or algorithm of your choice [39].

3.6 Timeline of CRISPR Screening

1. 2–4 months—cloning and initial testing of reporter construct transgenes.
2. 2 months—integration of the reporter construct and single-cell cloning.
3. 1–2 months—characterization of isolated clonal cell lines to select a reporter.
4. 2 months—preparation of lentiviral library and titering of virus (can be done concurrently with cell line creation).
5. 1–2 weeks—CRISPR screen.

6. 1 week—DNA preparation, library amplification, and quality control.
7. 2 weeks—sequencing and analysis.

4 Notes

1. Use proper personal protective equipment (PPE) and precautions when handling lentivirus.
2. Bleach all containers that have contacted with lentivirus with freshly diluted 10% bleach before disposal in biohazard waste.
3. Test the 15 mL conical tubes used by snap-freezing and thawing with buffer or media before using them to store lentivirus. Some types of conical tubes become brittle and crack after thawing.
4. Use secondary containers to thaw lentivirus in case of conical tube malfunction.
5. Do not refreeze thawed lentivirus. Since the amount of virus necessary per screen will not be known until after titering, storing smaller aliquots (~10 mL) avoids wasting virus stocks.
6. Do not handle lentiviral library or prepared lentivirus in the same workspace as your CRISPR screen library preparation. These are potent sources of DNA contamination that will efficiently amplify by PCR during preparation of the sequencing library.
7. If the cells are overgrown in the no puromycin control at the time of CellTiter-Glo analysis, you will not get accurate results.
8. When conducting the CRISPR screen, use a titer to infect approximately 30% of plated cells to avoid superinfections.
9. The protocol is written for 500× coverage of the lentiviral library for screens, but we have had success with a screen at 100× coverage [17]. When splitting cells throughout the screen, the total cell number should never drop below the number of cells necessary to maintain this coverage.
10. To ensure equal representation of the 500× initial infection, pool cells from all plates when splitting for the duration of the screen even though you will have an excess of cells that need to be discarded.
11. One-step PCR protocols can be used (e.g., [36], but we prefer this two-step PCR protocol. If too many of cycles are used during library preparation, duplication will muddle the analysis. The two-step protocol uses a conservative initial cutoff (18 cycles) to ensure that sufficient quantities of DNA are available to do cycle testing (Subheading 3.4, **steps 4–9**) and

subsequent preparative amplification. Admittedly, a disadvantage of this protocol is that the dilution of the PCR#1 into the PCR#2 reaction requires a few additional cycles.

12. Before amplification of the sorted sample, we ensure all reagents are working properly by first amplifying a single reaction of an input sample using the two-step procedure.
13. To ensure accurate cycle number determination, pool all ExTaq enzyme and use this pool for both the cycle test and PCR#2 steps.
14. It is common for labs to use gene-specific sgRNA plasmids found in the Brunello library. These provide a potent source of PCR contamination during library preparation. Therefore, do not work with the unamplified DNA in the same area in which you prepared DNA for the lentiviral library or generated used Brunello-based plasmids (e.g., LentiCRISPRv2) for gene-specific knockdowns. Similarly, the amplified products from PCR#1 or PCR#2 are potential sources of contamination.
15. For 500× coverage, ~40 million (M) reads per sample is appropriate. This estimate was generated by multiplying 76,411 Brunello sgRNAs by 500 to yield ~38.2 M reads.

Acknowledgments

We would like to thank Dr. John Schoggins and Dr. Joshua Mendell and their lab members for their invaluable guidance in setting up CRISPR screens. This work was supported by the National Institutes of Health R01 GM127311, U01 CA242115, and T32 GM007062.

References

1. Middleton R, Gao D, Thomas A, Singh B, Au A, Wong JJ, Bomane A, Cosson B, Eyras E, Rasko JE, Ritchie W (2017) IRFinder: assessing the impact of intron retention on mammalian gene expression. Genome Biol 18(1):51. https://doi.org/10.1186/s13059-017-1184-4
2. Gordon JM, Phizicky DV, Neugebauer KM (2021) Nuclear mechanisms of gene expression control: pre-mRNA splicing as a life or death decision. Curr Opin Genet Dev 67:67–76. https://doi.org/10.1016/j.gde.2020.11.002
3. Monteuuis G, Wong JJL, Bailey CG, Schmitz U, Rasko JEJ (2019) The changing paradigm of intron retention: regulation, ramifications and recipes. Nucleic Acids Res 47(22): 11497–11513. https://doi.org/10.1093/nar/gkz1068
4. Jacob AG, Smith CWJ (2017) Intron retention as a component of regulated gene expression programs. Hum Genet 136(9):1043–1057. https://doi.org/10.1007/s00439-017-1791-x
5. Grabski DF, Broseus L, Kumari B, Rekosh D, Hammarskjold ML, Ritchie W (2021) Intron retention and its impact on gene expression and protein diversity: a review and a practical guide. Wiley Interdiscip Rev RNA 12(1): e1631. https://doi.org/10.1002/wrna.1631
6. Wegener M, Muller-McNicoll M (2018) Nuclear retention of mRNAs—quality control, gene regulation and human disease. Semin Cell

Dev Biol 79:131–142. https://doi.org/10.1016/j.semcdb.2017.11.001

7. Wong JJ, Au AY, Ritchie W, Rasko JE (2016) Intron retention in mRNA: no longer nonsense: known and putative roles of intron retention in normal and disease biology. BioEssays 38(1):41–49. https://doi.org/10.1002/bies.201500117
8. Rekosh D, Hammarskjold ML (2018) Intron retention in viruses and cellular genes: detention, border controls and passports. Wiley Interdiscip Rev RNA 9(3):e1470. https://doi.org/10.1002/wrna.1470
9. Li Y, Bor YC, Misawa Y, Xue Y, Rekosh D, Hammarskjold ML (2006) An intron with a constitutive transport element is retained in a tap messenger RNA. Nature 443(7108): 234–237. https://doi.org/10.1038/nature05107
10. Wong JJ, Ritchie W, Ebner OA, Selbach M, Wong JW, Huang Y, Gao D, Pinello N, Gonzalez M, Baidya K, Thoeng A, Khoo TL, Bailey CG, Holst J, Rasko JE (2013) Orchestrated intron retention regulates normal granulocyte differentiation. Cell 154(3):583–595. https://doi.org/10.1016/j.cell.2013.06.052
11. Yap K, Lim ZQ, Khandelia P, Friedman B, Makeyev EV (2012) Coordinated regulation of neuronal mRNA steady-state levels through developmentally controlled intron retention. Genes Dev 26(11):1209–1223. https://doi.org/10.1101/gad.188037.112
12. Braunschweig U, Barbosa-Morais NL, Pan Q, Nachman EN, Alipanahi B, Gonatopoulos-Pournatzis T, Frey B, Irimia M, Blencowe BJ (2014) Widespread intron retention in mammals functionally tunes transcriptomes. Genome Res 24(11):1774–1786. https://doi.org/10.1101/gr.177790.114
13. Boutz PL, Bhutkar A, Sharp PA (2015) Detained introns are a novel, widespread class of post-transcriptionally spliced introns. Genes Dev 29(1):63–80. https://doi.org/10.1101/gad.247361.114
14. Pendleton KE, Park SK, Hunter OV, Bresson SM, Conrad NK (2018) Balance between MAT2A intron detention and splicing is determined cotranscriptionally. RNA 24(6): 778–786. https://doi.org/10.1261/rna.064899.117
15. Mauger O, Lemoine F, Scheiffele P (2016) Targeted intron retention and excision for rapid gene regulation in response to neuronal activity. Neuron 92(6):1266–1278. https://doi.org/10.1016/j.neuron.2016.11.032
16. Ninomiya K, Kataoka N, Hagiwara M (2011) Stress-responsive maturation of Clk1/4 pre-mRNAs promotes phosphorylation of SR splicing factor. J Cell Biol 195(1):27–40. https://doi.org/10.1083/jcb.201107093
17. Scarborough AM, Flaherty JN, Hunter OV, Liu K, Kumar A, Xing C, Tu BP, Conrad NK (2021) SAM homeostasis is regulated by CFIm-mediated splicing of MAT2A. Elife 10: e64930. https://doi.org/10.7554/eLife.64930
18. Horlbeck MA, Gilbert LA, Villalta JE, Adamson B, Pak RA, Chen Y, Fields AP, Park CY, Corn JE, Kampmann M, Weissman JS (2016) Compact and highly active next-generation libraries for CRISPR-mediated gene repression and activation. Elife 5: e19760. https://doi.org/10.7554/eLife.19760
19. Gilbert LA, Horlbeck MA, Adamson B, Villalta JE, Chen Y, Whitehead EH, Guimaraes C, Panning B, Ploegh HL, Bassik MC, Qi LS, Kampmann M, Weissman JS (2014) Genome-scale CRISPR-mediated control of gene repression and activation. Cell 159(3):647–661. https://doi.org/10.1016/j.cell.2014.09.029
20. Konermann S, Brigham MD, Trevino AE, Joung J, Abudayyeh OO, Barcena C, Hsu PD, Habib N, Gootenberg JS, Nishimasu H, Nureki O, Zhang F (2015) Genome-scale transcriptional activation by an engineered CRISPR-Cas9 complex. Nature 517(7536): 583–588. https://doi.org/10.1038/nature14136
21. Shalem O, Sanjana NE, Zhang F (2015) High-throughput functional genomics using CRISPR-Cas9. Nat Rev Genet 16(5): 299–311. https://doi.org/10.1038/nrg3899
22. Moore MJ, Proudfoot NJ (2009) Pre-mRNA processing reaches back to transcription and ahead to translation. Cell 136(4):688–700. https://doi.org/10.1016/j.cell.2009.02.001
23. Le Hir H, Nott A, Moore MJ (2003) How introns influence and enhance eukaryotic gene expression. Trends Biochem Sci 28(4): 215–220. https://doi.org/10.1016/S0968-0004(03)00052-5
24. Muller-McNicoll M, Neugebauer KM (2013) How cells get the message: dynamic assembly and function of mRNA-protein complexes. Nat Rev Genet 14(4):275–287. https://doi.org/10.1038/nrg3434
25. Donnelly MLL, Luke G, Mehrotra A, Li X, Hughes LE, Gani D, Ryan MD (2001) Analysis of the aphthovirus 2A/2B polyprotein 'cleavage' mechanism indicates not a proteolytic reaction, but a novel translational effect: a putative ribosomal 'skip'. J Gen Virol 82 (Pt 5):1013–1025. https://doi.org/10.1099/0022-1317-82-5-1013

26. Pendleton KE, Chen B, Liu K, Hunter OV, Xie Y, Tu BP, Conrad NK (2017) The U6 snRNA m(6)a methyltransferase METTL16 regulates SAM Synthetase intron retention. Cell 169 (5):824-835:e814. https://doi.org/10.1016/j.cell.2017.05.003
27. Park SK, Zhou X, Pendleton KE, Hunter OV, Kohler JJ, O'Donnell KA, Conrad NK (2017) A conserved splicing silencer dynamically regulates O-GlcNAc transferase intron retention and O-GlcNAc homeostasis. Cell Rep 20(5): 1088–1099. https://doi.org/10.1016/j.celrep.2017.07.017
28. Roschke AV, Stover K, Tonon G, Schaffer AA, Kirsch IR (2002) Stable karyotypes in epithelial cancer cell lines despite high rates of ongoing structural and numerical chromosomal instability. Neoplasia 4(1):19–31. https://doi.org/10.1038/sj.neo.7900197
29. Oceguera-Yanez F, Kim SI, Matsumoto T, Tan GW, Xiang L, Hatani T, Kondo T, Ikeya M, Yoshida Y, Inoue H, Woltjen K (2016) Engineering the AAVS1 locus for consistent and scalable transgene expression in human iPSCs and their differentiated derivatives. Methods 101:43–55. https://doi.org/10.1016/j.ymeth.2015.12.012
30. Sanjana NE, Shalem O, Zhang F (2014) Improved vectors and genome-wide libraries for CRISPR screening. Nat Methods 11(8): 783–784. https://doi.org/10.1038/nmeth.3047
31. Hockemeyer D, Soldner F, Beard C, Gao Q, Mitalipova M, DeKelver RC, Katibah GE, Amora R, Boydston EA, Zeitler B, Meng X, Miller JC, Zhang L, Rebar EJ, Gregory PD, Urnov FD, Jaenisch R (2009) Efficient targeting of expressed and silent genes in human ESCs and iPSCs using zinc-finger nucleases. Nat Biotechnol 27(9):851–857. https://doi.org/10.1038/nbt.1562
32. Huang Y, Wimler KM, Carmichael GG (1999) Intronless mRNA transport elements may affect multiple steps of pre-mRNA processing. EMBO J 18(6):1642–1652. https://doi.org/10.1093/emboj/18.6.1642
33. Doench JG, Fusi N, Sullender M, Hegde M, Vaimberg EW, Donovan KF, Smith I, Tothova Z, Wilen C, Orchard R, Virgin HW, Listgarten J, Root DE (2016) Optimized sgRNA design to maximize activity and minimize off-target effects of CRISPR-Cas9. Nat Biotechnol 34(2):184–191. https://doi.org/10.1038/nbt.3437
34. Golden RJ, Chen B, Li T, Braun J, Manjunath H, Chen X, Wu J, Schmid V, Chang TC, Kopp F, Ramirez-Martinez A, Tagliabracci VS, Chen ZJ, Xie Y, Mendell JT (2017) An Argonaute phosphorylation cycle promotes microRNA-mediated silencing. Nature 542(7640):197–202. https://doi.org/10.1038/nature21025
35. Largaespada DA, Collier LS (2008) Transposon-mediated mutagenesis in somatic cells: identification of transposon-genomic DNA junctions. Methods Mol Biol 435. https://doi.org/10.1007/978-1-59745-232-8_7
36. Richardson RB, Ohlson MB, Eitson JL, Kumar A, McDougal MB, Boys IN, Mar KB, De La Cruz-Rivera PC, Douglas C, Konopka G, Xing C, Schoggins JW (2018) A CRISPR screen identifies IFI6 as an ER-resident interferon effector that blocks flavivirus replication. Nat Microbiol 3(11): 1214–1223. https://doi.org/10.1038/s41564-018-0244-1
37. Shalem O (2014) Genome-scale CRISPR-Cas9 knockout screening in human cells. Science 343:84–87. https://doi.org/10.1126/science.1247005
38. Broad Institute PCR of sgRNAs for Illumina sequencing protocol. https://media.addgene.org/cms/filer_public/61/16/611619f4-0926-4a07-b5c7-e286a8ecf7f5/broadgpp-sequencing-protocol.pdf
39. Li W, Xu H, Xiao T, Cong L, Love MI, Zhang F, Irizarry RA, Liu JS, Brown M, Liu XS (2014) MAGeCK enables robust identification of essential genes from genome-scale CRISPR/Cas9 knockout screens. Genome Biol 15(12):554. https://doi.org/10.1186/s13059-014-0554-4

Chapter 17

Tethered Function Assays to Elucidate the Role of RNA-Binding Proteins

Andrea B. Eberle and Oliver Mühlemann

Abstract

The fate of each RNA molecule is strongly determined by RNA-binding proteins (RBPs) which accompany transcripts from its synthesis to its degradation. To elucidate the effect of a specific RBP on bound RNA, it can be artificially recruited to a specific site on a reporter mRNA that can be followed by a variety of methods. In this so-called tethering assay, the protein of interest (POI) is fused to the coat protein of the MS2 bacteriophage and expressed in your favorite cells together with a reporter gene containing MS2 binding sites. The MS2 binding sites are recognized by the MS2 coat protein (MS2CP) with high affinity and specificity and by doing so, the POI is tethered to the reporter RNA. Here, we describe how with the help of this assay the human cytoplasmic poly(A) binding protein is recruited to a mini-μ RNA reporter, thereby influencing the stability of the reporter transcript.

Key words RNA-binding protein, Tethering assay, Tethered protein, MS2 coat protein (MS2CP), MS2 binding sites, Poly(A) binding protein, Nonsense-mediated mRNA decay (NMD)

1 Introduction

RNA-binding proteins (RBPs) are essential for all steps in gene expression, ranging from synthesis and processing of transcripts in the nucleus to transport to the cytoplasm and localization within the cell. Furthermore, stability/turnover of transcripts and translation of mRNAs are heavily influenced and regulated by RBPs. Proteome-wide studies in different organisms showed that the number of RBPs was previously underestimated and it is now believed that more than 1500 RBPs are encoded in the human genome (reviewed in [1–3]). Equally important to the identification of these numerous RBPs is the careful elucidation of the function of each RBP, which may depend where on the mRNA it binds. Here, the tethered function assay makes a valuable contribution.

Peter Scheiffele and Oriane Mauger (eds.), *Alternative Splicing: Methods and Protocols*,
Methods in Molecular Biology, vol. 2537, https://doi.org/10.1007/978-1-0716-2521-7_17,

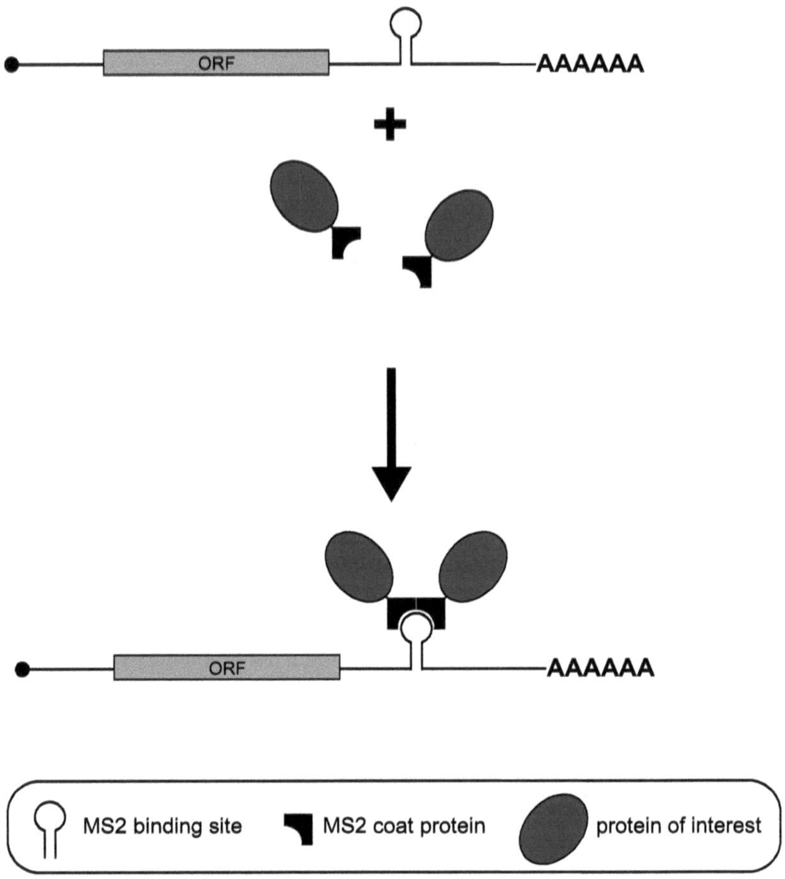

Fig. 1 A schematic overview of a tethered function assay where the protein of interest (POI) is tethered to the 3′UTR of a reporter mRNA via interaction of the MS2 coat protein (MS2CP) fused to the POI and the MS2 binding site located in the reporter transcript (for more details *see* introduction)

In a tethering assay experiment, an RBP is attached to a reporter RNA independently of its RNA-binding specificity and the impact of this RBP recruitment to the reporter RNA is examined. This allows to investigate proteins with unknown RNA substrates or even with no direct RNA binding capability (e.g., proteins indirectly associated with RNA molecule via protein-protein interaction). Two elements are needed in the tethering approach (Fig. 1): First, a specific RNA motif that serves as a high affinity binding site is introduced into a reporter gene (e.g., MS2 binding site in Fig. 1). The reporter gene is chosen in a way that the desired properties of the transcribed RNA can be easily followed (e.g., translation efficiency, RNA stability, processing or localization). Second, the protein of interest (POI) is expressed as fusion protein together with a moiety that binds the cognate binding site expressed on the reporter RNA (e.g., MS2 coat protein (MS2CP) in Fig. 1). If both parts are present in the cell, the POI is tethered with high affinity and specificity to the reporter RNA and its effect on the reporter can be determined. The strength of a tethering experiment is that the impact of one protein is investigated in relation to one step in gene expression in a well-defined and controllable approach.

The focus of this chapter is on MS2 tethering, which was pioneered 1998 by Wickens's group [4] and used in numerous studies elucidating important functions of RBPs in RNA metabolism (for examples see Table 1). Similar tethering systems have been developed by using the λN peptide or the PP7 coat protein instead of the MS2CP to attach the POI to high affinity binding motifs in the reporter RNA, called BoxB binding sites for the λN peptide and PP7 hairpins for the PP7 coat protein. The PP7 tethering system is often used in combination with the MS2 system to investigate single mRNA molecules by microscopy [5, 6]. The λN peptide, the PP7 approach or other less commonly used tethering systems are not further discussed here, but concisely summarized in [7, 8].

In this chapter, the four steps of an MS2 tethering assay will be discussed to successfully investigate the effect of a POI on a specific readout of the reporter mRNA. Exemplarily, the protocol shows how the recruitment of the human cytoplasmic poly(A) binding protein (PABPC1) downstream of a stop codon in a reporter transcript leads to the stabilization of the mRNA molecule.

1.1 Designing and Cloning of Reporter Genes Containing MS2 Binding Sites

One of the most important considerations before starting a tethering experiment is choosing a suitable reporter gene, which strongly depends on the desired downstream application. For instance, to record translation efficiency Renilla luciferase is a good choice as luminescence detection is a fast and easy proxy for translation efficiency. In addition, it is also important that *cis*-acting features important for processing and stability of the reporter transcripts are characterized in detail. In our example, the abundance of a reporter mRNA with and without a tethered protein will be analyzed using reverse transcription followed by quantitative PCR (RT-qPCR) to characterize RNA stabilizing proteins. To this end, a version of an immunoglobulin μ (mini-μ) reporter gene harboring a premature translation termination codon (PTC) is used that has been studied extensively in the field of nonsense-mediated mRNA decay (NMD), a post-transcriptional mRNA surveillance pathway [9, 10] (Fig. 2a).

The localization of the MS2 binding sites in the reporter RNA is another critical aspect in the experiment design. Many examples show that the tethering position influences the outcome of the experiment and therefore testing different positions might be favorable. The RBP Staufen (STAU1) for instance elicits NMD when tethered to the 3′UTR, while attachment to the 5′UTR stimulates translation [11, 12]. It was even demonstrated that recruitment of proteins to the 5′UTR of reporter genes can lead to an inhibition of

Table 1
Short selection of MS2 tethering experiments

Analysis of		Protein of interest tethered	Comment	Reference
Abundance of mRNA	*Increased*	PABPC1 (poly(A) binding protein cytoplasmic 1)	PABPC1 tethering near a premature stop codon (3′ of PTC) suppresses NMD leading to increased reporter mRNA levels	[21]
	Reduced	UPF1, SMG6 (two proteins involved in NMD)	Tethering of NMD factors (e.g., UPF1 or SMG6) to 3′UTR of reporter mRNAs elicit NMD leading to the decay of the mRNA	[16, 15]
Translation efficiency	*Activation*	METTL3 (m6A methyl-transferase like 3)	METTL3 enhances translation when tethered close to a stop codon. Firefly luciferase is measured to determine translation efficiency	[22]
	Repression	TNRC6 (trinucleotide repeat containing 6; shown to interact with argonaute proteins)	Tethering of TNRC6b or TNRC6c to the 3′UTR of a luciferase RNA exerts a strong translation inhibition. Studies with mutant versions of TNRC6b indicate that ago-binding is not required for translation silencing	[23]
Alternative splicing		SRSF1	Tethering of SRSF1 to an identified *cis*-element in exon 4 in DOK7 pre-mRNA activates the intron-distal 5′ splice site and suppresses the intron-proximal 5′SS of DOK7 intron 4	[24]
RNA localization		GFP, YFP, mCherry	Tethering of fluorescent proteins is used to visualize localization of mRNAs in living cells	Numerous studies, discussed in [25]
	Transport	She3p	Tethering of She3p to lacZ reporter mRNA leads to the transport of the reporter mRNA to the bud tip in *S. cerevisiae*	[26]
	Site of transcription	–	Active site of transcription of a reporter gene was detected by RNA FISH with a Cy5-labeled probe that hybridizes to the MS binding sites of the reporter mRNA	[27]

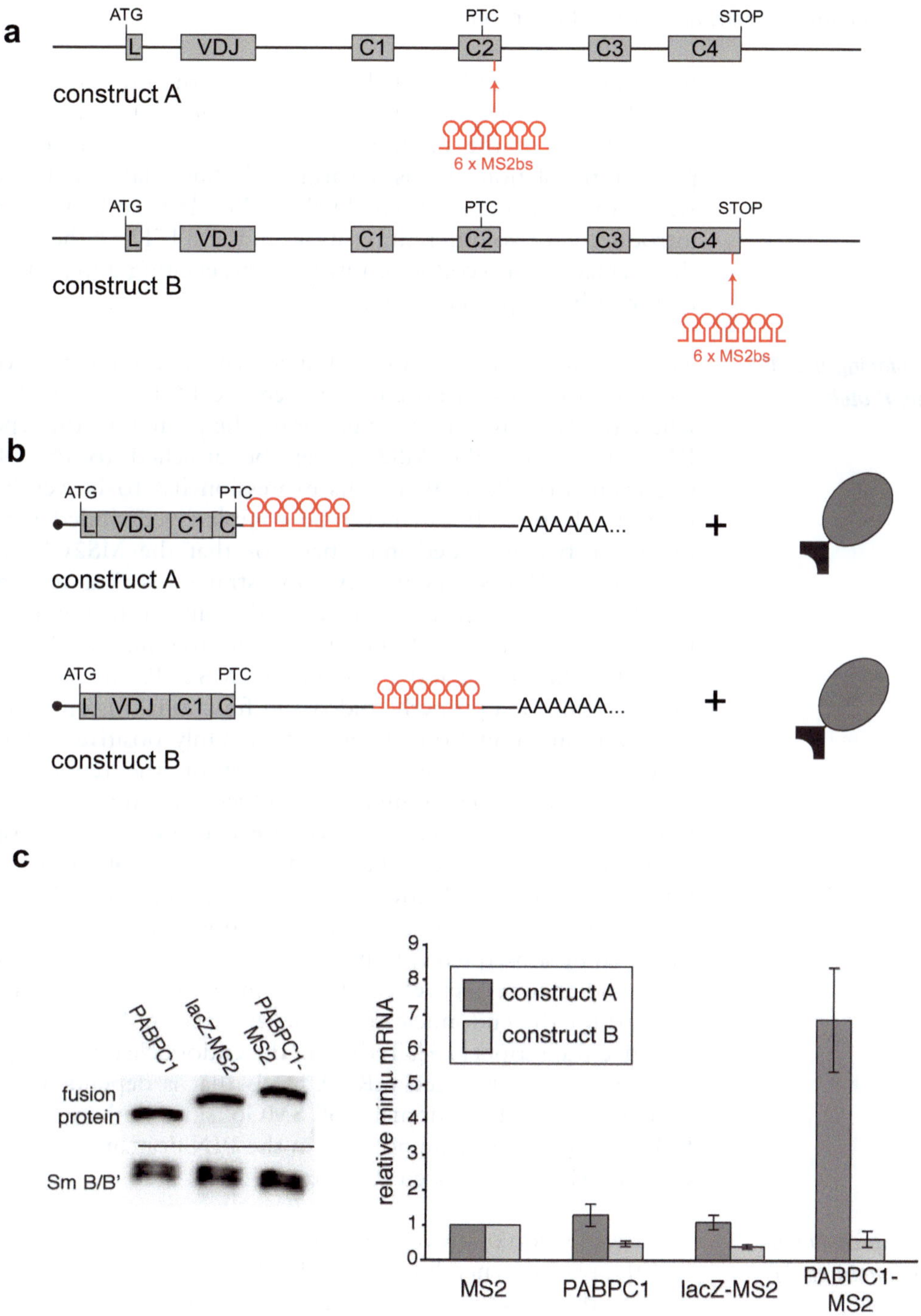

Fig. 2 Detailed workflow of the tethered function assays to elucidate the role of PABPC1 in mRNA abundance. (**a**) Insertion of six MS2 binding sites into a reporter plasmid (pβ mini-μ ter310) at two different positions (close to the PTC in construct A and further away in construct B). (**b**) Expression of these reporter plasmids in eukaryotic cells together with PABPC1 fused to MS2CP. (**c**) Left panel: Expression of PABPC1-MS2-HA fusion protein and the controls (PABBC1-HA without MS2CP or lacZ-MS2-HA) is analyzed by western blotting, SmB/B′ serves as loading control. Right panel: Relative mini-μ mRNA levels of construct A (dark bars) and construct B (light bars), normalized to cotransfected GPx1, is measured by RT-qPCR showing that tethering PABPC1 close to a termination codon leads to stabilization of the reporter mRNA (data taken from [21])

translation independent of the function of the protein [13]. The 3′UTR of reporter genes often represents a good starting choice for the addition of the binding sites, since this position does not prevent translation and is a natural binding platform for many RBPs. In the example outlined below, the MS2 binding sites were introduced at two different positions in the 3′UTR: either close to the premature stop codon or further away, eliciting different effects by the tethered proteins (Fig. 2a).

1.2 Cloning the MS2 Fusion Protein

The next step is to prepare a plasmid which contains the coding sequence of a fusion protein between the POI and the MS2CP, where the latter is crucial for tethering the protein to the reporter RNA. Generally, the MS2CP can be attached to the N- or C-terminus of the POI, but its expression has to be verified by western blotting. It can never be completely excluded that the fusion protein is folded improperly or that the MS2CP moiety disturbs the POI's function. A good strategy is thus to add the MS2CP to either terminus of the POI from the beginning. Furthermore, the addition of a GC linker helps to minimize the risk for sterical clashes between the POI and the MS2CP. The requirement of expression of a properly folded and functional fusion protein is an important limitation of the system. Only positive results are functionally meaningful while no effect on the reporter RNA could always be due to improperly folded and thus inactive proteins. In addition, several controls are necessary to monitor the quality of the experiment that are discussed in Subheading 3.2. If tethering of the POI affects the reporter, further insights into important domains or amino acid residues of the POI can be obtained by subsequently tethering mutant proteins to the reporter RNA and analyze their effect, for example, as shown for PABPC1 [14] or for the endonuclease SMG6 involved in degrading NMD sensitive transcripts [15]. Tethering the endonuclease to a reporter mRNA led to a reduced mRNA levels that is dependent on the catalytic active PIN domain of SMG6 since tethering mutant SMG6 (three crucial amino acids in the PIN domain are mutated) did not exhibited reduced RNA levels [15].

1.3 Expression of Both Constructs in Cells

The transfection of the reporter plasmid with the MS2 binding sites and the plasmid encoding the POI-MS2CP fusion protein to cells leads to the tethering of the POI to the reporter RNA by virtue of the MS2CP binding to the MS2 binding site. Expression of the reporter and effector constructs need to be carefully monitored to minimize indirect effects deriving from their overexpression in the cells.

Furthermore, elegant experiments can be conducted by combining knockdown experiments with the tethering approach, thereby investigating the interplay between two proteins. This approach was for example extensively and successfully used for exploring the NMD pathway: The endonuclease SMG6 was tethered to a reporter RNA, leading to rapid degradation of this transcript as mentioned above. However, if the SMG6 tethering was combined with a knockdown of the NMD key factor UPF1, the level of the reporter RNA was not reduced and SMG6-mediated decay of the reporter mRNA impaired, demonstrating that the presence of UPF1 is required for SMG6 to exert its endonuclease activity [15].

1.4 Functional Analysis of the Protein of Interest

The functional study at the end of the experiment lies of course in the suspected function of the RBP and the reporter gene chosen in the beginning. Different downstream measurements can be performed to analyze the impact of the POI to RNA processing/ splicing, stability, mRNA localization or translation (for examples *see* Table 1 and references therein): Processing of the reporter RNA can be assessed by northern blotting or analytical PCR, subcellular localization of RNA molecules can be observed by microscopy or by fractionation, and translation efficiency can be recorded by checking the protein abundance by luminescence (for luciferase), fluorescence (for GFP) or western blotting. In the example shown here, the levels of RNA molecules are determined by RT-qPCR addressing the question whether PABPC1 affects the abundances of transcripts.

2 Materials

2.1 Cloning of Plasmids

1. Oligonucleotides used for cloning are listed in Table 2.
2. DNA template for PCR.
   (a) to amplify MS2 binding sites, for example, β2-3MS2 containing six MS2 binding sites ([16] and *see* **Note 1**).
   (b) to amplify your POI, for example, pETNHis-hPABPC1 (obtained by Elmar Wahle, University of Halle, Germany).
3. Reporter plasmid where the MS2 binding sites will be introduced, for example, pβ mini-μ ter310 [9].
4. Expression plasmid encoding MS2CP, for example, pCMV-H2b-MS2-HA (*see* **Note 2**).
5. 10× buffer for high-fidelity polymerase (e.g., *pfu*Ultra high-fidelity reaction buffer, Agilent).
6. dNTP stock (10 mM each).
7. High-fidelity polymerase (e.g., *pfu*Ultra high-fidelity DNA polymerase, Agilent).
8. Tubes for PCR reaction and PCR machine.

Table 2
Oligonucleotides used for cloning and qPCR

Name	Sequence (5′–3′)	Notes
For cloning		
MS2 bs_for	CTAAGTTGTACAACCAAACTGGGTC TAGCTCTAG	To amplify MS2 binding sites (+*BsrGI* site)
MS2 bs_rev	ATTTAGTGTACACTATAGAATAGGGCCC TCTAG	To amplify MS2 binding sites (+*BsrGI* site)
PABPC1_for	CATGGTACCACCATGAACCCCAG TGCCCCCAGCTACCCCATGGCCTCGC TCTACGTG	To amplify PABPC1 (*KpnI* site)
PABPC1_rev	CATGGATCCGGAACAG TTGGAACACCGGTGG	To amplify PABPC1 (*BamHI* site)
For MS2-HA control_for	CACCATG	Annealed oligonucleotides for cloning of MS2-HA (*see* **Note 13**)
For MS2-HA control_rev	GATCCATGGTGGTAC	Annealed oligonucleotides for cloning of MS2-HA (*see* **Note 13**)
For PABPC1-HA control_for	GATCCATACCCATATGATGTTCCAGA TTACGCTTCACTCGAATGAGC	Annealed oligonucleotides for cloning of PABPC1-HA (*see* **Note 14**)
For PABPC1-HA control_rev	GGCCGCTCATTCGAGTGAAGCGTAATC TGGAACATCATATGGGTATG	Annealed oligonucleotides for cloning of PABPC1-HA (*see* **Note 14**)
lacZ_for	AAGGTACCACCATGGTGCGCTG TTCGCATTATCC	To amplify part of lacZ (*KpnI* site)
lacZ_rev	TTGGATCCGG TTTTTGACACCAGACCAACTG	To amplify part of lacZ (*BamHI* site)
For qPCR		
Mini μ assay:	Forward: GTCTCACCTTCTTGAAGAACGTGTC Reverse: GGGATGGTGAAGGTTAGGATGTC Taqman probe: FAM-CACATGTGCTGCCAGTCCCTCCAC -BHQ	To quantify mini μ mRNA: TaqMan probe lies over exon–exon junction (C2-C3)
GPx1 assay:	Forward: TGGTGGTGCTCGGTTTCC Reverse: GACATACTTGAGGGAA TTCAGAATCTC Taqman probe: FAM-CCATTCTCCTGATGTCCGAACTGA TTGC-TAMRA	To quantify GPx1 mRNA to normalize: TaqMan probe lies over exon–exon junction

9. Reagents and equipment for agarose gel: agarose, TAE buffer (40 mM Tris, 20 mM acetic acid, 1 mM EDTA), ethidium bromide, loading buffer, DNA ladder.
10. Restriction enzyme of interest and corresponding buffer, for example, *Bsr*GI, *Bam*HI, *Not*I, *Kpn*I.
11. DNA purification kit (e.g., Wizard SV Gel and PCR Clean-Up kit, Promega).
12. Buffer for Antarctic phosphatase (50 mM Bis-Tris–Propane–HCl, 1 mM $MgCl_2$, 0.1 mM $ZnCl_2$, pH 6).
13. Antarctic phosphatase.
14. 10× buffer for T4 DNA ligase.
15. T4 DNA ligase.
16. 100 mM NaCl.
17. 10× buffer for T4 polynucleotide kinase (PNK).
18. 10 mM ATP.
19. T4 PNK.
20. Chemically competent *E. coli* XL1 blue.
21. Ampicillin (Amp) or Kanamycin (Kan).
22. Lysogeny broth (LB) medium.
23. LB agar plates.
24. Plasmid Miniprep kit.
25. NanoDrop instrument (UV spectrophotometer).

2.2 Expression of Constructs in Cells

1. Human cervix epithelioid carcinoma cell line (HeLa cells).
2. 37 °C incubator, 5% CO_2.
3. Dulbecco's modified Eagle's medium (DMEM).
4. DMEM +/+: DMEM supplemented with 10% heat-inactivated fetal calf serum (FCS), 100 U/mL penicillin, and 100 μg/mL streptomycin.
5. 6-well plates.
6. Transfection reagent (e.g., DreamFect, OZ Biosciences).
7. Plasmid to transfect for normalization, for example, pCMV-rGPx1 (*see* **Note 3**).
8. PBS buffer: 137 mM NaCl, 2.7 mM KCl, 10 mM Na_2HPO_4, 2 mM KH_2PO_4.
9. T/E: 0.05% trypsin–EDTA.
10. Neubauer counting chamber or cell counter.
11. SDS-PAGE loading buffer.
12. Lysis buffer for RNA analysis (from RNA isolation kit) or TRI-mix (*see* **Note 4**).

2.3 Functional Analysis of the Protein of Interest

1. RNase-free water: filtered or DEPC-treated.
2. RNA isolation kit (e.g., Absolutely RNA Miniprep Kit, Agilent) or reagent for TRI-mix extraction method.
3. Reagents for DNase treatment (e.g., TURBO DNA-free™ kit, Ambion).
4. Reagents for reverse transcription (RT): 150 ng/μL random hexamers, RT-buffer, 10 mM dNTP, RNase inhibitor, reverse transcriptase.
5. Reagents and equipment for qPCR: qPCR machine, qPCR tubes, 2× master mix for qPCR (e.g., Brilliant III Ultra-Fast qPCR Master Mix, Agilent).
6. Oligonucleotides and Taqman probes for qPCR are listed in Table 2.
7. General reagents and equipment for SDS-PAGE and western blotting.
8. Antibodies to detect tagged fusion proteins and a protein to normalize, e.g., anti-HA and anti-SmB/B′.
9. For analytical PCR: suitable oligonucleotides, Taq Polymerase, PCR buffer, dNTPs.

3 Methods

3.1 Designing and Cloning of Reporter Genes Containing MS2 Binding Sites

Before starting the experiment, decide which reporter gene is suitable for the desired readout and where the MS2 binding sites should be located in the reporter gene (see introduction for further discussion). Additionally, the number of the MS2 binding sites should be carefully evaluated. It was shown that two MS2 stem loops are sufficient for successful recruitment of a protein to the reporter gene [4, 16–18], but often more than two MS2 binding sites, up to 24 binding sites, are used, eventually increasing the detected effect of the RBP on the reporter RNA [19]. However, it has to be assessed that by adding a long sequence with strong secondary structure, processing or stability of the reporter transcript is not impaired. Additionally, 24 repeats of the MS2 binding site may cause problems to efficiently and correctly replicate such a plasmid in bacteria, often resulting in low plasmid yield and skipping of MS2 binding sites. If a cassette consisting of ≥12 binding site is needed, for example, to reach sufficient sensitivity for single molecule live cell imaging experiments [19] recombinase-deficient *E. coli* strains are recommended for plasmid amplification and for each plasmid preparation, the number of MS2 binding sites should be confirmed. Here, six MS2 binding sites were inserted ~70 nts downstream of a PTC (ter310) in the C2 exon of the pβ mini-μ ter310 plasmid (construct A, Fig. 2a) or further away, ~45 nts upstream of the physiological termination codon in exon C4 (construct B, Fig. 2a). In particular, the MS2 binding sites were amplified by PCR and introduced into the *BsrGI* restriction site located in the C2 exon (downstream of ter310) and in the C4 exon.

1. Find suitable restriction sites to introduce the MS2 binding sites into the reporter gene of choice and design oligonucleotides accordingly (*see* **Note 5**):

   Forward oligonucleotides:

   CTAAGTTGTACA<u>ACCAAACTGGGTCTAGCTCTAG</u>-3'

   Reverse oligonucleotides:

   ATTTAGTGTACA<u>CTATAGAATAGGGCCCTCTAG</u>-3'

   TGTACA: *BsrGI* restriction site.

   bold: additional nucleotides for efficient digestion.

   <u>underlined:</u> complementary to template plasmid to amplify MS2 binding sites (sequences for MS2 binding sites *see* **Note 6**).

2. Perform PCR to amplify the six MS2 binding sites with *BsrGI* overhangs at both sides. As template, pcβwt β2-3MS2 containing six MS2 binding sites is used [16]. Alternatively, any plasmid with MS2 binding sites can serve as template (for suitable plasmids from Addgene *see* **Note 1**). Pipette the following reagents in PCR tube: 1× PCR reaction buffer, 0.2 mM of each dNTP, 0.4 μM of each primer, 50 ng template DNA, and 2.5 U high-fidelity DNA polymerase.

3. Run the PCR according to the following scheme: 1 × (95 °C—2 min); 35 × (95 °C—30 s, 55 °C—30 s, 72 °C—30 s); 1 × (72 °C—5 min).

4. Analyze 5 μL of the PCR reaction on an 1.5% agarose gel (*see* **Note 7**).

5. Digest the PCR reaction with *BsrGI* enzyme in the recommended buffer at 37 °C. Afterward, heat-inactivate the enzyme at 80 °C for 20 min.

6. Purify the amplified and digested PCR fragment by a PCR purification kit (*see* **Note 8**).

7. Digest reporter pβ mini-μ ter310 plasmid with *BsrGI*. Since two *BsrGI* sites are present in pβ mini-μ ter310 a partial digestion is necessary to get a linearized plasmid that is cut at only one *BsrGI* site. For that prepare several restriction reactions and incubate each tube for a different time span at 37 °C (*see* **Note 9**).

8. Run an aliquot of each digestion reaction on an analytical agarose gel and decide for the best digestion condition to receive plasmids cut at one *BsrGI* site. For comparison load undigested plasmid on the agarose gel.

9. Repeat the partial digestion with the decided condition and run the whole reaction on an agarose gel.
10. Purify the linear plasmid by cutting out the desired band from the agarose gel and follow the protocol of the purification kit.
11. Dephosphorylate linearized vector to reduce religation of the vector: Incubate purified plasmid with 5 U Antarctic phosphatase and corresponding buffer for 30 min at 37 °C. Heat-inactivate phosphatase at 80 °C for 2 min.
12. Insert the digested and purified PCR fragment either into the *BsrGI* site in exon C2 (for construct A) or into *BsrGI* site in exon C4 (for construct B) of pβ mini-μ ter 310 plasmid. For that, add the linearized and dephosphorylated pβ mini-μ ter 310 plasmid (from **step 11**) and the digested PCR fragment (from **step 6**) in a tube with T4 DNA ligase and the corresponding buffer (including ATP) in a molar ratio of 1:3 and incubate overnight at 16 °C (*see* **Note 10**).
13. Next day, transform the ligation reaction to chemically competent *E. coli* XL1 blue and plate them on LB plates containing ampicillin (Amp).
14. Analyze the colonies from the plate: Inoculate single colonies in 4 mL LB/Amp and cultivate them in a shaker (200 rpm) over night at 37 °C.
15. Isolate the plasmids by a Miniprep kit the next day.
16. Find correct plasmids by analysis restriction digestions on agarose gel. Here, the minipreps are digested with *XbaI* (cutting once in MS2 binding sites and once in mini-μ ter310 backbone) so that the orientation of the insert can also be determined.
17. Send plasmids for sequencing to confirm that they are correct.
18. The new constructs (construct A and construct B) have to be tested to confirm that the insertion of the MS2 binding sites does not disturb the stability and processing of the expressed RNA of the reporter plasmid. Splicing of the reporter RNA is analyzed by analytical RT-PCR as described in Subheading 3.4. To monitor the stability of the newly cloned plasmids, the RNA abundances of construct A and construct B can be measured by RT-qPCR and compared to the original plasmid. For that, transfect the original plasmid pβ mini-μ ter310t or pβ mini-μ ter310 construct A or pβ mini-μ ter310 construct B in HeLa cells and analyzed the RNA levels 48 hours post transfection by RT-qPCR (for details *see* Subheadings 3.3 and 3.4).

3.2 Cloning the MS2 Fusion Protein

Expression plasmids encoding a fusion protein between the POI and the MS2CP are needed as second piece for a functional tethering approach (Figs. 1 and 2c: PABPC1-MS2). The MS2CP of 14 kDa should be appended to the N- and C-terminus of the POI to test possible interference of the MS2CP moiety to protein's function (data not shown). In addition, several controls are needed to convincingly show that the effect detected relies on the POI's recruitment to the reporter RNA: (1) MS2CP alone (Fig. 2c: MS2), (2) POI without the MS2CP moiety (Fig. 2c: PABPC1), and (3) an unrelated protein of similar size as the POI fused with MS2CP (Fig. 2c: LacZ-MS2). Additionally, mutant versions of the POI are interesting to include as discussed in the introduction.

In all constructs, an HA-tag is added to monitor the expression of the proteins by western blotting. In this example, the MS2CP is fused to the C-terminus of the POI. However, attaching the MS2CP moiety to the N-terminus was also tested and no difference was found between the two positions of the MS2 protein.

3.2.1 Cloning of pCMV-PABPC1-MS2-HA

The starting plasmid for the generation of the MS2 fusion plasmids was a plasmid containing the open reading frame of histone H2b fused to the sequence encoding the MS2CP and an HA-tag (pCMV-H2b-MS2-HA). H2b is flanked by *Kpn*I and *Bam*HI and can therefore be easily replaced with any sequence of a POI that is amplified by PCR with flanking *Kpn*I and *Bam*HI sites (*see* **Note 11** for sequence information). Any other plasmid containing the sequence of the MS2CP can be used as a starting point for cloning or as template for PCR amplification (*see* **Note 2**).

1. Design forward and reverse oligos to amplify the open reading frame of your POI and include the desired restriction sites at the end. Here, the primers used to amplify the open reading frame of PABPC1 contain a *Kpn*I (forward oligonucleotide) and *Bam*HI (reverse oligonucleotide) restriction site and a few additional nucleotides at the end to enable efficient binding and cutting of the respective restriction enzymes (oligonucleotides are listed in Table 2).
2. Run PCR to amplify the open reading frame of the POI with *Kpn*I and *Bam*HI overhangs at the sides. As template pETNHis-hPABPC1 is used (obtained by Elmar Wahle, University of Halle, Germany). For that, pipette 1× PCR reaction buffer, 0.2 mM of each dNTP, 0.4 μM of each primer, 50 ng template DNA and 2.5 U high-fidelity DNA polymerase to a PCR tube and run the reaction in a PCR cycler according to the T_m of the oligonucleotides, length of the amplicon, and properties of the polymerase.
3. Digest the amplicon with *Kpn*I and *Bam*HI at 37 °C.
4. Run the digested amplicon on an agarose gel and purify it via a purification kit (*see* **Note 8**).

5. Digest the vector pCMV-H2b-MS2-HA with *Kpn*I and *Bam*HI at 37 °C to cut out the H2b insert of about 400 bp.
6. Add phosphatase and corresponding buffer to the restriction reaction to dephosphorylate the vector. Incubate the reaction for 30 min at 37 °C.
7. Run the digested and dephosphorylated reaction on an 1% agarose gel. Cut out vector backbone band and purify via a purification kit (*see* **Note 12**).
8. Ligate vector backbone with PCR amplified and digested PABPC1 (molar ratio 1:3) over night at 16 °C to generate pCMV-PABPC1-MS2-HA (as done before in Subheading 3.1, **step 12**).
9. Transform the ligation reaction to competent *E. coli* XL1 blue and plate them on LB plates containing kanamycin (Kan).
10. Analyze the colonies by restriction analysis and sequencing of the isolated plasmids (as described above in Subheading 3.1, **step 14**).

3.2.2 Cloning of Controls: MS2CP Alone (pCMV-MS2-HA)

1. Use the digested and purified vector backbone from **step** 7 (H2b sequence removed).
2. Design two oligonucleotides that can anneal and close the digested vector backbone: 5′-CACCATG-3′ and 5-′-GATCCATGGTGGTAC-3′. The Kozak sequence is included to increase translation initiation (*see* **Note 13** for clarification).
3. Anneal the two oligonucleotides: Mix 10 μL of each oligonucleotides (10 μM each) and add 20 μL 100 mM NaCl (final concentration: 50 mM). Incubate the tube at 95 °C for 5 min, slowly cool to room temperature.
4. Phosphorylate the annealed oligos with 10 U T4 polynucleotide kinase (PNK) in 1 × PNK reaction buffer and 1 mM ATP at 37 °C for 30 min. Heat-inactivate the enzyme by incubating at 65 °C for 20 min.
5. Ligate the annealed and phosphorylated oligonucleotides with the digested vector backbone and continue as described above for cloning of pCMV-PABPC1-MS2-HA (**steps 9** and **10**).

3.2.3 Cloning of Controls: PABPC1 Without MS2CP (pCMV-PABPC1-HA)

The cloning strategy of pCMV-PABPC1-HA is the same as for pCMV-MS2-HA. pCMV-PABPC1-MS2-HA is digested with *Bam*HI and *Not*I to remove MS2-HA. The complementary oligonucleotides (5′-GATCCATACCCATATGATGTTCCAGATTAC GCTTCACTCGAATGAGC-3′ and 5′-GGCCGCTCATTCGA GTGAAGCGTAATCTGGAACATCATATGGGTATG-3′) are annealed, phosphorylated, and ligated into the vector backbone (cut by *Bam*HI and *Not*I) bringing back the HA-tag, but not the MS2 sequence (*see* **Note 14**).

3.2.4 Cloning of Controls: Unrelated Protein with MS2CP (pCMV-lacZ-MS2-HA)

Follow the protocol described for cloning of pCMV-PABPC1-MS2-HA, but design primers to amplify a sequence encoding a control protein with no function in RNA metabolism. In our example here, a C-terminal fragment of lacZ (1910 bp) of similar size than PABPC1 was amplified by PCR, including suitable restriction sites at both ends (*Kpn*I and *Bam*HI).

3.3 Expression of Constructs in Cells

After confirming the cloned plasmids by sequencing, they are transfected to eukaryotic cells leading to POI's tethering to the MS2 binding sites of the reporter RNA (Fig. 2b). Here, we use HeLa cells cultivated in Dulbecco's modified Eagle's medium (DMEM), supplemented with 10% heat-inactivated fetal calf serum (FCS), 100 U/mL penicillin, and 100 μg/mL streptomycin.

1. Seed 2×10^5 HeLa cells in 6-well plates and incubate overnight at 37 °C, 5% CO_2.
2. Transfect the plasmids the next day. Titration experiments for finding optimal plasmid concentrations have been performed before (*see* **Note 15**). Set up transfection reaction accordingly.
3. Prepare the following four plasmid mixtures for **mini-μ construct A** in 100 μL DMEM −/−: 100 ng mini-μ construct A (6 MS2 binding sites in C2 exon, *see* Subheading 3.1); 100 ng pCMV-rGPx1 (for normalization, *see* **Note 3**); 300 ng MS2-HA or PABPC1-HA or lacZ-MS2-HA or PABPC1-MS2-HA (*see* Subheading 3.2).
4. Prepare the following 4 plasmid mixtures for **mini-μ construct B** in 100 μL DMEM −/−: 100 ng mini-μ construct B (6 MS2 binding sites in C4 exon, *see* Subheading 3.1); 100 ng pCMV-rGPx1(for normalization, *see* **Note 3**); 300 ng MS2-HA or PABPC1-HA or lacZ-MS2-HA or PABPC1-MS2-HA (*see* Subheading 3.2).
5. Prepare the transfection reagent mixture in a tube: 4 μL transfection reagent DreamFect and 96 μL DMEM −/− per transfection (*see* **Note 16**).
6. Mix 100 μL of the transfection reagent mixture with 100 μL plasmid mixture and incubate for 20 min at room temperature.
7. In between wash cells once with PBS and add 1 mL DMEM −/−.
8. After 20 min, add each 200 μL DreamFect-plasmid mix dropwise to the cells in the 6-well plate.
9. Change medium 3–4 h later and add DMEM +/+.
10. Change the medium once again on the following day.
11. Harvest the cells 48–52 post transfection: Wash cells once with PBS, add a few drops T/E, and incubate for 5–10 min at 37 °C.

Resuspend cells in DMEM +/+, put them on ice, and count them by cell counter or with the help of a Neubauer counting chamber.

12. For Western Blotting, add 2×10^5 cells to a tube, centrifuge for 5 min at 200 rcf (4 °C) and remove supernatant. The cells are lysed in SDS-PAGE loading buffer and stored at −20 °C until usage (continue in Subheading 3.4, Western Blot).
13. For RNA analysis, spin down the residual cells (5 min, 200 rcf, 4 °C), remove supernatant and resuspend pellet in 1 mL TRI-mix or Lysis Buffer of an RNA Isolation kit (*see* **Note 4**).

3.4 Functional Analysis of the Protein of Interest

Finally, the effect of the tethered protein to the reporter transcript can be analyzed. Here, the **abundance of the reporter RNA** in the different conditions is measured **by RT-qPCR**:

1. Isolate total RNA by a suitable kit (e.g., Absolutely RNA Miniprep kit) or by TRI-mix (*see* **Note 4**).
2. Degrade residual DNA by DNase treatment (e.g., Turbo DNA-free kit or add DNase to the column when using RNA isolation kit).
3. Measure RNA concentration by nanodrop.
4. Reverse transcribe 1 μg RNA to cDNA. To control possible DNA contamination, include RT control samples in which the reaction is performed as depicted below, but substitute the reverse transcriptase with H_2O. A specific RT protocol follows: add 1 μg RNA, H_2O and 300 ng random hexamer in a tube (total volume 41 μL). Incubate for 5 min at 65 °C to remove secondary structures. Cool down at room temperature for 10 min. Add RT buffer, 0.4 mM dNTPs, 40 U RNase inhibitor, 50 U reverse transcriptase to the tube (total volume 50 μL). Incubate for 60 min at 42 °C. For inactivation of the RT incubate at 85 °C for 10 min. Add 75 μL DEPC-treated H_2O to the tube (→8 ng/μL).
5. Run a quantitative PCR (qPCR, real-time PCR) with 40 ng of reverse transcribed RNA per reaction using assays to quantify mini-μ mRNA and GPx1 mRNA levels (TaqMan probe [200 nM], forward and reverse oligo [800 nM each]). Include in the same run H_2O as nontemplate control and RT control samples to detect any DNA contamination.
6. Analyze the effect of the tethered POI in the RT-qPCR as in Fig. 2c.

Two important controls have to be conducted to convincingly show the effect of the tethered protein. First, the **expression of the fusion proteins** in HeLa cells has to be detected **by western blotting**:

7. Run whole cell lysates corresponding to 2×10^5 HeLa cells (*see* Subheading 3.3, **step 12**) per lane on a 10% SDS-PAGE.
8. Transfer proteins to a nitrocellulose membrane.
9. After blocking the membrane with 5% milk powder dissolved in TBS–Tween for 30–60 min, add the first antibody (α-HA and α-SmB/B′ antibody diluted in 5% milk–TBS–tween) to the membrane and incubate overnight in the cold room.
10. Wash the membrane three times for 5 min with TBS–Tween and incubate membrane with corresponding secondary antibody for 1 h at room temperature.
11. Wash the membrane with TBS–Tween and detect proteins with standard method of the laboratory. SmB/B′ serves as loading control (*see* Fig. 2c for results and **Note 17**).

Second, the **correct processing of the reporter transcripts** expressed HeLa cells has to be examined **by analytical RT-PCR**:

12. Use the cDNA prepared in **steps 1–4** or prepare cDNA using oligo(dT) as primer (*see* **Note 18**).
13. Amplify reverse transcribed material corresponding to 200 ng by PCR using Taq polymerase, 0.4 μM of forward and reverse oligonucleotides in 1× PCR buffer (including dNTPs). For checking splicing of the reporter RNA, use primers flanking the insertion site of the MS2 binding sites. For construct A, primers complementary to a sequence in C1 exon (forward) and C4 exon (reverse) are used; for construct B, primers anneal to C2 exon (forward) and the 3′ UTR (after C4 exon, reverse).
14. Analyze end-point RT-PCR amplicons on a 1.25% agarose gel to confirm that splicing is not impaired.

4 Notes

1. The following Addgene plasmids can serve as templates to amplify MS2 binding sites:
   (a) pCI-TPI_WT-xrRNA-4MS2-4H (#108370, Niels Gehring).
   (b) pSL-MS2-6X (#27118, Robert Singer).
   (c) pSL-MS2-12X (#27119, Robert Singer).
2. The following Addgene plasmids contain the sequence of the MS2CP:
   (a) MS2_GFP (#61764, Daniel Larson).
   (b) phage UBC NLS-HA-2XMCP-tagRFPt (#64541, Jeffrey Chao).

3. An unrelated normalizer plasmid, which codes for an mRNA that is not present in HeLa cells, is used to compensate for varying transfection efficiencies in different samples.
4. RNA can be isolated by column-based methods such as the silica-based purification or by Guanidium Thiocyanate-Phenol-Chloroform extraction. For the latter extraction method, a TRI-mix consisting of water-saturated phenol, chloroform and a chaotropic denaturing agent called guanidinium thiocyanate is necessary (e.g., TRIreagent (Sigma-Aldrich), TRIzol (Invitrogen)). The protocol for producing a "homemade" TRImix can be found in [20].
5. Newer cloning strategies that are independent on restriction sites (e.g., In-Fusion® Cloning (Takara), Gibson Assembly® (New England Biolabs), GeneArt Seamless Cloning (Invitrogen)) as well as gene synthesis technology are of course applicable for all the cloning procedures.
6. Sequence of amplicon containing the six MS2 binding sites, flanked by *Bsr*GI sites:

Tgtacaaccaaactgggtctagctctagctgtagaaaacatgaggat**c**acccatgtctgctggacgactgt
agaaaacatgaggat**c**acccatgtctgctgtctagctgtagaaaacatgaggat**c**acccatgtctgctgga
cgactgtagaaaacatgaggat**c**acccatgtctgctgtctagctgtagaaaacatgaggat**c**acccatgtc
tgctggacgactgtagaaaacatgaggat**c**acccatgtctgctgtctagagggccctattctatag**tGTAC**
A

Underlined: complementary to oligonucleotides used in PCR.

21 nt RNA stem loops are marked in gray; each stem loop contains a mutation (**c in bold** instead of U in original stem loop) leading to a higher affinity for MS2CP [7].

7. A single DNA band of 297 bp should be present on the agarose gel if the indicated primers and template is used.
8. Purify the digested PCR amplicon directly by a PCR purification kit if a single band is detected on the agarose gel. If more than one band is detected, run the digested PCR product on a new preparative agarose gel, cut out the expected band and extract the amplicon from the agarose gel piece according to the protocol of the purification kit.
9. Partial digestions can be performed in different ways and have to be empirically determined for each restriction enzyme. It is recommended to change only one parameter, either time or amount of enzyme, per experiment. If the enzyme concentration is changed, use for instance a 1:10 dilution so that in the end reaction 20 U, 2 U, 0.2 U, 0.02 U, and 0.002 U of the

enzyme is used and incubated at 37 °C for 20 min. Stop the reaction immediately by addition of EDTA (10 mM final concentration) or by heating it up to 80 °C for 20 min. Otherwise keep the amount of enzyme constant and prepare five restriction reactions that are incubated at 37 °C but stop the reaction at different time points, e.g., after 5, 10, 15, 30, 60 min. Another common strategy that can be combined with the conditions above is the addition of ethidium bromide to the reaction to inhibit efficient second cutting of the restriction enzyme.

10. To control the preparation of the vector backbone (digestion and dephosphorylation), it is recommended to perform a non-insert control. To do so, add the same amount of vector to the ligation buffer and the ligase, but substitute the insert with H_2O.

11. A part of the sequence of the pCMV-H2b-MS2-HA plasmid depicting the three restriction sites *Kpn*I, *Bam*HI, and *Not*I:

ggtaccgccaccatgctatgccagagccagcgaagtctgctcccgccccgaaaagggctccaagaag

gcggtgactaaggcgcagaagaaaggcggcaagaagcgcaagcgcagccgcaaggagagctattccat

ctatgtgtacaaggttctgaagcaggtccaccctgacaccggcatttcgtccaaggccatgggcatca

tgaattcgtttgtgaacgacattttcgagcgcatcgcaggtgaggcttcccgcctggcgcattacaac

aagcgctcgaccatcacctccagggagatccagacggccgtgcgcctgctgctgcctggggagttggc

caagcacgccgtgtccgagggtactaaggccatcaccaagtacaccagcgctaagGATCCAATGGCTT

CTAACTTTACTCAGTTCGTTCTCGTCGACAATGGCGGAACTGGCGACGTGACTGTCGCCCCAAGCAAC

TTCGCTAACGGGATCGCTGAATGGATCAGCTCTAACTCGCGTTCACAGGCTTACAAAGTAACCTGTAG

CGTTCGTCAGAGCTCTGCGCAGAATCGCAAATACACCATCAAAGTCGAGGTGCCTAAAGGCGCCTGGC

GTTCGTACTTAAATATGGAACTAACCATTCCAATTTTCGCCACGAATTCCGACTGCGAGCTTATTGTT

AAGGCAATGCAAGGTCTCCTAAAAGATGGAAACCCGATTCCCTCAGCAATCGCAGCAAACTCCGGCAT

CTACGCG**TACCCATATGATGTTCCAGATTACGCTTCACTCGAATGA**GCGGCCGC

Underlined: H2b sequence.

Bold, gray: MS2CP sequence.

Bold, black: HA-tag sequence.

12. If the indicated plasmid and primers are used, the expected bands on an agarose gel are 4,382 bp for vector backbone (desired) and 390 bp for the H2b insert.

13. The oligonucleotides contain part of the Kozak consensus sequence (GCCRCC**ATG**G; start codon in bold) and by annealing the two oligonucleotides, the necessary sticky ends of the restriction enzymes are generated (***Kpn*I** and ***Bam*HI** site):

5´-**C**ACCAT**G**-3´

3´-**CATGG**TGGTA**CCTAG**-5´

14. The annealed oligonucleotides have sticky ***Bam*HI** and ***Not*I** ends. In addition, the sequence of the **HA-tag (bold)** is introduced in the oligos:

5´-**GATCC**A**TACCCATATGATGTTCCAGATTACGCTTCACTCGAATGA**GC-3´

3´-

GT**ATGGGTATACTACAAGGTCTAATGCGAAGTGAGCTTACT**CGCCGG-5´

15. Before starting the real experiments, it is strongly recommended to identify the optimal amount of the reporter plasmid (containing the MS2 binding sites) for transfection (suggested range: 50–500 ng/well in 6-well plates). If a condition with reasonable readout is found (assessable by, e.g., C_t value in RT-qPCR), the plasmid expressing the MS2 fusion protein should be also titrated (suggested range 300–1500 ng/well in 6-well plates).
16. To minimize differences in transfection efficiencies, a master mix for the transfection reagent mixture should be prepared. For 8 transfection reactions, mix 34 μL DreamFect (4 μL × 8.5) with 816 μL DMEM −/− (96 μL × 8.5) as master mix.
17. MS2CP has a molecular weight of 14 kDa and is not shown on the western blot. The calculated molecular weight of PABPC1-HA, LacZ-MS2-HA, and PABPC1-MS2-HA are 71 kDa, 87 kDa, and 85 kDa, respectively.
18. To enrich the synthesis of cDNA deriving from mRNA, the RT reaction can be performed with oligo(dT) as primer instead of random hexamers (use 200 nM 5′-d(T)$_{30}$VN-3′).

Acknowledgments

The research of A.B.E. and O.M. is supported by the NCCR RNA & Disease funded by the Swiss National Science Foundation (SNSF), by the SNSF grant 310030B-182831 to O.M., and by the canton of Bern.

References

1. Gerstberger S, Hafner M, Tuschl T (2014) A census of human RNA-binding proteins. Nat Rev Genet 15(12):829–845. https://doi.org/10.1038/nrg3813
2. Hentze MW, Castello A, Schwarzl T, Preiss T (2018) A brave new world of RNA-binding proteins. Nat Rev Mol Cell Biol 19(5): 327–341. https://doi.org/10.1038/nrm.2017.130
3. Singh G, Pratt G, Yeo GW, Moore MJ (2015) The clothes make the mRNA: past and present trends in mRNP fashion. Annu Rev Biochem 84:325–354. https://doi.org/10.1146/annurev-biochem-080111-092106
4. Coller JM, Gray NK, Wickens MP (1998) mRNA stabilization by poly(a) binding protein is independent of poly(A) and requires translation. Genes Dev 12(20):3226–3235. https://doi.org/10.1101/gad.12.20.3226
5. Hocine S, Raymond P, Zenklusen D, Chao JA, Singer RH (2013) Single-molecule analysis of gene expression using two-color RNA labeling in live yeast. Nat Methods 10(2):119–121. https://doi.org/10.1038/nmeth.2305
6. Wu B, Chen J, Singer RH (2014) Background free imaging of single mRNAs in live cells using split fluorescent proteins. Sci Rep 4:3615. https://doi.org/10.1038/srep03615
7. Bos TJ, Nussbacher JK, Aigner S, Yeo GW (2016) Tethered function assays as tools to elucidate the molecular roles of RNA-binding proteins. Adv Exp Med Biol 907:61–88. https://doi.org/10.1007/978-3-319-29073-7_3
8. Coller J, Wickens M (2007) Tethered function assays: an adaptable approach to study RNA regulatory proteins. Methods Enzymol 429: 299–321. https://doi.org/10.1016/S0076-6879(07)29014-7
9. Buhler M, Paillusson A, Muhlemann O (2004) Efficient downregulation of immunoglobulin mu mRNA with premature translation-termination codons requires the 5′-half of the VDJ exon. Nucleic Acids Res 32(11): 3304–3315. https://doi.org/10.1093/nar/gkh651
10. Buhler M, Steiner S, Mohn F, Paillusson A, Muhlemann O (2006) EJC-independent degradation of nonsense immunoglobulin-mu mRNA depends on 3′ UTR length. Nat Struct Mol Biol 13(5):462–464. https://doi.org/10.1038/nsmb1081
11. Dugre-Brisson S, Elvira G, Boulay K, Chatel-Chaix L, Mouland AJ, DesGroseillers L (2005) Interaction of Staufen1 with the 5′ end of mRNA facilitates translation of these RNAs. Nucleic Acids Res 33(15):4797–4812. https://doi.org/10.1093/nar/gki794
12. Kim YK, Furic L, Desgroseillers L, Maquat LE (2005) Mammalian Staufen1 recruits Upf1 to specific mRNA 3'UTRs so as to elicit mRNA decay. Cell 120(2):195–208. https://doi.org/10.1016/j.cell.2004.11.050
13. Stripecke R, Oliveira CC, McCarthy JE, Hentze MW (1994) Proteins binding to 5′ untranslated region sites: a general mechanism for translational regulation of mRNAs in human and yeast cells. Mol Cell Biol 14(9): 5898–5909. https://doi.org/10.1128/mcb.14.9.5898
14. Joncourt R, Eberle AB, Rufener SC, Muhlemann O (2014) Eukaryotic initiation factor 4G suppresses nonsense-mediated mRNA decay by two genetically separable mechanisms. PLoS One 9(8):e104391. https://doi.org/10.1371/journal.pone.0104391
15. Nicholson P, Josi C, Kurosawa H, Yamashita A, Muhlemann O (2014) A novel phosphorylation-independent interaction between SMG6 and UPF1 is essential for human NMD. Nucleic Acids Res 42(14): 9217–9235. https://doi.org/10.1093/nar/gku645
16. Lykke-Andersen J, Shu MD, Steitz JA (2000) Human Upf proteins target an mRNA for nonsense-mediated decay when bound downstream of a termination codon. Cell 103(7): 1121–1131. https://doi.org/10.1016/s0092-8674(00)00214-2
17. Minshall N, Thom G, Standart N (2001) A conserved role of a DEAD box helicase in mRNA masking. RNA 7(12):1728–1742. https://doi.org/10.1017/s135583820101158x
18. Ruiz-Echevarria MJ, Peltz SW (2000) The RNA binding protein Pub1 modulates the stability of transcripts containing upstream open reading frames. Cell 101(7):741–751. https://doi.org/10.1016/s0092-8674(00)80886-7
19. de Turris V, Nicholson P, Orozco RZ, Singer RH, Muhlemann O (2011) Cotranscriptional effect of a premature termination codon revealed by live-cell imaging. RNA 17(12): 2094–2107. https://doi.org/10.1261/rna.02918111
20. Nicholson P, Joncourt R, Muhlemann O (2012) Analysis of nonsense-mediated mRNA decay in mammalian cells. Curr Protoc Cell Biol. Chapter 27:Unit27 24. https://doi.org/10.1002/0471143030.cb2704s55

21. Eberle AB, Stalder L, Mathys H, Orozco RZ, Muhlemann O (2008) Posttranscriptional gene regulation by spatial rearrangement of the 3′ untranslated region. PLoS Biol 6(4): e92. https://doi.org/10.1371/journal.pbio.0060092
22. Choe J, Lin S, Zhang W, Liu Q, Wang L, Ramirez-Moya J, Du P, Kim W, Tang S, Sliz P, Santisteban P, George RE, Richards WG, Wong KK, Locker N, Slack FJ, Gregory RI (2018) mRNA circularization by METTL3-eIF3h enhances translation and promotes oncogenesis. Nature 561(7724):556–560. https://doi.org/10.1038/s41586-018-0538-8
23. Baillat D, Shiekhattar R (2009) Functional dissection of the human TNRC6 (GW182-related) family of proteins. Mol Cell Biol 29(15):4144–4155. https://doi.org/10.1128/MCB.00380-09
24. Ahsan KB, Masuda A, Rahman MA, Takeda JI, Nazim M, Ohkawara B, Ito M, Ohno K (2017) SRSF1 suppresses selection of intron-distal 5′ splice site of DOK7 intron 4 to generate functional full-length DOK-7 protein. Sci Rep 7(1):10446. https://doi.org/10.1038/s41598-017-11036-z
25. Pichon X, Robert MC, Bertrand E, Singer RH, Tutucci E (2020) New generations of MS2 variants and MCP fusions to detect single mRNAs in living eukaryotic cells. Methods Mol Biol 2166:121–144. https://doi.org/10.1007/978-1-0716-0712-1_7
26. Long RM, Gu W, Lorimer E, Singer RH, Chartrand P (2000) She2p is a novel RNA-binding protein that recruits the Myo4p-She3p complex to ASH1 mRNA. EMBO J 19(23): 6592–6601. https://doi.org/10.1093/emboj/19.23.6592
27. Sebbag-Sznajder N, Brody Y, Hochberg-Laufer H, Shav-Tal Y, Sperling J, Sperling R (2020) Dynamic Supraspliceosomes are assembled on different transcripts regardless of their intron number and splicing state. Front Genet 11:409. https://doi.org/10.3389/fgene.2020.00409

Chapter 18

Probing Liquid–Liquid Phase Separation of RNA-Binding Proteins In Vitro and In Vivo

Stephanie Heinrich and Maria Hondele

Abstract

Biomolecular condensates and the concept of liquid–liquid phase separation (LLPS) have transformed cell biology in recent years. Condensates organize cellular content and compartmentalize biochemical reactions, in particular many processes involving RNA. This protocol is aimed at readers new to the LLPS field who want to test their protein or cellular structure of interest. We describe the basic principles of liquid–liquid phase separation, and outline initial approaches—both in vitro and in yeast cells—for the characterization of a candidate cellular condensate. First, we focus on strategies to purify phase-separating proteins and to reconstitute condensates from recombinant proteins in vitro for observation by light microscopy. Second, we describe in vivo experiments (including fluorescence recovery after photobleaching (FRAP) microscopy and 1,6-Hexanediol treatment) to test whether a subcellular structure displays liquid-like behavior in cells.

Key words RNA, RNA binding proteins, Liquid–liquid phase separation, Biomolecular condensates, Protein purification, Light microscopy, FRAP, 1,6-Hexanediol

1 Introduction LLPS and LLPS in Splicing

1.1 Liquid–Liquid Phase Separation Is a Major Organizing Principle of Cellular Biochemistry

Cells use various mechanisms to organize their content. Eukaryotic cells contain intracellular membranes that form specialized organelles to segregate complex biological processes, for example transcription in the nucleus from translation in the cytoplasm. However, in the past decade it has become evident that all cells—from prokaryotes to any kind of cell in the human body—rely in addition heavily on self-organizing principles and in particular a phenomenon called liquid–liquid phase separation (LLPS) to dynamically and reversibly compartmentalize biochemical reactions or to concentrate biomolecules [1, 2]. The resulting "biomolecular condensates" can be tiny—containing only some dozen molecules—or manifest as "membraneless organelles" well visible

Peter Scheiffele and Oriane Mauger (eds.), *Alternative Splicing: Methods and Protocols*,
Methods in Molecular Biology, vol. 2537, https://doi.org/10.1007/978-1-0716-2521-7_18,

under the light microscope. Prominent examples of membraneless organelles are for example nuclear speckles, the nucleolus or cytoplasmic stress granules.

Organization of biochemical processes and structures by phase separation or condensation has been implicated in many biological processes, including chromatin structure, signal transduction cascades, formation of centrosomes and meiotic spindles, as well as many steps of gene expression and RNA metabolism [3–7]. However, in many cases the molecular function or cellular advantage of condensation is not well understood. With the concept of phase separation and membraneless organelles gaining increasing attention, many laboratories are setting out to study such phenomena and structures. This protocol aims to give the reader an overview of the basic molecular principles of phase separation they should pay attention to, and outlines initial approaches—both in vitro and in cells—to study whether a structure or assembly of interest behaves like a condensate.

1.2 Molecular Basis and Cellular Importance of LLPS

What is the molecular basis of phase separation? In simple words, condensation or phase separation means that some proteins and nucleic acids—beyond a certain "critical concentration"—prefer to interact with each other instead of remaining soluble in the cytoplasm. Condensation is based on multivalent, weak interactions between proteins, or between proteins and other polymers such as nucleic acids [8]. The interactions need to be multivalent—with a valency (i.e., the number of interaction options) of at least three [9]—to form branched meshworks that manifest as "droplets," and weak to keep the meshwork reversible and dynamic. The resulting condensates behave like the content of a "lava lamp": they can undergo fusion and fission events with subsequent relaxation to a spherical shape. The components of the droplet phase are in constant exchange with their environment: they switch binding partners within the condensate, or transition between the droplet and the soluble phase [10].

LLPS-promoting protein–protein interactions can involve folded domains, but prominently also intrinsically disordered regions (IDRs) containing different types of sequence stretches that facilitate weak interactions. Among these, prion-like or low-complexity domains (LCDs) rich in amino acids like glutamine, asparagine or serine were the first to be characterized. Sequences containing tyrosine/phenylalanine and arginine/lysine that allow for the formation of cation–π or π–π interactions (involving the π electron systems of the aromatic rings in the amino acid side chains) have recently gained much attention because they drive phase separation of "FET-type" proteins. Also blocks of negative and positive charge can facilitate condensation [11].

How these sequence stretches need to be arranged to facilitate phase separation is described by the "stickers and spacers" model. Sticker and spacer stretches alternate along a protein amino acid sequence. The stickers "glue" the different protein partners together: each sticker represents a weak interaction that facilitates reversibility and exchange of its binding partners, but in sum they are strong enough to keep the condensate together. The stickers are connected and separated by (often) solubility-promoting "spacer" stretches. If the stickers cluster too much, the local interaction becomes relatively strong and irreversible, and the assemblies aggregate [12].

Phase separation is a concentration-dependent phenomenon: at low concentrations, the system exists in a "mixed" phase of soluble components. Once the concentration of a phase-separation driving component reaches a specific, though context-dependent "critical concentration" [13], the system transitions to a "de-mixed" or droplet phase where condensates nucleate and grow, or form by a mechanism called "spinodal decomposition" [14, 15]. The saturation threshold depends on the interaction chemistry of the phase-separating component(s) and on environmental conditions such as pH, salt, temperature, and other interacting proteins. With increasing concentration, once nucleation has started, the concentration of the phase-separating component remains constant both in the soluble and in the droplet phase, only the size of the droplet increases [16].

The weak protein–protein and protein–nucleic interactions acid driving LLPS are easily diminished or enhanced by diverse post-translational modifications (phosphorylation, methylation, glycosylation, ribosylation, acetylation) or changes in intracellular pH (e.g., in stress situations) or even temperature which can enhance or disturb the interaction potential of individual amino acids [17–19]. This constitutes at least to some extent the molecular basis why some condensates form or dissolve so quickly in response to triggers like stress signaling cascades [20]. Furthermore, intracellular crowding or binding to proteins that shield or create binding sites modulate the condensation behavior [21, 22].

Condensates can mature and "age," presumably because initially only some of all the potential interactions are formed, and over time the system equilibrates to a more stable state. This results in a transition from an initially liquid-like to a more viscous, gel-like or glass-like state [23, 24], which can also have a specific function [25], and ultimately can transform into aggregate-like or even amyloid-like structures. Cells constantly counteract these solidifications with various energy-consuming processes. In consequence, cellular condensates are rarely at thermodynamic equilibrium—which in turn allows them to perform work and regulate the flux of information and substrates [26]. Malfunction of the mechanisms counteracting aggregation has been implicated in numerous diseases, for example, neurodegenerative pathologies or Alzheimer's disease [27–29].

1.3 LLPS in Splicing

Evidence is accumulating that all steps of gene expression are heavily linked to membraneless compartmentalization, which is suggested to accelerate reaction rates, confer directionality or to orchestrate different gene expression events in space and time [30, 31]. Already during transcription initiation, RNA Polymerase II (Pol II), the Mediator complex and transcriptional cofactors form condensates, especially at super enhancer regions [31]. Changes in the phosphorylation pattern of the unstructured Pol II—C terminus (CTD) trigger release from these initiation condensates and subsequent formation of "elongation condensates" composed of Pol II and elongation factors, a process that accelerates the transcription process. These elongation condensates also contain splicing factors such as SRSF2 [32]. The phosphorylated CTD interacts with many other spliceosome components, thereby initiating cotranscriptional splicing [33]. Inversely, dephosphorylation of the CTD can increase posttranscriptional splicing by reducing the formation of phase-separated condensates during the cotranscriptional splicing process [31].

Intriguingly, many splicing factors contain LCDs, for example the arginine–serine dipeptide repeats in the SR family of RNA binding proteins [34, 35]. Another example is AKAP95, a nuclear protein that binds to introns of pre-mRNAs, and whose phase separation propensity is required to regulate splicing [36]. In addition, phase-separated compartments called nuclear speckles are storage sites for several splicing factors [37] and have been reported to occur in the vicinity of highly transcribed genes [38]. Lastly, RNA itself has been shown to phase-separate [39], and it can be chemically modified and thereby trigger interactions with RNA-binding molecules, which in turn increases multivalency and condensate formation [40]. For example, the N^6-methyladenosine (m^6A) modification enhances the phase separation propensity of RNA [41], and also its nuclear "reader" protein YTHDC contains LCDs, which may recruit the RNA into specific nuclear structures to facilitate, for example, splicing [42].

1.4 Characterization of Condensation Phenomena

When setting out to study a liquid–liquid phase separation phenomenon, we would recommend to perform complementary in vivo and in vitro experiments and to probe and quantify these thoroughly to support the hypothesis. Condensation can easily arise as an artifact: as every crystallographer knows, proteins tend to phase-separate if mixed at high concentrations with the right buffer, but this does not necessarily mean that those "droplets" are a biologically relevant phenomenon. In consequence—in vitro but also in cells (i.e., in overexpression experiments)—it should be considered whether one is working at a biologically relevant (sometimes local) concentration range.

If phase separation is observed in vitro, with recombinant proteins, it is crucial to verify that the protein under study forms condensates also in cells. Some of these cellular structures might only contain some tens of molecules, sometimes called "nanocondensates," and thus be too tiny to be visible as large "membraneless organelles"; still, they can be revealed by superresolution microscopy [43] or methods like FC(C)S (*see* **Note 28**).

In cells, the characterization of condensates is intrinsically challenging [44]: first, accumulation of proteins at specific structures or "foci" can look very similar to a membraneless organelle, but might not represent a phase-separated structure. And second, formation of a membraneless organelle is the result of many interconnected processes within a crowded cellular environment. We therefore caution that the isolated view on a supposed cellular condensate requires careful interpretation of the experimental results. Apart from static properties like subcellular accumulation and (co)-localization of candidate proteins, a candidate condensate should be examined in live cells with respect to its dynamic behavior, for example fusion of different condensates and relaxation to spherical shape, recovery after photobleaching, material properties and formation/dissolution kinetics (*see* **Note 28**).

Furthermore, for both in vitro and in vivo/cellular experiments, it should be considered if the protein of interest has interaction partners (proteins, small molecules, specific nucleic acid binding motifs) that could influence a potential phase separation behavior.

2 Materials

In the following we describe experiments—in vitro with recombinant proteins, and in yeast cells—centered around the yeast DEAD-box ATPase Dhh1, a component of cytoplasmic processing bodies (P-bodies, PB) as a model phase-separating protein. In the Notes section, we describe in detail how to design expression constructs and purification protocols for your protein of interest (*see* **Notes 1–11**), and highlight considerations for in vitro (*see* **Notes 12–27**) and in vivo imaging experiments (*see* **Notes 28–36**).

2.1 Materials Required for Purification of Dhh1-mCherry

1. Protein expression plasmids for recombinant expression of Dhh1-mCherry (protein of interest) and Dhh1^{core}-mCherry (condensation-deficient control/comparison protein) in *E. coli* (available on request [45]).
2. *E. coli* protein expression strain (e.g., BL21 star).
3. LB medium.

4. 1000× Ampicillin stock solution (100 mg/mL).
5. Rich, buffered bacterial growth medium such as terrific broth (TB).
6. 1 M IPTG in H_2O.
7. Two 5 L Erlenmeyer flasks.
8. Temperature-controlled shaker for bacterial growth (required temperatures: 37 °C and 18 °C).
9. 1 L centrifugation bottles and cooled centrifuge for 1 L bottles.
10. 50 mL reaction ("Falcon") tubes.
11. Ice bucket and laboratory ice for cooling.
12. Sonicator with appropriate tip, or other cell lysis device.
13. 500 mL lysis buffer: 500 mM NaCl, 25 mM Tris pH 7.5, 10% glycerol, 10 mM imidazole.
14. Protease inhibitors (e.g., Roche Complete EDTA-free).
15. 100 mL elution buffer: 500 mM NaCl, 25 mM Tris pH 7.5, 10% glycerol, 500 mM imidazole.
16. 1.5 L storage buffer: 200 mM NaCl, 25 mM Tris pH 7.5, 10% glycerol, 2 mM DTT (add fresh before use).
17. 1.5 mL and 2 mL reaction tubes.
18. Rotor (e.g., SS-34) and suitable centrifugation bottles for ~40″000 × *g* speed.
19. Ni^{2+} affinity Sepharose (~10 mL slurry).
20. Flow-through columns for protein purification.
21. Dialysis membrane of suitable MWCO.
22. Protein concentration devices, for example, 30 kDa cutoff for the Dhh1 proteins.
23. Automated chromatography system.
24. Size exclusion chromatography column, for example, Superdex 200.
25. 10% acrylamide gels, 4× loading buffer, and Coomassie stain for protein analysis.
26. Liquid nitrogen and PCR 8-tube "strips" for snap freezing.

2.2 Materials Required for an In Vitro Reconstitution Experiment with Recombinant Proteins

1. Protein stock solutions of the protein of interest (e.g., Dhh1) and a control protein (e.g., $Dhh1^{core}$, in the same "storage buffer"), plus some of the storage buffer for dilutions. Typical protein stock solutions are ~500 μM.
2. Low-salt buffers (LSBs): LSB50 = 50 mM KCl, LSB200 = 200 mM KCl. (prepare 10 mL; they can be stored at room temperature for future experiments).
3. Buffer solutions: 1 M Hepes pH 6.8 and 7.2.

4. 1 mg/mL polyuridine (polyU, Sigma) in H2O.
5. 100 mM ATP/100 mM $MgCl_2$ (ATP needs to be pH-adjusted when dissolved).
6. 10 mg/mL BSA as a crowding agent, and to reduce unspecific binding.
7. Optional: 50% PEG solution (PEG6000 or PEG8000) in H2O.
8. RNase inhibitor.
9. Filter tips to prevent RNAse contamination when taking RNA solutions from their stock.
10. Filter all protein and buffer solutions before use (0.22 μm filters) to remove dust particles.

2.3 Materials for Yeast Cell Growth and Hexanediol Treatment

1. Yeast strain KWY8050 (*leu2–3112 trp1–1 can1–100 ura3–1 ade2–1 his3–11,15 DHH1-yEGFP-URA3*; allows to visualize the formation of P-bodies, available upon request) or strain of choice, *see* **Note 31**.
2. SCD media for culturing: Synthetic complete (SC) broth containing 2% dextrose (Formedium). For plates, add 2% Bacto agar.
3. SC without Glucose for acute PB induction (mix Yeast Nitrogen Base w/o Amino acids (Formedium) with Kaiser Complete SC mixture (Formedium) according to manufacturer's instructions).
4. Glass culture tubes.
5. Incubator and rotating wheel (25 °C).
6. 1,6-Hexanediol (50% stock solution in ddH_2O).
7. Concanavalin A (1 mg/mL in ddH_2O).

2.4 General Imaging Requirements

1. Imaging dish, for example, 8-well imaging slide or 384-multiwell dish.
2. Centrifuge with microscopy plate adapters.
3. Microscopes.
   - For in vitro reconstitution experiments: widefield or confocal light microscope with at least a 40× objective.
   - For hexanediol treatment: widefield or confocal light microscope (microscope able to perform time-lapse imaging experiments, for example,, for example, temperature-controlled Nikon Eclipse Ti-E with a Spectra X LED lamp using an Apochromat VC 100× objective NA 1.4 (Nikon)).
   - For FRAP imaging: confocal microscope able to perform FRAP bleaching (e.g., Leica SP8 Laser Scanning Confocal Microscope, with a PlanApo 63× 1.4 NA oil immersion objective and a conventional photomultiplier tube (PMT) detector).

2.5 Time Considerations

1. Production of recombinant proteins: 3 days for cell growth and harvesting; 2 days for purification.
2. In vitro reconstitution experiment: starting with purified proteins, 3–4 h for setting up and the first round of imaging.
3. In vivo imaging experiments: 3 days from glycerol stock; time-lapse imaging and FRAP imaging varies depending on sampling time.

3 Methods

3.1 Protein Purification under Condensate-Preventing Conditions (Fig. 1)

1. mCherry-tagged Dhh1 constructs (full-length Dhh1 and the phase-separation deficient "tailless" ATPase core Dhh1^{core} containing residues [48–425] [45] were cloned into a pETMCN-based expression vector with a N-terminal 6xHis tag, removable by TEV cleavage, and a C-terminal mCherry tag. These plasmids are available upon request. For design of your own expression plasmid (*see* **Notes 1–11**).

   Dhh1 = pKW3631/pMH124; after TEV cleavage: MW = 86.7 kDa, Abs 0.1% = 0.764, extinction coefficient = 66,240

   Dhh1^{core} = pKW4063/pMH126; after TEV cleavage: MW = 72.3 kDa, Abs 0.1% = 0.757, extinction coefficient = 54,780,
2. For protein expression, 10 ng plasmid were transformed into competent BL21star cells by heat shock. After a brief recovery on ice, the transformation reaction is used to inoculate a preculture of 25 mL LB with 1× Ampicillin. The preculture is grown over night shaking at 37 °C and should be dense the next morning.
3. The next day, 1 L of TB medium is prepared with 1× Ampicillin and inoculated with 10 mL of the preculture; keep the rest of the preculture at 4 °C. The large culture is grown shaking (180–200 rpm) at 37 °C until OD_{600} = ~0.7. Protein production is initiated by addition of IPTG to a final concentration of 250 μM; at the same time, the temperature of the shaker is reduced to 18 °C and cultures are kept shaking overnight. If cells over-grew and OD_{600} is larger than 1 before induction, the large culture should be discarded and restarted with fresh medium and the rest of the preculture.
4. The next day, cells are harvested in 1 L bottles at 4500 × *g*, 10 °C for 10 min. The supernatant is discarded. From now on, keep all reagents on ice.
5. The cell pellet is resuspended in 15 mL lysis buffer supplemented with protease inhibitors by pipetting (e.g., with a 10 mL pipette) or (upright) in the bacterial growth shaker at 150 rpm

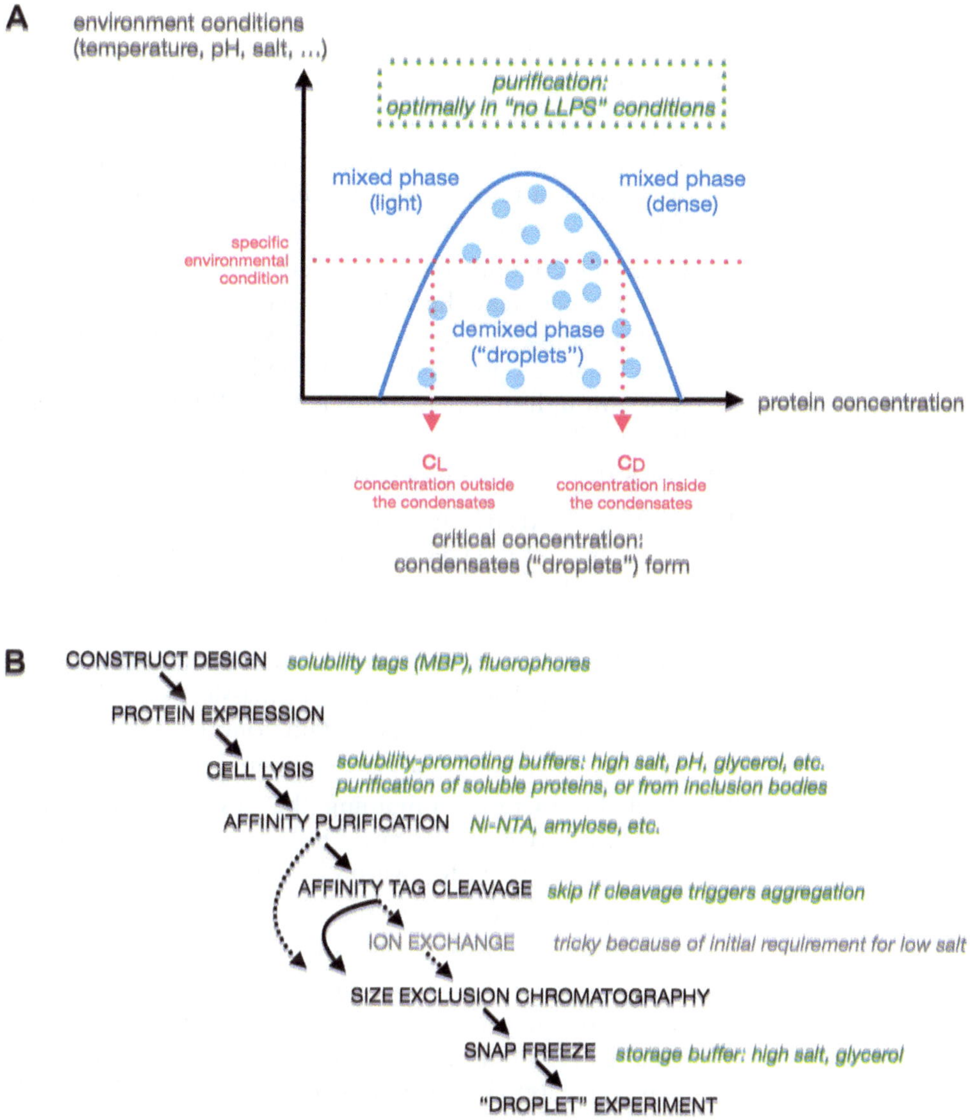

Fig. 1 Protein purification for in vitro *condensate experiments.* (**a**) Liquid–liquid phase separation phase diagram, *see* [15] for a detailed explanation. Proteins should optimally be purified in conditions where they do not phase separate. (**b**) Schematic flow of a typical protein purification, with considerations for phase-separating proteins in green/gray

for 5–10 min; avoid extensive foaming by, for example, vortexing. Transfer the resuspended pellet solution to a 50 mL Falcon tube.

6. Lyse cells by sonication (ask someone familiar with the machine for optimal settings) or method of your choice. For an analytic gel (see below), take a 100 μL sample into an 1.5 mL reaction tube; spin 5 min at full speed in a tabletop centrifuge. Transfer 50 μL of the supernatant to a new 1.5 mL reaction tube

("SN"), discard the rest of the supernatant; dissolve the pellet in 100 μL lysis buffer ("P").

7. Spin lysate at ~40′000 × g, 10 °C for 45 min. Filter supernatant into a fresh Falcon tube.
8. Transfer 5 mL of Ni^{2+} slurry to a flow-through column, attached to a retort stand. Equilibrate beads by washing with 20 mL water and then 20 mL lysis buffer.
9. Close the column bottom outlet, add the lysate supernatant, close the top, take the column from the retort stand and incubate (rotating) for about 20 min. Re-attach the column to the retort stand, open column top and bottom, and collect the flow-through in a glass beaker; save 100 μL in an Eppendorf tube ("FT").
10. Wash the beads 4 times with 20 mL lysis buffer.
11. Elute in 2 mL fraction with elution buffer; number the individual fractions. Typically, 10 elution fractions should be sufficient to elute all protein. The Ni^{2+} beads can be regenerated according to the manufacturer's instructions.
12. Run 10 μL of SN, P, FT and from each elution fraction on an acrylamide gel and stain with Coomassie. Since the SN, P, and FT fractions were kept at constant volume, the protein amount in these lanes can be directly compared.
13. Pool the fractions containing the desired protein, add TEV protease (1:100 molar ratio TEV–target protein) and dialyze against storage buffer overnight in the cold room.
14. The next day, equilibrate the size exclusion column with storage buffer with freshly added DTT.
15. Transfer the dialyzed protein solution to 2 mL reaction tubes and spin 5 min full-speed in a tabletop centrifuge to remove aggregates. Pool and concentrate down to 2 mL using the Amicon Ultra concentrators by spinning at 10 °C in 15-min intervals; "rinse" the concentrator membrane between spins by inverting the tube, or by flushing the membrane with a 1 mL pipette using the protein solution to be concentrated.
16. Run the concentrated protein solution on the equilibrated size exclusion column.
17. Run 10 μL of the protein-containing fractions on an acrylamide gel. Pool fractions containing clean protein and concentrate to ~500 μM as described above.
18. Snap-freeze 25 μL aliquots in PCR strips in liquid nitrogen; store at −80 °C (e.g., in labeled plastic bag in paper boxes).

3.2 In Vitro Reconstitution of Biomolecular Condensates

1. For an initial screen, we would recommend to use a matrix scheme. In the following, rather extensive example, we use a protein of interest—for example, Dhh1—plus a control protein—for example, Dhh1^{core}—at two different concentrations plus a buffer control, which makes 5 "protein solutions" (variable A: A1 to A5). These protein solutions are mixed with buffers of two different salt concentrations (LSB50 or LSB200; variable B) and pH (1 M HEPES pH 6.4 or pH 7.2; variable C), with or without additives like RNA (variable D). The result would be an $A \times B \times C \times D = 5 \times 2 \times 2 \times 2 = 40$ conditions matrix.
2. Variable A: for both Dhh1 and Dhh1^{core}, we use the concentrated protein stock (~ 500 μM) and a 1:5 predilution with storage buffer (~100 μM) (*see* **Notes 14** and **15**).
3. All other variable components (B, C, D), ATP/$MgCl_2$, and BSA: prepare master mixes for each condition to be tested (for A = 5 protein solutions, make 5 + 1 = six-fold the volume you need for an individual well to have some spare for pipetting). To take the polyU or RNA solution and the RNase inhibitor from the stock, use filter tips (*see* **Notes 16–23**).
4. The final reactions should contain per well the following.
   - 1 μL concentrated protein (protein of interest or control) or 1:5 predilution (protein of interest or control), or buffer [A = 5].
   - 15 μL LSB50 or LSB200 [B = 2].
   - 1 μL 1 M HEPES pH 6.4 or 1 M HEPES pH 7.2 [C = 2].
   - 1 μL 1 mg/mL polyU (final concentration 50 ng/μL) or 10 mM Tris pH 8 (control); use filter tip to take from stock solution [D = 2].
   - 1 μL 10 mg/mL BSA as crowding agent.
   - 1 μL 100 mM ATP/$MgCl_2$.
   - 0.05 μL RNase inhibitor.
5. A sixfold master mix "B1-C1-D1" would for example contain [B1] (6*15=) 90 μL LSB50, [C1] (6*1=) 6 μL HEPES pH 6.4, [D1] (6*1=) 6 μL 1 mg/mL polyU, (6*1=) 6 μL 10 mg/mL BSA, (6*1=) 6 μL 100 mM ATP/$MgCl_2$ and (6*0.05=) 0.3 μL RNase inhibitor. The master mix "B1-C2-D1" would contain HEPES pH 7.2 instead of 6.4; master mix "B2-C1-D1" would be prepared with LSB200.
6. It helps to "logically" arrange the conditions in the 384-multiwell plate, for example the protein solution variable A1 to A5 on the y axis and the B-C-D combinations on the x axis.

7. During pipetting, especially when working with unlabeled proteins, it is important to avoid air bubbles which may look like condensates in bright field microscopy.
8. Pipet 1 μL of protein solution about 1 mm below the upper rim of the corner of the well. For multiprotein systems, this is also where we mix the different protein solutions.
9. Add the master mix containing all other components to the protein droplet and mix carefully by pipetting up and down once or twice.
10. Incubate for 10–20 min at room temperature.
11. Let settle by gravity onto the glass surface; alternatively, spin down the imaging plate at 50 rpm for 30 seconds, or take up the mixture with the pipette and transfer to the bottom of the plate.
12. At the microscope (*see* **Notes 12**, **13**, and **24–27** for imaging suggestions), find settings for bright field and mCherry fluorescence that capture your condensates without overexposure; image all wells with the same settings. Acquire several images for each well since droplets might not be equally distributed over the well surface. Droplets will mature over time, so optimally imaging of all conditions should be performed as fast as possible using an automated script.
13. We suggest reimaging the plate after incubation at room temperature or 4 °C for 1 or 2 h (colder temperatures typically help with condensate formation).
14. If after 2 h still no droplets are visible, crowding agents like PEG6000 or dextran can be added to a final concentration of 10—20% from a 50% stock solution in H_2O, also on top of a preexisting, already imaged setup.
15. Evaluation: Several laboratories have developed software and scripts that quantify the number, size, shape, and fluorescence intensity of in vitro condensates using common image analysis programs like FIJI/ImageJ. Furthermore, one can measure the partitioning coefficient—that is, what intensity of the protein is outside versus inside a condensate/droplet.

3.3 Yeast Cell Growth and Imaging Dish Preparation

To induce P-bodies in Dhh1-GFP expressing cells

1. Inoculate selected yeast strain (e.g., Dhh1-GFP, *see* **Note 31**) as overnight culture in 3 mL SCD medium at 25 °C.
2. Dilute cells the next morning to OD_{600} 0.15 in 3 mL SCD and grow to OD_{600} 0.6–0.8.
3. Coat imaging dish with Concanavalin A by applying the stock solution to the well to cover the bottom surface for 5 min.
4. Remove Concanavalin A from well and let well air-dry.

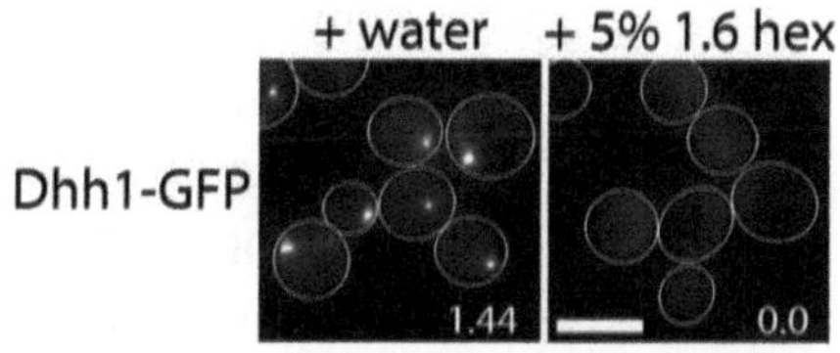

Fig. 2 1,6-Hexanediol treatment of glucose-starved cells expressing Dhh1-GFP. Log phase cells were shifted in SC medium without glucose for 30 min and subsequently treated with either 5% 1,6-Hexanediol or water for 15 min before imaging. Shown is a maximum projection of the central eight slices of a Z-stack image with slice spacing of 0.4 μm. Numbers indicate PBs per cell. Scale bar: 5 μm

5. When cells reach an OD_{600} 0.6–0.8 wash in SC without glucose.
   (a) Spin down for 5 min at 3000 × *g*.
   (b) Remove supernatant.
   (c) Resuspend cells in 3 mL SC without glucose.
   (d) Repeat 3 times.
6. Incubate for 30 min at 25 °C to induce P-body formation.
7. Continue with 1,6-Hexanediol treatment or FRAP measurement (*see* Subheading 3.4).

3.4 Probing the Dynamic Behavior of Condensates

1. Hexanediol, end-point assay (*see* **Notes 28** and **29** *for caveats and alternative approaches*).
   (a) Transfer 900 μL of PB-induced yeast cell culture to a 1.5 mL tube.
   (b) Add 100 μL of either 50% 1,6-Hexanediol stock solution or ddH_2O to cells (final concentration 5%).
   (c) Incubate for 15 min.
   (d) Subject 200 μL of each treatment to a Concanavalin A coated imaging dish.
   (e) Spin down for 1 min at 100 rcf.
   (f) Image at widefield or confocal microscope with appropriate laser intensity settings (Fig. 2).
2. Hexanediol, time-lapse imaging.
   (a) Prepare 2× stock solution 1,6-Hexanediol in SC without glucose (i.e., 10% in SC -Glc).
   (b) Subject 200 μL of PB-induced yeast cell culture to a concanavalin A coated imaging dish.
   (c) Spin down for 1 min at 100 rcf.

(d) Remove 100 μL of supernatant.

(e) Place imaging dish into microscope, select a region and record a pretreatment image.

(f) Add 100 μL 2× 1,6-Hexanediol in SC -Glc, and carefully mix to not disturb cells on bottom of the well.

(g) Immediately afterwards start time lapse imaging with appropriate laser intensity settings and time intervals (typical time interval for PBs: image every minute for 30 min; either image single plane or perform Z stack imaging with slice spacing of 0.4 μm). Perform time-lapse imaging with laser intensities and imaging intervals that minimize photobleaching and phototoxicity.

3. Fluorescence Recovery After Photobleaching (FRAP) (*see* **Notes 28, 30–36**).

(a) Subject 200 μL of PB-induced yeast cell culture to a Concanavalin A coated imaging dish.

(b) Spin down for 1 min at 100 rcf.

(c) In FRAP imaging software (e.g., Leica LAS AF SP8 software):

- Select a cell.
- Select PB and draw the bleach region around it (region of interest; ROI, standardized size, for size selection *see* **Notes 32** and **33**).
- Subject ROI to 5–10 pulses of an argon laser at 488 nm.
- Collect images from a single plane with a 2.92 nm pin-hole at 500 ms intervals for 50 s postbleach (*see* **Note 34**).

4. Evaluation.

(a) Hexanediol-dependent condensate modulation: Use publicly available particle tracking tools (e.g., TrackMate in ImageJ/FIJI or Diatrack) to count particles and to track them over time.

(b) Deriving kinetic data from FRAP measurements (*see* **Note 35**): Subtract extracellular background (I_{bg}) from the intensity of the bleached region (I_{bl}), and bleach- correct the values by normalizing for total cell intensity (I_{total}) resulting in $(I_{bl} - I_{bg})/(I_{total} - I_{bg})$ [46]. Generate recovery curves by normalizing the intensities for prebleach and postbleach values and fit the curves with an exponential function: $f(t) = A \cdot [1 - \exp.(-r \cdot t)]$ with f(t) = normalized fluorescence intensity, A = nonrecovering pool, r = time constant of recovery, t = time, with the method "NonlinearLeastSquares." Alternative models can also be

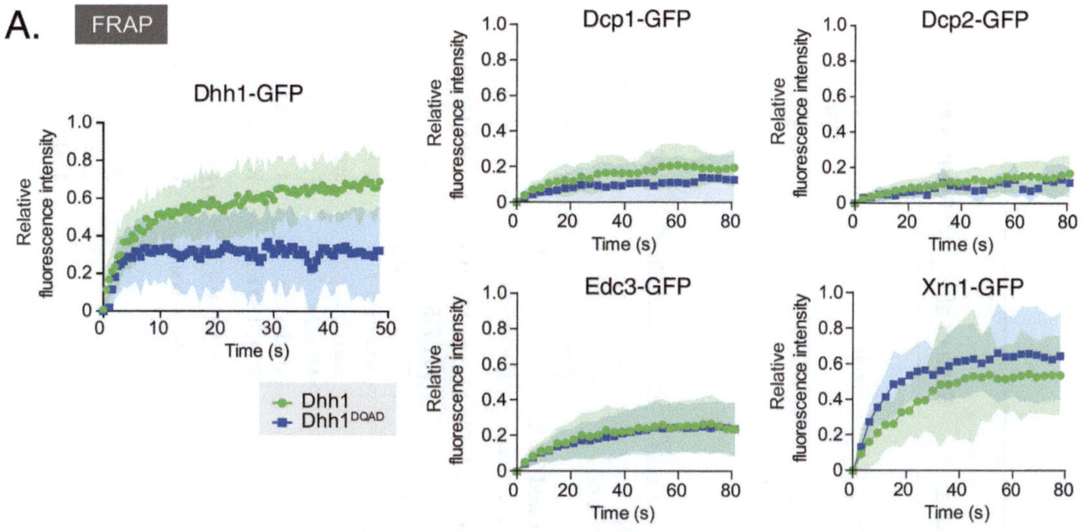

Fig. 3 FRAP example. Example of FRAP measurements of different PB components 20 min after acute glucose starvation (Image from [47])

employed depending on the protein of choice, *see* **Note 36**. Figure 3 shows example curves from >8PBs, typically ~12 PBs.

4 Notes

Considerations for the Design, Purification, and Testing of Recombinant Proteins

1. In vitro reconstitution experiments require purified protein, which is mostly recombinantly purified from bacteria or insect cells. It is important to keep in mind that phase-separating proteins evolved not to be soluble, and therefore standard purification protocols might not be the road to success. When surveying the literature, it becomes evident that many laboratories have developed similar strategies to increase solubility, based on chemical factors like purification in high-salt buffers, and by reducing the overall duration of purification (Table 1, [48–53])—aggregation is in the end a kinetic phenomenon.
2. To identify candidate proteins or protein regions, much progress has been made in the development of algorithms that predict sequences driving "homotypic" condensation, especially for sequences in intrinsically disordered protein regions [54–58].
3. A typical expression construct for a phase-separating protein contains a cleavable affinity and possibly solubility tag for purification, and often a fluorophore for visualization, for example,

Table 1
Examples of purification protocols for phase-separating proteins, as described in [48–53]

‘	Boehning et al.	Klein et al.		Guillen-Boixet et al.	Wang et al.	Saha et al.	Putnam et al.
Construct	6xHis_MBP _TEV_ hRPB1_CTD	6xHis_BFP /mEGFP/ mCherry_SR SF2	6xHis_BFP /mEGFP/ mCherry_+ NPM1, Med1, Brd4, Fib1, HP1alpha	6xHis-3C- GFP- G3BP1- MBP	Many FUS, hnRNPs, TAF15 constructs: N-terminal His-MBP- 3C-	PGL-3- 6xHis- mEGFP	MBP-TEV- PGL-3, MBP-TEV- MEG-3
Expression	E. coli (BL21 RIL); 3–4 h at 37 °C	*E. coli* (LOBSTR); 20 h 16 °C		Insect cells / baculo expression system	SF9 insect cells/ baculo expression system; harvested 72 h postinfection	SF9 insect cells/ baculo expression system; harvested 72 h postinfection	*E. coli* (Rosetta); 16 h 16 °C
Lysis buffer	LB300 = 20 mM HEPES pH 7.4, 300 mM NaCl, 30 mM imidazole, 10% glycerol, 1 mM DTT	*Denaturating buffer*: 50 mM Tris pH 7.5, 300 mM NaCl, 8 M urea	50 mM Tris pH 7.5, 500 mM NaCl	50 mM Tris–HCl pH 7.5, 1000 mM KCl, 2 mM EDTA, 2 mM DTT	50 mM Tris–HCl pH 7.4, 1000 mM KCl, 5% glycerol, 10 mM imidazole	25 mM HEPES pH 7.5, 300 mM KCl, 300 mM L-arginine, 10 mM imidazole, 1 mM DTT	25 mM HEPES pH 7.4, 500 mM NaCl, 400 mM L-arginine, 20% (v/v) glycerol, 1 mM DTT, RNase A
Affinity purification	*Ni-NTA*: Load in LB300, wash with HSB1000 (1000 mM NaCl, 20 mM HEPES pH 7.4, 30 mM	*Ni-NTA:* Wash/ elute with lysis buffer	*Ni-NTA:* Wash/elute with lysis buffer	*Amylose resin:* Wash/elute with lysis buffer	*Ni-NTA*: Wash/ elute with lysis buffer	*Ni-NTA*: Wash/ elute in lysis buffer	*MBP-trap:* Wash with 25 mM HEPES pH 7.4, 40% glycerol, 1000 mM

	imidazole, 10% glycerol, 1 mM DTT), elute with LB300						NaCl; elute with additional 20 mM maltose
	Amylose resin: Wash with HSB1000, elute with LB300 + maltose	Stepwise dialysis into 4 M, / 2 M, / 0 M urea, 500 mM NaCl, 1 mM DTT + 10% glycerol for 0 M urea	Dialysis into final storage buffer	*Ni-NTA*; wash/ elute with lysis buffer	His-MBP tag cleaved with 3C prescission protease at room temperature for 4 h		Dilute with 0 mM NaCl buffer to final 400 mM NaCl (rest constant); TEV cleavage 2–4 h at RT
				Tag cleavage (3C prescission) during overnight dialysis in lysis buffer	*Amylose resin:* to remove uncleared protein (collect flow-through)	*Desalting column:* change to 50 mM KCl, rest constant	
Ion exchange						*HiTrapQ anion-exchange column*; linear salt gradient to 1000 mM KCl	Dilute to 200 mM NaCl; immediately load onto *HeparinTrap* column, elute with B1000 salt gradient
Size exclusion	*Superdex 200* in storage buffer			*Superdex 200* in storage buffer	*Superdex 200* in storage buffer	*Superdex 200* in storage buffer	

(continued)

Table 1 (continued)

'	Boehning et al.	Klein et al.		Guillen-Boixet et al.	Wang et al.	Saha et al.	Putnam et al.
Final storage buffer	20 mM HEPES, pH 7.4, 300 mM NaCl, 10% glycerol, 1 mM TCEP	50 mM Tris pH 7.5, 500 mM NaCl, 1 mM DTT, 10% glycerol	50 mM Tris 7.5, 125 mM NaCl (500 mM for HP1alpha), 10% glycerol, 1 mM DTT	50 mM Tris/HCl, pH 7.5, 300 mM KCl, 1 mM DTT	50 mM Tris–HCl, 500 mM KCl, 1 mM DTT, 5% glycerol	25 mM HEPES pH 7.5, 300 mM KCl, 1 mM DTT	25 mM HEPES pH 7.4, 40% glycerol, 1 mM DTT, 500/800 mM NaCl

[HisTag]-MBP-[protease-cleavage-site]-PROTEIN-OF-INTEREST-[linker]-mCherry. It is important to design control constructs that have exactly the same affinity and fluorescence tags, and are purified alongside the protein of interest, to help distinguish artifacts from condensation phenomena.

4. For affinity purification, any commonly used tag like 6xHis can be used. It is worth noting that affinity tags like HIS, GST (glutathione S-transferase, a dimeric protein) or TwinStrep (our observations) can substantially enhance condensation, by adding additional interaction sites and thus increasing valency. Cleavage of the purification tag before the experiment is highly recommended.

5. Aggregation-prone proteins tend to accumulate in inclusion bodies or in the pellet fraction after cell lysis and centrifugation. Purification from inclusion bodies for phase separation has been described, mostly for largely unstructured proteins [48, 59]. Alternatively, fusion tags like maltose binding protein (MBP) or thioredoxin can bring a protein back into the soluble fraction. However, they may later on prevent condensation of the overall protein construct during the experiment. We recommend to include a protease cleavage site in the linker region between the purification/solubility tags and the final protein, to remove the enhancing or inhibiting influence of the tags. If the protein is aggregating upon MBP cleavage during the purification process, one can try to store it while still fused to the MBP tag and add the protease to the final experimental setup in the microscopy well.

6. Although not strictly required, fluorescent protein tags are commonly fused to the protein of interest for visualization of condensates, especially when studying multicomponent systems. The fluorophore can be added at the N- or the C-terminus, which can make a difference in terms of protein solubility and phase separation behavior (especially if the condensation-promoting sequences are concentrated at one of those termini). Many fluorescent proteins—for example GFP—are natural dimers, which will increase the valency of the construct and can thus strongly influence phase separation. Commonly used, better behaved alternatives are mCherry or the monomeric GFP variant A206K. Alternatively, a protein can be labeled with SNAP tag or chemical labeling technologies. However, also these chemistries can modulate the phase separation behavior (our observation), but since the dyes are very bright, it is sufficient and advisable to spike only a small amount into a larger pool of unlabeled protein for the LLPS experiment.

7. In our hands, protein expression works best in "slow" expression conditions like overnight at 18 or 25 °C, while others, for example, isolated LCDs, seem to express better when expressed for 3–4 h at 37 °C (higher temperatures might counteract cold temperature-induced phase separation/aggregation). In general, we recommend the use of rich and buffered expression media like TB.
8. For the purification procedure, it is important to work at conditions where the protein is least likely to condensate or precipitate, since such condensates can quickly convert into irreversible aggregates. Thus, it is recommended to work swiftly, with relatively high salt buffers—which break most of the protein–protein and protein–nucleic acid interactions that lead to condensation—and possibly also at room temperature, although this may accelerate protein degradation.
9. Many laboratories include ~10% glycerol in their buffers and work at about 300–500 mM sodium chloride. Other additives—for example arginine for sequences where phase separation relies primarily on cation–π or π–π interactions—can be helpful as well. Since pH influences the charge and thus the solubility of your protein, it is worth testing a range from, for example, pH 6 to pH 9 for purification.
10. While gel filtration chromatography is commonly used to separate the (soluble) protein of interest from aggregates and contaminations, ion exchange chromatography is little used (*see* Fig. 1 and Table 1 [48–53])—likely because most proteins start to condensate or aggregate when transferred to the low salt buffers required for ion exchange separation protocols.
11. For storage, it is advisable to concentrate the protein in a solubility-promoting buffer, for example, with high salt and glycerol, and freeze small aliquots. From those stocks, one can directly dilute a small volume manyfold into the final experimental buffer to trigger phase separation. We commonly snap-freeze in liquid nitrogen ~50× concentrated protein stock solutions as 20–50 μL aliquots in PCR tubes.
12. In vitro condensates are most commonly observed by widefield or confocal microscopy. Condensates of unlabeled protein and nucleic acid condensates are well visible in bright-field/phase-contrast microscopy. For one component systems this should be the method of choice since any fluorescent tag will alter the protein's biochemical properties. Recent technical developments in quantitative phase imaging even enable measurements of protein concentration within condensates from such bright field images [60]. For two or more component systems, or for confocal analysis of condensate ultrastructures, it is recommended to use fluorescence microscopy. Widefield microscopy

allows for fast screening and imaging of multiple conditions, but is mostly used to image condensates on the microscopy well/glass surface. Confocal microscopy is better suited for quantification of signal intensity and to obtain information about ultrastructural details like "hollowness," since it removes out-of-focus light.

13. For imaging, 8- to 384-well plates are usually preferred over "squeezing" the droplet solution between two coverslips, since interaction with the two glass surfaces can strongly influence the condensation behavior. 384-well plates can be used with volumes as low as 10–12 μL, which is an advantage if the yield of protein is limiting. Condensates interact to some extent with surfaces, which can be acknowledged, or counteracted by coating the well surfaces by silanization [61] (Caveat: this can dissolve the plastic walls of multiwell plates) or coating with BSA, but this might affect crowding and thus condensate formation. The type of coating that is suitable for a condensate needs to be tested individually, therefore uncoated glass surfaces are recommended for initial experiments.
14. Protein stock solution: typically, at least 500 μM in storage buffer. Take an aliquot from −80 °C and thaw quickly in your hand. Spin before use (3 min 12'000 × *g* in a tabletop centrifuge) to precipitate and remove aggregates, then remeasure the concentration. Optimally, one reaches at least a tenfold dilution of the storage buffer in the final experimental setup. Snap freezing in liquid nitrogen and small aliquots is in most cases successful; some proteins might survive freezing at −20 and repetitive thawing cycles, others have to be purified freshly for every experiment and will not survive freezing.
15. We typically predilute the protein with storage buffer so we can add the same volume to the final experiment for all conditions. For example, from a 500 μM protein stock solution, prepare a 1:5 predilution with storage buffer (now 100 μM). Then dilute 1:20 into the experimental master mix, which gives a final experimental concentration of 25 μM (from the 500 μM stock) and 5 μM (from 1:5 predilution = 100 μM stock).
16. The setup described is quite complex, but it gives a good overview of screening options. For an initial test, one can also simply mix the protein stock solution with the low salt "LSB" buffers. In general, any component added will influence phase separation, so for optimal quantification one should add as few components as possible.
17. Most proteins condense at low salt conditions and LSB50 should be successful. If the tendency to form condensates is strong, LSB50 might result in aggregation, therefore also LSB200 should be tested. In case there is aggregation even

with LSB200, higher salt concentrations—or different salts like NaCl—can be tested.

18. For proteins undergoing phase-separation based on pi–pi interactions, NaCl might not strongly affect the phase diagram. 1,6-Hexanediol, urea or temperature can have a large influence in these cases.

19. Buffer solutions, for example, 1 M HEPES with variable pH: We typically start out with two solutions of different pH (e.g., pH 6.4 and pH 7.2). We are aware that HEPES is not buffering well below 6.8, but prefer to use one buffer "type" in all experiments. The laboratory of Paolo Arosio would recommend to use Bis-Tris for pH 5–5.5, Tris for pH 6–9 and CAPSO for pH 9.5–10. Once conditions where condensates form are identified, we perform a finer screen with intermittent pH values. Other substances such as phosphate buffers can also be tested. We observed in some cases that they prevent protein aggregation.

20. Many biomolecular condensates form around RNA molecules. For in vitro setups, the RNA can be obtained from diverse sources: isolated from cells (total RNA or polyA-enriched RNA), commercially purchased as homomultimers (polyU, polyG etc) or in vitro transcribed from specific templates, if there is indication for specific binding motifs. One should be careful not to add excess RNA or nucleic acid since they add charge, and furthermore they can "outcompete" or saturate binding sites on the proteins and in consequence break the multivalency network. Especially when working with cellular or in vitro transcribed RNAs, we recommend to add commercial RNase inhibitors (at low concentration) and to use filter tips.

21. Other additives that might stimulate phase separation are, for example, ATP for ATP-binding proteins. We use a 100 mM stock solution supplemented with 100 mM $MgCl_2$, and dilute it to a concentration of 2–5 mM in the final reaction setup. ATP powder is very acidic and should be dissolved in a pH-buffered solution such as HEPES buffer, or pH-adjusted.

22. Crowding agents like PEG, BSA, or Ficoll will promote any kind of phase separation, but should be used with care. PEG at high concentrations for example is also used to precipitate DNA or RNA and might thus make them appear condensate-like. BSA on the other hand was proposed to shield unspecific protein–protein interactions and to distinguish real condensates from artifacts.

23. Some condensates and proteins do not well tolerate centrifugation or pipetting. If the sample seems partially aggregated, it may help to minimize any sample handling, especially after mixing the protein stock solution with the master mix.

24. Phase-separated droplets typically appear as round spheres that settle onto the glass surface where they grow and undergo fusion events. For unlabeled proteins/nucleic acids, dyes like thioflavin T (ThT) (final concentration 0.02–0.1 mM from a 2 mM stock solution) [62], which stains β-sheet rich amyloid-like aggregates but also many phase-separated, liquid-like structures, can be added to distinguish condensates from air bubbles. If the condensates do not fuse, they might be in a gel-like material state already. If the structures look more like "coral reefs," they are probably not condensates but aggregates, which result from stronger interactions and often partial unfolding of the proteins.
25. In contrast to aggregates, which are quite irreversible, condensates—especially if freshly formed—are often easily dissolved by addition of high salt or large volumes of buffer (dilution lowers the protein concentration below the saturation threshold), increased temperature, addition of alcohols like 1,6-Hexanediol (final concentration 5–10%, see below), or RNase in case they contain RNA.
26. As an alternative to light microscopy, light scattering has successfully been used to detect condensation [63, 64].
27. To start probing the material properties of a condensate, fluorescence-based imaging techniques such as FRAP (see below) are a good starting point. Alternatives include the inclusion of nanoparticle tracers, optical tweezers, or microfluidic setups [65, 66].

Considerations for In Vivo Condensate Imaging

28. We focus here on addressing the dynamic nature of condensates, where one can probe condensate properties such as viscosity, exchange rates between dense and dilute phase, and how they change in various cellular conditions. Besides 1,6-Hexanediol and FRAP, which we discuss in this protocol, methods like FLIM (fluorescence lifetime imaging), FRET (Forster resonance energy transfer), or FCS (fluorescence correlation spectroscopy) have been used to investigate condensate liquidity through molecular mobility [15, 65]. In combination with genetic or chemical perturbation experiments, these methods can reveal factors or conditions that tune condensate material properties.
29. General use of Hexanediol: Studies suggest that 1,6-Hexanediol treatment influences a range of cellular structures and compartments [45, 67–69]. In addition, hexanediol-mediated LLPS modulation seems to affect only a subset of interactions (hydrophobic) that drive condensation, while other types of interactions (e.g., electrostatic) are not sensitive to 1,6-Hexanediol [69, 70]. Lastly, 1,6-Hexanediol has recently been

reported to affect the activity of several kinases and phosphatases [71]. These enzymes have been described to be involved in condensate formation or dissolution [72], making a careful interpretation of experimental results imperative.

30. General use of FRAP for LLPS measurements: Fluorescence recovery depends on a multitude of parameters, such as the concentration of the molecule of interest, its diffusion coefficient and its interaction with other factors that might affect its diffusion properties. In addition, one needs to take technical aspects such as the type of microscope into careful consideration. Changes in any one of these parameters can influence the rate of recovery and the results need to be carefully evaluated [44].
31. To perform FRAP measurements, fluorescent labeling of the protein of interest is necessary. In yeast, the endogenous copy of a protein is typically tagged with a fluorescent protein like GFP, preferably as the monomeric A206K variant. An optional but helpful step is to include a second reference label of a component of the same condensate with a spectrally compatible fluorophore (e.g., mCherry or mKate) to monitor the behavior of the condensate structure during and after bleaching.
32. Bleaching can be performed on the full droplet, but half-droplet bleaching gives additional information on the mobility of molecules within the droplet. It is important to consider the size of the bleached area relative to the size of the droplet as it can dramatically affect the recovery rate and also be indicative of diffusion or binding as dominant recovery factors [44].
33. The bleached area should not be defined too small as it can introduce technical artefacts (through limited number of pixels and high signal variability). The signal from the bleach region should be significantly brighter than the background signal.
34. Condensates can move within the cell over the time course of the experiment. We recommend choosing a time frame that allows the acquisition of recovery kinetics and keeps the bleached area in focus (this is easier to control when only a part of a droplet is bleached rather than the full droplet).
35. FRAP can detect if there are substructures within a droplet as these regions would recover at different rates. However, curve fitting needs to be interpreted carefully [73].
36. Several other models for curve fitting exist. The model choice is important, for example, for extraction of diffusion coefficients [74].

Acknowledgments

We would like to thank Miriam Linsenmeier, Daan Overwijn, and Bastian Kohl for critical reading of the manuscript, and the Weis and Hondele lab members for helpful discussions. S.H. would like to acknowledge funding from an EMBO long-term fellowship (ALTF 290-2014, EMBOCOFUND2012, GA-2012-600394). This work was supported by the Swiss National Science Foundation (PCEFP3_187052 to M.H.) and a European Research Council Starter Grant (ERC-ST2020 950262 to M.H.).

References

1. Banani SF, Lee HO, Hyman AA et al (2017) Biomolecular condensates: organizers of cellular biochemistry. Nat Rev Mol Cell Biol 18: 285–298
2. Shin Y, Brangwynne CP (2017) Liquid phase condensation in cell physiology and disease. Science 357:eaaf4382
3. Lafontaine DLJ, Riback JA, Bascetin R et al (2021) The nucleolus as a multiphase liquid condensate. Nat Rev Mol Cell Biol 22: 165–182
4. Larson AG, Narlikar GJ (2018) The role of phase separation in heterochromatin formation, function, and regulation. Biochemistry 57:2540–2548
5. Nandana V, Schrader JM (2021) Roles of liquid-liquid phase separation in bacterial RNA metabolism. Curr Opin Microbiol 61: 91–98
6. Razin SV, Ulianov SV (2020) Divide and rule: phase separation in eukaryotic genome functioning. Cell 9:2480
7. So C, Cheng S, Schuh M (2021) Phase separation during germline development. Trends Cell Biol 31:254–268
8. Roden C, Gladfelter AS (2020) RNA contributions to the form and function of biomolecular condensates. Nat Rev Mol Cell Biol 22(3): 183–195
9. Sanders DW, Kedersha N, Lee DSW et al (2020) Competing protein-RNA interaction networks control multiphase intracellular organization. Cell 181:306–324.e28
10. Choi J-M, Holehouse AS, Pappu RV (2020) Physical principles underlying the complex biology of intracellular phase transitions. Annu Rev Biophys 49:107–133
11. Martin EW, Holehouse AS (2020) Intrinsically disordered protein regions and phase separation: sequence determinants of assembly or lack thereof. Emerg Top Life Sci 4:307–329
12. Martin EW, Holehouse AS, Peran I et al (2020) Valence and patterning of aromatic residues determine the phase behavior of prion-like domains. Science 367:694–699
13. Riback JA, Zhu L, Ferrolino MC et al (2020) Composition-dependent thermodynamics of intracellular phase separation. Nature 581: 209–214
14. Bracha D, Walls MT, Wei M-T et al (2018) Mapping local and global liquid phase behavior in living cells using photo-oligomerizable seeds. Cell 175:1467–1480.e13
15. Alberti S, Gladfelter A, Mittag T (2019) Considerations and challenges in studying liquid-liquid phase separation and biomolecular condensates. Cell 176:419–434
16. Berry J, Brangwynne CP, Haataja M (2018) Physical principles of intracellular organization via active and passive phase transitions. Rep Prog Phys 81:046601
17. Nosella ML, Forman-Kay JD (2021) Phosphorylation-dependent regulation of messenger RNA transcription, processing and translation within biomolecular condensates. Curr Opin Cell Biol 69:30–40
18. Söding J, Zwicker D, Sohrabi-Jahromi S et al (2020) Mechanisms for active regulation of biomolecular condensates. Trends Cell Biol 30:4–14
19. Garcia-Cabau C, Salvatella X (2021) Regulation of biomolecular condensate dynamics by signaling. Curr Opin Cell Biol 69:111–119
20. Yoo H, Triandafillou C, Drummond DA (2019) Cellular sensing by phase separation: using the process, not just the products. J Biol Chem 294:7151–7159
21. Gomes E, Shorter J (2019) The molecular language of membraneless organelles. J Biol Chem 294:7115–7127
22. Bratek-Skicki A, Pancsa R, Meszaros B et al (2020) A guide to regulation of the formation

of biomolecular condensates. FEBS J 287: 1924–1935

23. Jawerth L, Fischer-Friedrich E, Saha S et al (2020) Protein condensates as aging Maxwell fluids. Science 370:1317–1323
24. Linsenmeier M, Hondele M, Grigolato F et al (2021) Dynamic arrest and aging of biomolecular condensates are regulated by low-complexity domains, RNA and biochemical activity. 2021.02.26.433003
25. Woodruff JB, Hyman AA, Boke E (2018) Organization and function of non-dynamic biomolecular condensates. Trends Biochem Sci 43:81–94
26. Hondele M, Heinrich S, De Los Rios P et al (2020) Membraneless organelles: phasing out of equilibrium. Emerg Top Life Sci 4:331–342
27. Boija A, Klein IA, Young RA (2021) Biomolecular condensates and cancer. Cancer Cell 39: 174–192
28. Alberti S, Hyman AA (2021) Biomolecular condensates at the nexus of cellular stress, protein aggregation disease and ageing. Nat Rev Mol Cell Biol 22:196–213
29. Cai D, Liu Z, Lippincott-Schwartz J (2021) Biomolecular condensates and their links to cancer progression. Trends Biochem Sci 46(7):535–549
30. Hnisz D, Shrinivas K, Young RA et al (2017) A phase separation model for transcriptional control. Cell 169:13–23
31. Peng L, Li E-M, Xu L-Y (2020) From start to end: phase separation and transcriptional regulation. Biochim Biophys Acta 1863: 194641
32. Guo YE, Manteiga JC, Henninger JE et al (2019) Pol II phosphorylation regulates a switch between transcriptional and splicing condensates. Nature 572:543–548
33. Nojima T, Rebelo K, Gomes T et al (2018) RNA polymerase II phosphorylated on CTD serine 5 interacts with the spliceosome during co-transcriptional splicing. Mol Cell 72: 369–379.e4
34. Galganski L, Urbanek MO, Krzyzosiak WJ (2017) Nuclear speckles: molecular organization, biological function and role in disease. Nucleic Acids Res 45:10350–10368
35. Saitoh N, Spahr CS, Patterson SD et al (2004) Proteomic analysis of Interchromatin granule clusters. MBoC 15:3876–3890
36. Li W, Hu J, Shi B et al (2020) Biophysical properties of AKAP95 protein condensates regulate splicing and tumorigenesis. Nat Cell Biol 22:960–972
37. Maita H, Nakagawa S (2020) What is the switch for coupling transcription and splicing? RNA Polymerase II C-terminal domain phosphorylation, phase separation and beyond. Wiley Interdiscip Rev RNA 11:e1574
38. Quinodoz SA, Ollikainen N, Tabak B et al (2018) Higher-order inter-chromosomal hubs shape 3D genome organization in the nucleus. Cell 174:744–757.e24
39. Jain A, Vale RD (2017) RNA phase transitions in repeat expansion disorders. Nature 546: 243–247
40. Gao Y, Pei G, Li D et al (2019) Multivalent m 6 A motifs promote phase separation of YTHDF proteins. Cell Res 29:767–769
41. Ries RJ, Zaccara S, Klein P et al (2019) M 6 A enhances the phase separation potential of mRNA. Nature 571:424
42. Zaccara S, Ries RJ, Jaffrey SR (2019) Reading, writing and erasing mRNA methylation. Nat Rev Mol Cell Biol 20:608–624
43. Cho W-K, Spille J-H, Hecht M et al (2018) Mediator and RNA polymerase II clusters associate in transcription-dependent condensates. Science 361:412–415
44. McSwiggen DT, Mir M, Darzacq X et al (2019) Evaluating phase separation in live cells: diagnosis, caveats, and functional consequences. Genes Dev 33:1619–1634
45. Hondele M, Sachdev R, Heinrich S et al (2019) DEAD-box ATPases are global regulators of phase-separated organelles. Nature 573(7772):144–148
46. Bancaud A, Huet S, Rabut G et al (2010) Fluorescence perturbation techniques to study mobility and molecular dynamics of proteins in live cells: FRAP, photoactivation, photoconversion, and FLIP. Cold Spring Harb Protoc 2010:pdb.top90
47. Mugler CF, Hondele M, Heinrich S et al (2016) ATPase activity of the DEAD-box protein Dhh1 controls processing body formation. eLife 5:e18746
48. Putnam A, Cassani M, Smith J et al (2019) A gel phase promotes condensation of liquid P granules in C. elegans embryos. Nat Struct Mol Biol 26:220–226
49. Boehning M, Dugast-Darzacq C, Rankovic M et al (2018) RNA polymerase II clustering through carboxy-terminal domain phase separation. Nat Struct Mol Biol 25:833–840
50. Klein IA, Boija A, Afeyan LK et al (2020) Partitioning of cancer therapeutics in nuclear condensates. Science 368:1386–1392

51. Guillén-Boixet J, Kopach A, Holehouse AS et al (2020) RNA-induced conformational switching and clustering of G3BP drive stress granule assembly by condensation. Cell 181: 346–361.e17
52. Wang J, Choi J-M, Holehouse AS et al (2018) A molecular grammar governing the driving forces for phase separation of prion-like RNA binding proteins. Cell 174:688–699.e16
53. Saha S, Weber CA, Nousch M et al (2016) Polar positioning of phase-separated liquid compartments in cells regulated by an mRNA competition mechanism. Cell 166: 1572–1584.e16
54. Saar KL, Morgunov AS, Qi R et al (2021) Learning the molecular grammar of protein condensates from sequence determinants and embeddings. Proc Natl Acad Sci U S A 118: e2019053118
55. Bolognesi B, Lorenzo Gotor N, Dhar R et al (2016) A concentration-dependent liquid phase separation can cause toxicity upon increased protein expression. Cell Rep 16: 222–231
56. Vernon RM, Chong PA, Tsang B et al (2018) Pi-Pi contacts are an overlooked protein feature relevant to phase separation. Elife 7:e31486
57. Li Q, Wang X, Dou Z et al (2020) Protein databases related to liquid-liquid phase separation. Int J Mol Sci 21:6796
58. Shen B, Chen Z, Yu C et al (2021) Computational screening of biological phase-separating proteins. Genomics Proteomics Bioinformatics 19(1):13–24
59. Küffner AM, Linsenmeier M, Grigolato F et al (2021) Sequestration within biomolecular condensates inhibits Aβ-42 amyloid formation. Chem Sci 12:4373–4382
60. McCall PM, Kim K, Fritsch AW et al (2020) Quantitative phase microscopy enables precise and efficient determination of biomolecular condensate composition. 2020.10.25.352823
61. Alberti S, Saha S, Woodruff JB et al (2018) A user's guide for phase separation assays with purified proteins. J Mol Biol 430:4806–4820
62. Biancalana M, Koide S (2010) Molecular mechanism of Thioflavin-T binding to amyloid fibrils. Biochim Biophys Acta 1804: 1405–1412
63. Van Lindt J, Bratek-Skicki A, Nguyen PN et al (2021) A generic approach to study the kinetics of liquid–liquid phase separation under near-native conditions. Commun Biol 4:1–8
64. Peran I, Martin EW, Mittag T (2020) Walking along a protein phase diagram to determine coexistence points by static light scattering. In: Kragelund BB, Skriver K (eds) Intrinsically disordered proteins: methods and protocols. Springer, New York, NY, pp 715–730
65. Bracha D, Walls MT, Brangwynne CP (2019) Probing and engineering liquid-phase organelles. Nat Biotechnol 37:1435–1445
66. Linsenmeier M, Kopp MRG, Stavrakis S et al (2021) Analysis of biomolecular condensates and protein phase separation with microfluidic technology. Biochim Biophys Acta Mol Cell Res 1868:118823
67. Wheeler JR, Matheny T, Jain S et al (2016) Distinct stages in stress granule assembly and disassembly. Elife 5:e18413
68. Onischenko E, Tang JH, Andersen KR et al (2017) Natively unfolded FG repeats stabilize the structure of the nuclear pore complex. Cell 171:904–917.e19
69. Kroschwald S, Maharana S, Simon A (2017) Hexanediol: a chemical probe to investigate the material properties of membrane-less compartments. Matters 3:e201702000010
70. Krainer G, Welsh TJ, Joseph JA et al (2021) Reentrant liquid condensate phase of proteins is stabilized by hydrophobic and non-ionic interactions. Nat Commun 12:1085
71. Düster R, Kaltheuner IH, Schmitz M et al (2021) 1,6-Hexanediol, commonly used to dissolve liquid–liquid phase separated condensates, directly impairs kinase and phosphatase activities. J Biol Chem 296:100260
72. Rai AK, Chen J-X, Selbach M et al (2018) Kinase-controlled phase transition of membraneless organelles in mitosis. Nature 559: 211–216
73. Sprague BL, McNally JG (2005) FRAP analysis of binding: proper and fitting. Trends Cell Biol 15:84–91
74. Taylor NO, Wei M-T, Stone HA et al (2019) Quantifying dynamics in phase-separated condensates using fluorescence recovery after Photobleaching. Biophys J 117:1285–1300

Chapter 19

Analysis of Oligonucleotide Biodistribution and Metabolization in Experimental Animals

François Halloy, Paulina Brönnimann, Jonathan Hall, and Daniel Schümperli

Abstract

We describe methods to follow the fate of oligonucleotides after their injection into experimental animals. The quantitation in various tissues, blood or bone marrow cells is possible by chemical ligation PCR. This method works independently of chemical modifications of the oligonucleotide and/or its conjugations to lipid or peptide moieties. Moreover, metabolization intermediates can be detected by mass spectrometry. Together with a readout assay for the biochemical or physiological effects, which will differ, depending on the particular purpose of the oligonucleotide, these methods allow for a comprehensive understanding of oligonucleotide behavior in a living organism.

Key words Oligonucleotide therapeutics, Oligonucleotide conjugates, Oligonucleotide biodistribution, Chemical ligation qPCR, Mass spectrometry

1 Introduction

Oligonucleotides (ON) are becoming increasingly important as therapeutic compounds, be it as antisense molecules that inhibit gene expression or correct aberrant splicing decisions, as double-stranded small interfering RNAs or as gapmers used for RNaseH-mediated degradation of a target RNA [1, 2]. In preclinical research, the functionality of such ON is first established by direct biochemical assays and cell culture experiments. The next step is to use an experimental animal model (most often mouse) for preclinical development. However, such a model, in many respects, is like a "black box." The biochemical or physiological outcome measure of a treatment is usually well defined and easy to record. More difficult is to assess pharmacokinetics, tissue distribution, and metabolization of a compound.

ON can be chemically modified or conjugated to different chemical moieties in order to improve properties such as stability

Peter Scheiffele and Oriane Mauger (eds.), *Alternative Splicing: Methods and Protocols*, Methods in Molecular Biology, vol. 2537, https://doi.org/10.1007/978-1-0716-2521-7_19,

in different environments, persistence in the bloodstream, targeting to different tissues, cellular uptake, and binding affinity for the target [3, 4]. However, while such modifications and conjugations are used in many studies, optimization often relies on a trial and error approach, using the final outcome measure as the only readout.

The purpose of this chapter is, first, to present a method by which ON can be detected and quantified in various tissues and/or the blood of mice injected with the compound. The method chosen—chemical ligation quantitative PCR (CL-qPCR)—functions independently of chemical modifications [5, 6] of the ON or conjugations to lipid or peptide moieties [7]. A second method—liquid chromatography mass spectrometry (LC-MS)—allows for the detection and identification of metabolization intermediates [8, 9]. This yields important insights into the fate of conjugated chemical moieties. The combination of these methods with a functional outcome assay (which will be specific for the particular purpose of the ON) may bring a deeper layer of analysis, for example, by helping researchers to design more stable conjugates or to design ones adapted to specific target tissues.

As illustration, we will use data from one of our recent publications where we used splice-switching oligonucleotides (SSO) in a mouse model for Erythropoietic Protoporphyria (EPP) [7]. EPP is a defect in heme metabolism, caused by a reduced production or function of ferrochelatase (FECH) [10]. This leads to an excessive accumulation of the enzyme's substrate protoporphyrin IX (PPIX), which is a complete heme molecule lacking the central ferrous cation. PPIX in erythrocytes reacts with visible light and releases oxygen radicals which can damage capillaries and the surrounding tissue. The result is an extremely painful sensitivity to daylight or intense artificial light. In severe cases, PPIX deposits can also affect the liver. Our SSOs target a single nucleotide polymorphism (SNP) upstream of exon 4 in the human *FECH* gene. This SNP is not pathogenic per se, but reduces FECH production by promoting the use of an aberrant splice site nearby [11]. Because this SNP is present in about 95% of all EPP patients in combination with a strong EPP loss of function mutation, it represents an attractive target for splice-switching therapies.

In what follows, we assume that the users will have one or more ON of known sequence which they want to test in a mouse or other animal disease model and that they have appropriate assays for the biochemical or physiological outcome measures.

2 Materials

2.1 General Points

1. Prior to these experiments, mice (or other experimental animals) will have been injected with ON in doses and by the injection regime of choice, under an authorization and in accordance with local animal protection laws. Animals injected by the same regime, but with an appropriate vehicle control (e.g., 0.9% saline, if your ON were dissolved in this solution), are also required for the downstream assays.
2. Prepare all solutions using ultrapure water and chemicals of analytical grade.

2.2 Isolation and Processing of Animal Tissues

1. Sterile 0.9% saline or Dulbecco's phosphate-buffered saline (DPBS).
2. Liquid nitrogen.
3. Styrofoam recipient with a flat bottom for liquid nitrogen.
4. Dry ice in an appropriate insulating container.
5. Sterile 1.5 mL Eppendorf tubes. These should be weight-tared prior to use.
6. Sterile plastic Petri dishes.
7. Sterilized dissection instruments (scalpels, forceps, scissors, etc.).
8. EDTA blood tubes.
9. Sterilized spatula (15–20 cm).
10. Sterilized mortar for tissue grinding (9 cm diameter).
11. Needle sonicator (VibraCell).
12. Clarity OTX Lysis-Loading buffer v2.0 (Phenomenex).

2.3 Analysis of Tissue Biodistribution by CL-qPCR

2.3.1 Primer Design and Synthesis

1. Therapeutic ON are commonly modified with phosphorothioate backbones and ribose modifications which hinder direct amplification by Taq polymerases. This problem is circumvented by the CL-qPCR protocol (Fig. 1). In a first step, two oligodeoxynucleotides (termed "ligators") which are each complementary to one half of the ON to be detected are synthesized. These will be chemically ligated to produce a DNA template for the second step, a real-time qPCR. The ligation reaction only occurs when the two substrate oligonucleotides are properly annealed to the target ON. Therefore, the concentrations of the ON to be detected and the ligated DNA strand are proportional to each other.
2. The PS ligator (numbered 1 in Fig. 1; *see* **Note 1**) and the BPS ligator (2; *see* **Note 2**), both consist of a template-binding section (colored in Fig. 1) and a noncomplementary overhang

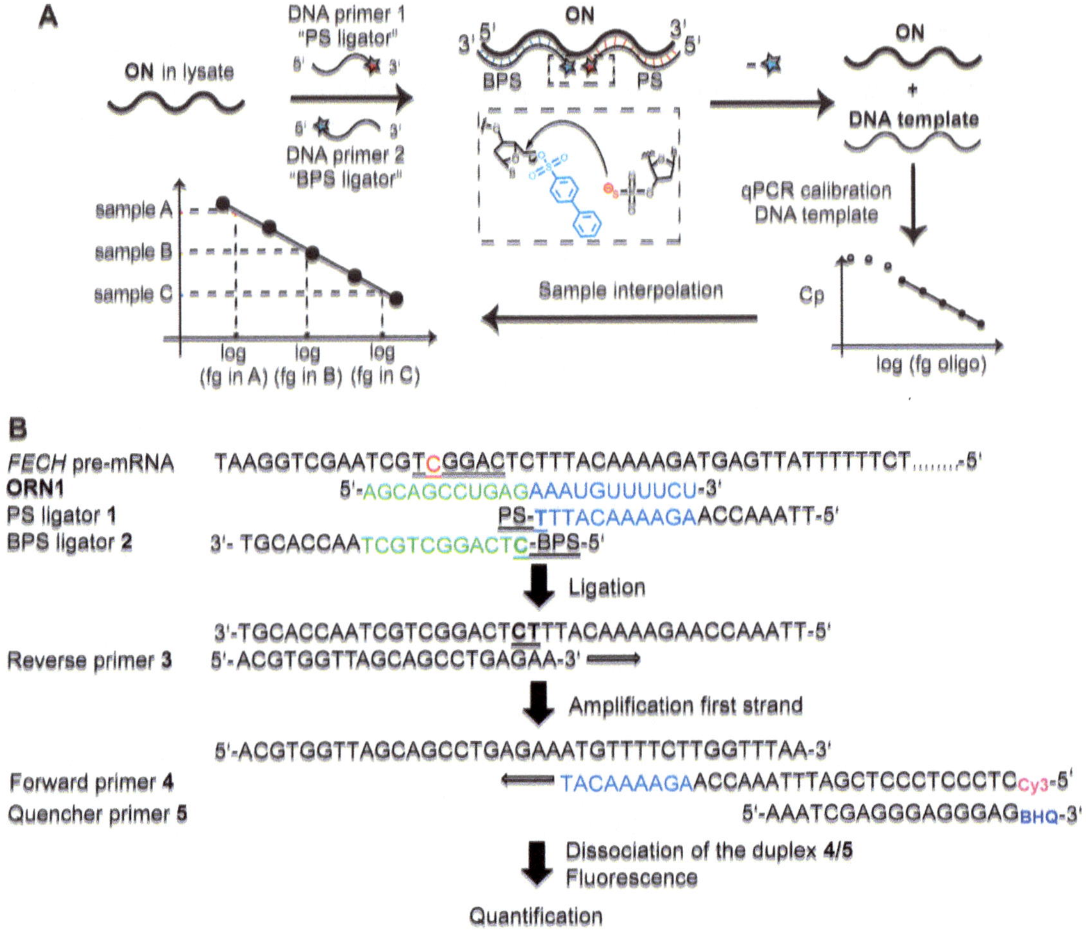

Fig. 1 Principle of Chemical Ligation-qPCR (Boos 2013, Ref. 5). (**a**) Graphical overview and (**b**) example for the ON ORN1 used in Ref. 7. Two end-modified DNA oligonucleotides, PS ligator (1) and BPS ligator (2), are added to a sample containing the target ON and incubated for 1 h at 33 °C. Hybridization of the two ligators to the target ON results in their chemical ligation and formation of a continuous DNA strand. This then serves as template for quantitative PCR using the primer and quencher oligonucleotides (3–5). (Part **a** modified from [7] under the terms of the Creative Commons Attribution License)

of 8 nucleotides (nt). When hybridized to the ON of interest, the two ligators abut each other. The 3′-terminal nucleotide of the PS ligator has a 3′ phosphorothioate modification (red); the BPS ligator is modified with a 5′ biphenylsulfonate group (blue). These two end-groups, when brought into close vicinity, undergo a bond-forming reaction that results in a phosphorothioate linkage at the junction of the ligated oligonucleotides.

3. The reverse PCR primer (numbered 3 in Fig. 1) is an unmodified oligodeoxynucleotide. Its sequence is fully complementary to the PS ligator and additionally contains 2 nt complementary

to the BPS ligator at its 3′ end. Thus, it will only prime synthesis on a correctly ligated DNA strand.

4. Two modified DNA oligonucleotides are required to complete the real-time PCR reaction. The forward primer (numbered 4 in Fig. 1) contains, from 3′ to 5′, those nucleotides of the BPS ligator that are not covered by the reverse primer, followed by a 14-nt overhang and a Cyanine-3 (Cy3) group at the 5′ end. The quencher oligo (numbered 5 in Fig. 1) is complementary to 16 nt at the 5′ end of the forward primer and carries a black hole quencher (BHQ) group at its 3′ end. During PCR, the quencher oligonucleotide will be dissociated from the forward primer which will allow the Cy3 group to exhibit fluorescence. *See* **Note 3** on the storage of these two oligos.

2.3.2 CL-qPCR

1. Clarity OTX Lysis-Loading buffer v2.0, Phenomenex.
2. Polyadenylic acid (PolyA), GE Healthcare, prepared at 1 mg/mL in water.
3. FastTaq Polymerase enzyme and associated buffer (Roche).
4. dNTPs.
5. Thermocycler.
6. LC480 qPCR instrument (Roche) or equivalent.

2.4 Metabolite Identification by LC-MS

1. Clarity OTX 96-well plate or 20 mg columns, Phenomenex.
2. Proteinase K (optional).
3. Methanol, analytical grade.
4. Equilibration buffer: 50 mM ammonium acetate, pH 5.5, 0.5% (m/v) Triton X-100.
5. Modified equilibration buffer: 50 mM ammonium acetate, pH 5.5, 0.5% (m/v) Triton X-100, 0.1% (m/v) cysteine.
6. Washing buffer: 50 mM ammonium acetate, pH 5.5, 50% (v/v) analytical grade acetonitrile.
7. Elution buffer: 100 mM ammonium bicarbonate, pH 10, 40% (v/v) analytical grade acetonitrile, 10% (v/v) analytic grade tetrahydrofuran.
8. Tris-EDTA (TE) buffer, pH 8.0.
9. Vacuum applying device for 96-well plates.
10. 1.5 mL screw-cap tubes.
11. Speedvac vacuum centrifuge.
12. 96-well plates for oligonucleotide recovery (Eppendorf Deep-well 96/2000 μL, or equivalent).
13. LC-MS system (Agilent 1200/6130) with a Waters UPLC OST C-18 column (2.1 × 50 mm, 1.7 μm) for oligonucleotide detection.

14. LC-MS buffers A and B. A: 0.4 M hexafluoroisopropanol, 15 mM trimethylamine, pH 8; B: analytical grade methanol.

3 Methods

3.1 General Points

1. Wear clean plastic gloves at all steps.
2. All samples (tissue lysates, qPCR plates and purified metabolite solutions) should be prepared and kept at 4 °C.
3. When processing multiple samples, the dissection instruments, mortars for tissue grinding, and needle for needle sonication must be cleaned thoroughly with 70% ethanol between consecutive samples to avoid any cross-contamination. Especially, any contamination between ON-treated samples and vehicle- or untreated samples can result in undesired amplification of control samples during the qPCR run.

3.2 Isolation and Processing of Animal Tissues

1. Euthanize the animals by the method approved by your local regulations.
2. Proceed with a standard dissection protocol. Open and pull back the skin on the front of the animal and rinse the underlying surface with sterile DPBS before opening the thoracic and abdominal cavities.
3. Blood: Collect by heart puncture and transfer to EDTA tubes. For further processing, continue with **step 9**.
4. Bone marrow: Collect bone marrow cells by cleaning the long bones and flushing them with DPBS following a known procedure (e.g., [12]). For further processing, continue with **step 9**.
5. Other organs: Collect the organs and tissues of choice and rinse them in ice-chilled DPBS. Briefly soak up the excess liquid.
6. Snap-freeze the organs/tissues in liquid nitrogen. Keep on dry ice for immediate processing or store for long term at −80 °C.
7. Precool a mortar, a pestle, sterile Eppendorf tubes and a spatula in a recipient (Styrofoam box or equivalent) filled with 1 cm of liquid nitrogen.
8. Grind the tissue to a fine powder and transfer the content into an Eppendorf tube (*see* **Note 4**). Keep on dry ice for immediate processing or store the ground tissue at −80 °C.
9. Use the OTX lysis buffer at 4 °C without warming up at room temperature. Determine the mass of ground tissue powder contained per tube and add 10 volumes (v/w) of OTX Lysis buffer, for example, 400 μL for 40 mg of ground tissue. For blood, use 1 volume (v/v) of OTX lysis buffer, for example, 200 μL for 200 μL of EDTA blood. Pipet up and down eight

times to homogenize, vortex for 30 s. For uncharged ON (e.g., morpholino) *see* **Note 5**.

10. Homogenize the suspension by needle sonication at 4 °C. We used three cycles at 15 s ON-5 s OFF and 40% amplitude (*see* **Note 6**).
11. Briefly centrifuge at 14,000 × *g* for 30 s to pellet debris. Keep the supernatant (hereafter called lysate) for immediate use in Subheading 3.3 or Subheading 3.4 or long-term storage at −80 °C.

3.3 Analysis of Tissue Biodistribution by CL-qPCR

1. Dilute the lysate at a ratio of 1/750 (v/v) in water.
2. For each tissue and each compound, generate a calibration curve. Proceed as follows: from a stock solution of the target ON of known concentration (we used a stock at 25 μM), generate 6 serial, ten-fold dilutions in the diluted lysate of a control animal.
3. Conduct the chemical ligation step for each animal of the study and each calibration sample. Include a water control. Mix composition for Chemical Ligation: 0.1 μL PS ligator (10 μM stock), 0.1 μL BPS ligator (10 μM stock), 1 μL Poly A (1 mg/mL stock), 1 μL 10× Buffer with $MgCl_2$, and 5.8 μL ultrapure water, mixed with 2 μL diluted lysate. Run the chemical ligation for 1 h at 33 °C (*see* **Note 7**).
4. Prepare a qPCR mix containing 0.15 μL forward primer (10 μM stock), 0.15 μL reverse primer (10 μM stock), 0.13 μL BHQ primer (28.28 μM stock), 0.15 μL dNTPs mix, 1 μL 10× Buffer with $MgCl_2$, 0.1 μL 10× FastTaq Polymerase and 6.32 μL ultrapure water and mix with 2 μL of the chemical ligation reaction mixture. We recommend to prepare technical triplicates for each sample.
5. Run the qPCR with the following program: 10 min/95 °C, 50 cycles (3 s/95 °C—30 s/55 °C—10 s/72 °C). Ramping speeds were set at 2.5 °C/s.
6. Calculate the cycle values Cp by using the software of the qPCR machine.
7. Determine the linear detection range, the PCR efficiency and the value of each experimental sample for the tissues to be investigated. Exclude from analysis samples falling outside of the linear detection range. This is exemplified in Fig. 2, where we conducted a pilot study with our lead ON **ORN1** to compare subcutaneous (*s.c.*) and intraperitoneal (*i.p.*) injections. In the nine tissues we selected, we found no statistical difference in **ORN1** bioaccumulation between *s.c.* (blue) and *i. p* (red) injected mice. Brain and blood samples fell outside of

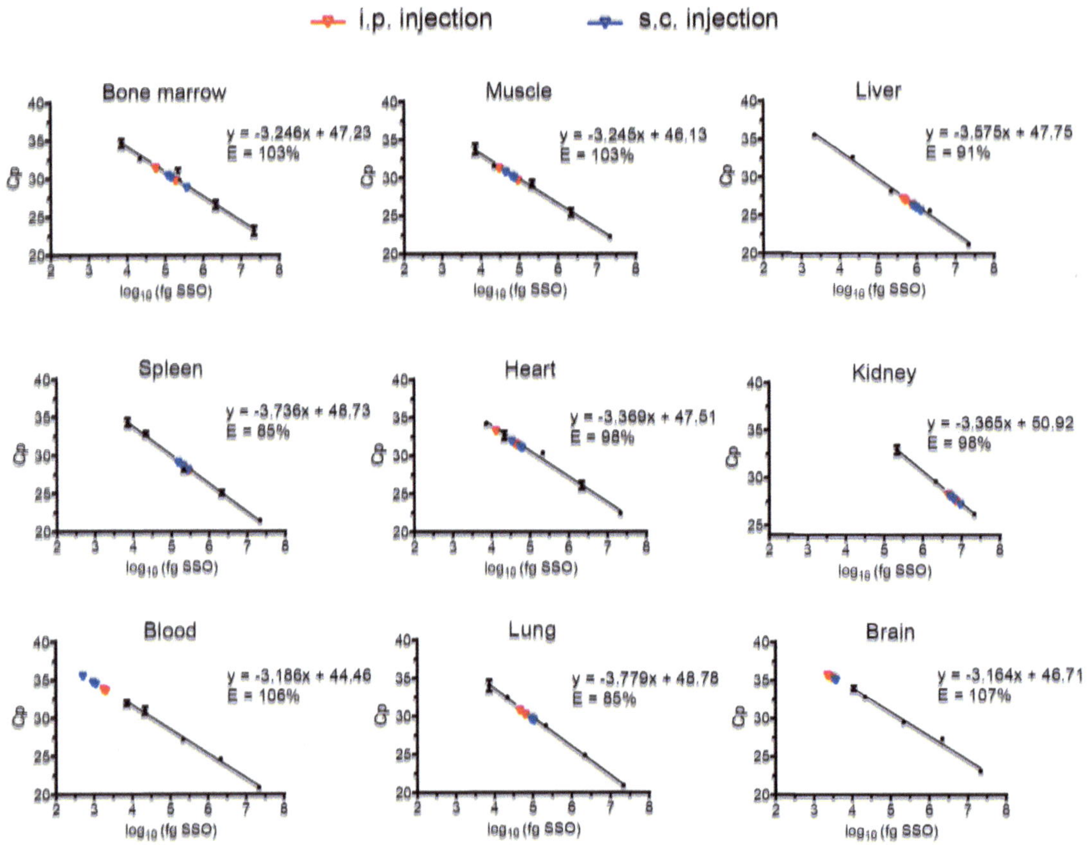

Fig. 2 Calibration range (defined by black data points) and comparison of intraperitoneal (i.p., red symbols) and subcutaneous (s.c., blue symbols) injections in selected mouse tissues. Animals were injected with four doses of oligonucleotide ORN1 at 50 mg/kg s.c. or i.p and sacrificed 24 h after the last injection. *Cp* cycle number at detection threshold (crossing point), *SSO* splice-switching oligonucleotide, *E* PCR efficiency

the linear detection range and, in the case of brain, were not distinguishable from the noninjected control (Cp ≥35 cycles).

8. Interpolate the ON concentration in tissues with help of the calibration curves and express in ng oligonucleotide per mg tissue (*see* **Note 8**). In an account published recently [7], we deployed the CL-qPCR to compare bioaccumulation of different conjugated versions of **ORN1** for optimized delivery to liver and bone marrow tissues (Fig. 3).

3.4 Metabolite Identification by LC-MS

1. Process the lysate. For extremely lipophilic conjugates (e.g., cholesterol), we recommend to compare tissue processing with or without the proteinase K digestion step as per manufacturer's instructions (*see* **Note 9**). Prepare for each compound a control solution by spiking a known amount of oligonucleotide stock into the lysate of a control animal. We recommend

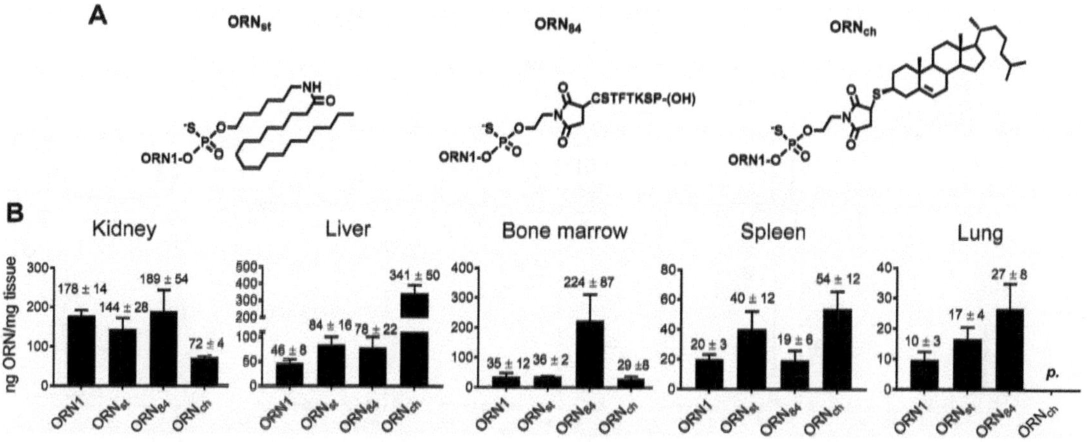

Fig. 3 Head-to-head comparison of SSO conjugates for improved tissue delivery in EPP. (**a**) Three conjugated versions of the parent oligonucleotide ORN1 were investigated for biodistribution studies. The stearic (ORN_{st}) and cholesteryl (ORN_{ch}) conjugates were selected for improved binding to plasma proteins [21, 22]. The peptide conjugate ORN_{84} was selected for CD84 binding in bone marrow cells [23]. Animals were injected over 2 weeks with four doses of ORN1, ORN_{st}, ORN_{84}, or ORN_{ch} at 50 mg/kg s.c., and sacrificed 24 h after the last injection. (**b**) Accumulation of oligonucleotides in a set of tissues expressed in ng ORN1/mg tissue as determined by the CL-qPCR. (Modified from [7] under the terms of the Creative Commons Attribution License)

for such controls a final oligonucleotide concentration of 10–50 μM for accurate detection in LC-MS.

2. Condition the desired number of wells of the Clarity OTX 96-well plate with 1 mL methanol (*see* **Note 10**).
3. Equilibrate using 1 mL of modified equilibration buffer.
4. Load lysates on the Clarity OTX 96-well plate and discard the flow-through. Dry the column before proceeding to the next step.
5. Wash (*see* **Note 11**) in the following order: 1 mL modified equilibration buffer, 2 × 1.5 mL washing buffer, 2 × 1.5 mL equilibration buffer. Dry the column before proceeding to the next step.
6. Add 1 mL of Elution buffer. Apply brief vacuum for 2 s to soak the column and wait 1 min before proceeding to elution into a 96 well plate. This step is prone to cross-contamination, and vacuum should be applied in brief pulses to avoid spilling of the eluates into adjacent wells.
7. Transfer the eluates into screw-cap tubes and concentrate them in a Speedvac centrifuge.
8. Resuspend in 100 μL of TE buffer, pH 8.0.
9. Briefly centrifuge (5,000 × *g*, 30 s) to pellet insoluble particles.

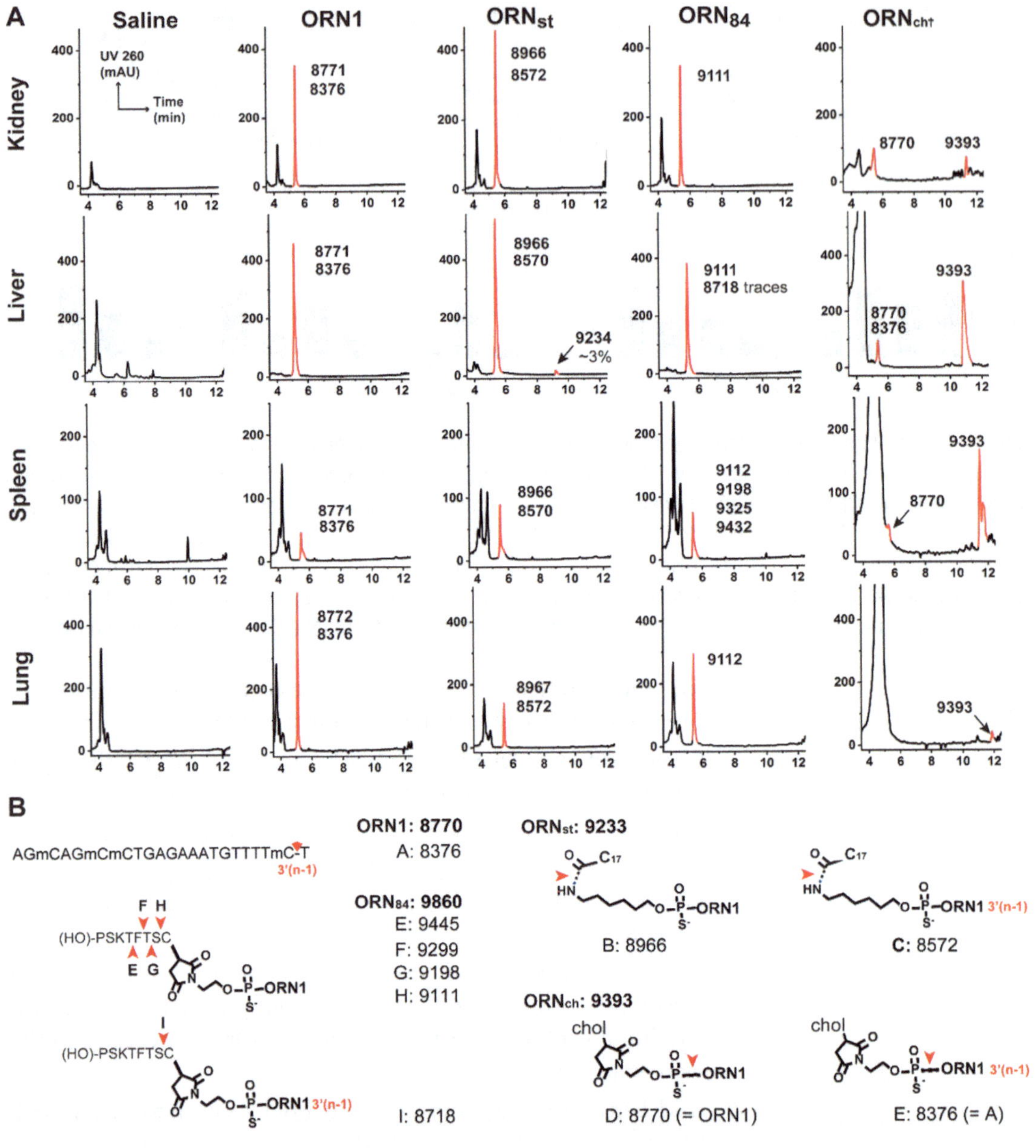

Fig. 4 Metabolism of SSO conjugates in selected mouse tissues after repeated injections. (**a**) Animals were injected and sacrificed as detailed in Fig. 3. Oligonucleotides were extracted from the tissue matrix using the Phenomenex Clarity OTX kit. Tissues from mice treated with **ORN$_{ch}$** were subjected to a proteinase K digestion before processing. Detected oligonucleotide species are marked in red, next to their associated masses determined by LC–MS. (**b**) Calculated molecular weights of parent SSOs and their inferred metabolites from LC–MS data shown in (**a**) (C17 is the stearoyl chain; chol is the thiocholesteryl group). (Modified from [7] under the terms of the Creative Commons Attribution License)

10. Inject the supernatant into the LC-MS apparatus using a standard gradient for each compound. We used a gradient of 10% to 95% buffer B in buffer A at 0.3 mL/min and 65 °C.
11. The spectrum of saline (or other solvent)-treated animals provides a reference spectrum with peaks derived from endogenous molecular species. Identify peak(s) specific to samples coming from ON-treated animals, record their molecular weight and the mass spectrum.
12. Metabolism may occur on (1) the ON moiety and, if applicable, on (2) the conjugate moiety. (1) Single-stranded oligonucleotides that are fully modified with phosphorothioate linkages and 2′-*O*-alkyl groups (e.g., methoxy, methoxyethyl) undergo mainly exonuclease degradation from the 3′-end, leading to detection of truncated sequences (n-1, n-2, etc.) [15], whereas siRNAs are primarily degraded by exonucleases from the 5′-end [16]. Finally, gapmers undergo degradation by endonucleases and exonucleases [17]. (2) Various metabolic reactions can be expected on a conjugated group. Amide and ester bonds can be cleaved enzymatically, whereas thiosuccinimide linkers are prone to loss of the conjugated group or hydrolysis and ring opening [18, 19]. In the case of peptide conjugates, significant peptidolysis is to be expected if the peptide sequence is not protected by modification (e.g., by pegylation or by the incorporation of unnatural amino-acids) [20]. We propose to assign metabolites and provide the qualitative analysis such as in Fig. 4 (*see* **Notes 12** and **13**).

4 Notes

1. Some steps in the synthesis of the PS primer for CL-qPCR have to be carefully controlled. We observed that a partial loss of the terminal 3′-phosphorothioate group could occur upon (1) the harsh conditions of oligonucleotide deprotection and (2) prolonged drying times in the lyophilizer. In such cases, an impurity of molecular weight M-16 g/mol is observed in the LC-MS chromatogram, whereby separation from the intact product is not possible. Therefore, we recommend a rapid purification process and storage of the final primer in aliquoted batches in water supplemented with 50 μM TCEP. Freeze-thaw cycles of stock samples should be avoided, and we recommend not to use batches older than 6 months. LC-MS quality-control injections were performed regularly.
2. The chemical stability of the BPS moiety has to be monitored carefully during ammonia deprotection of the primer. We did not notice degradation of the 5′ BPS group upon storage at −20 °C for up to two years.

3. Cyanine-3 (Cy3) and Black-Hole Quencher (BHQ) primers must be protected from light (e.g., with tubes wrapped in aluminum) to prevent photobleaching, which adversely affects performance in the RT-qPCR assay.
4. We recommend to prepare working aliquots containing 30–60 mg tissue per tube. Aliquots can be used to generate internal replicates for uptake and metabolism studies, but also for other assays, for example, lysis for RNA or protein work.
5. The solid-phase extraction procedure with OTX lysis buffer requires the ON to be negatively charged, and may not work correctly if using morpholino ON. For uncharged ON, a different extraction method (e.g., hybridization based [9] or phenol–chloroform based [13]) could be used.
6. Extended sonication times may lead to sample over-heating and degradation of some ON conjugate moieties, particularly when cleavable linkers are used. Compatibility and stability of ON conjugates toward needle sonication can be assessed by spiking the compound into a tissue lysate taken from a PBS-treated mouse, sonicating, and processing the lysate for ON cleanup and LC-MS injection. A bead beater homogenizer (Qiagen TissueLyser II, or equivalent) can be used alternatively to needle sonication.
7. The temperature of the chemical ligation reaction may impact ON detection. For long ON (i.e., more than 16 nt) a reaction temperature of 33 °C gave good quality data. Shorter ON sequences may require use of lower temperatures, for example 20 °C or 25 °C, as the corresponding ligator sequences will also be shorter and have lower melting temperatures (T_m). Similarly, when comparing ON with a common sequence, but of different chemistries (e.g., 2′-*O*-methyl, LNA), we observed variations in PCR efficiencies that we assumed to be derived from changes in hybridization affinities. For both of these reasons, the CL-qPCR protocol may require testing several CL temperatures until satisfactory PCR efficiencies are obtained in the qPCR step.
8. We have investigated 18S genomic DNA as an alternative to tissue weight for normalization [14]. A good correlation between these methods was obtained for several tissues and **ORN1** (Fig. 5). This can be applied for the investigation of samples of specific cell—or subtissue—populations where weighing of tissue may be difficult.
9. We limited the use of proteinase K pretreatment of the tissues to avoid any work-up related enzymatic degradations, for instance in the case of an oligonucleotide-peptide conjugate. In preliminary experiments, we observed that the amide bond of $\mathbf{ORN_{st}}$ remained unaltered after proteinase K treatment.

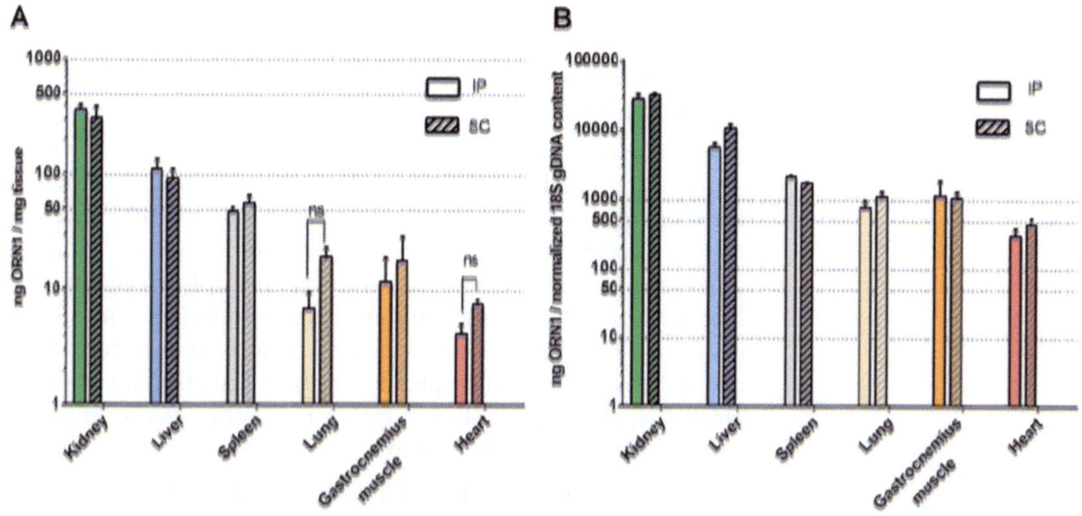

C

Tissue to liver bioaccumulation ratios	Tissue weight normalisation		18S gDNA normalisation	
	i.p.	s.c.	i.p.	s.c.
Kidney	3.28	3.36	4.97	3.04
Liver	1	1	1	1
Spleen	0.43	0.61	0.37	0.16
Heart	0.04	0.08	0.05	0.04
Muscle	0.11	0.2	0.2	0.1
Lung	0.06	0.21	0.13	0.1

Fig. 5 Comparison of ON content normalization to tissue weight or to genomic DNA content. Mice were injected with ORN1 and sacrificed as detailed in Fig. 2, and tissues were processed for oligonucleotide quantification by CL-qPCR. (**a**) ON accumulation expressed in ng ORN1/mg tissue showed kidney and liver to be the main sites of accumulation, followed by spleen, lung, gastrocnemius muscle and heart. No significant differences were observed in comparing i.p. and s.c. injections. (**b**) ON accumulation expressed in ng ORN1/18S genomic DNA (gDNA) content, using a TaqMan assay for mouse 18S gDNA [14]. A similar ranking of the tissues is observed. (**c**) Accumulation ratios relative to liver led to similar rankings when using tissue weight or 18S gDNA for normalization

However, we found proteinase K pretreatment to be necessary in the case of $\mathbf{ORN_{ch}}$ in the liver, where no extraction from ground tissue was obtained otherwise. Therefore, proteinase K treatment was done for all tissues derived from $\mathbf{ORN_{ch}}$-treated animals. While we do not have explanation for this, we believe it is due to the high hydrophobicity of $\mathbf{ORN_{ch}}$ in comparison to other conjugates. This observation should be kept in mind in the case of oligonucleotides conjugated to similarly lipophilic moieties.

10. We recommend the use of Clarity OTX 96-well plates when processing large numbers of samples. For individual 20 mg columns, however, some of the aforementioned steps may have to be adapted.
11. We observed a major improvement of the LC-MS spectra by replacing potassium phosphate by ammonium acetate in the buffer (not shown). Use of potassium phosphate buffer yielded noisy spectra, associated with multiple M-16 adducts corresponding to desulfurization at phosphorothioate linkages. While this might have been interpreted as a metabolic event, we obtained clean spectra with ammonium acetate. We assume that this issue is specific to phosphorothioate oligonucleotides, although it should be kept in mind whenever potassium phosphate buffer is used.
12. As described above, the OTX Clarity technique (solid-phase extraction based) allows for a qualitative determination of metabolism. The technique could be upgraded to quantification by spiking into the tissue lysate an internal control of known concentration and retention time.
13. We used a reverse-phase (RP) column for metabolite identification by LC-MS. In the case of lipophilic conjugates, the loss of a lipophilic moiety causes a significant shift in retention time on the RP column, which eases the identification of metabolites (*see* $\mathbf{ORN_{st}}$ and $\mathbf{ORN_{ch}}$ in Fig. 4). When working with unconjugated or polypeptidic conjugates, it may be prudent to use an ion-exchange (IE) column, as RP columns do not typically provide resolution of truncated oligonucleotides (i.e., n-1, n-2, etc.) or truncated-peptide fragments for area-under-the-curve determinations (see **ORN1** and $\mathbf{ORN_{84}}$ in Fig. 4).

Acknowledgments

We acknowledge financial support by the NCCR RNA and Disease of the Swiss National Science Foundation and by ETH Zürich to J.H.

References

1. Stein CA, Castanotto D (2017) FDA-approved oligonucleotide therapies in 2017. Mol Ther 25(5):1069–1075. https://doi.org/10.1016/j.ymthe.2017.03.023
2. Levin AA (2019) Treating disease at the RNA level with oligonucleotides. N Engl J Med 380(1):57–70. https://doi.org/10.1056/NEJMra1705346
3. Crooke ST, Wang S, Vickers TA, Shen W, Liang X-h (2017) Cellular uptake and trafficking of antisense oligonucleotides. Nat Biotechnol 35(3):230–237. https://doi.org/10.1038/nbt.3779
4. Crooke ST, Witztum JL, Bennett CF, Baker BF (2018) RNA-targeted therapeutics. Cell Metab 27(4):714–739. https://doi.org/10.1016/j.cmet.2018.03.004

5. Boos JA, Kirk DW, Piccolotto M-L, Zuercher W, Gfeller S, Neuner P, Dattler A, Wishart WL, Von Arx F, Beverly M, Christensen J, Litherland K, van de Kerkhof E, Swart PJ, Faller T, Beyerbach A, Morrissey D, Hunziker J, Beuvink I (2013) Whole-body scanning PCR; a highly sensitive method to study the biodistribution of mRNAs, noncoding RNAs and therapeutic oligonucleotides. Nucleic Acids Res 41(15):e145. https://doi.org/10.1093/nar/gkt515
6. Brunschweiger A, Gebert LFR, Lucic M, Pradere U, Jahns H, Berk C, Hunziker J, Hall J (2016) Site-specific conjugation of drug-like fragments to an antimiR scaffold as a strategy to target miRNAs inside RISC. Chem Commun 52(1):156–159. https://doi.org/10.1039/C5CC07478A
7. Halloy F, Iyer PS, Cwiek P, Ghidini A, Barman-Aksozen J, Wildner-Verhey van Wijk N, Theocharides APA, Minder EI, Schneider-Yin X, Schumperli D, Hall J (2020) Delivery of oligonucleotides to bone marrow to modulate ferrochelatase splicing in a mouse model of erythropoietic protoporphyria. Nucleic Acids Res 48(9):4658–4671. https://doi.org/10.1093/nar/gkaa229
8. Kim J, Basiri B, Hassan C, Punt C, van der Hage E, den Besten C, Bartlett MG (2019) Metabolite profiling of the antisense oligonucleotide Eluforsen using liquid chromatography-mass spectrometry. Mol Ther Nucleic Acids 17:714–725. https://doi.org/10.1016/j.omtn.2019.07.006
9. Sips L, Ediage EN, Ingelse B, Verhaeghe T, Dillen L (2019) LC–MS quantification of oligonucleotides in biological matrices with SPE or hybridization extraction. Bioanalysis 11(21):1941–1954. https://doi.org/10.4155/bio-2019-0117
10. Lecha M, Puy H, Deybach J-C (2009) Erythropoietic protoporphyria. Orphanet J Rare Dis 4(1):1–10. https://doi.org/10.1186/1750-1172-4-19
11. Gouya L, Puy H, Robreau A-M, Bourgeois M, Lamoril J, Da Silva V, Grandchamp B, Deybach J-C (2002) The penetrance of dominant erythropoietic protoporphyria is modulated by expression of wildtype FECH. Nat Genet 30(1):27–28. http://www.nature.com/ng/journal/v30/n1/suppinfo/ng809_S1.html
12. Amend SR, Valkenburg KC, Pienta KJ (2016) Murine hind limb long bone dissection and bone marrow isolation. J Vis Exp 110:53936. https://doi.org/10.3791/53936
13. Turnpenny P, Rawal J, Schardt T, Lamoratta S, Mueller H, Weber M, Brady K (2011) Quantitation of locked nucleic acid antisense oligonucleotides in mouse tissue using a liquid-liquid extraction LC-MS/MS analytical approach. Bioanalysis 3(17):1911–1921. https://doi.org/10.4155/bio.11.100
14. Boos JA, Beuvink I (2016) Whole-body scanning PCR, a tool for the visualization of the in vivo biodistribution pattern of endogenous and exogenous oligonucleotides in rodents. In: Medarova Z (ed) RNA imaging: methods and protocols. Springer New York, New York, NY, pp 99–111. https://doi.org/10.1007/978-1-4939-3148-4_8
15. Yu RZ, Geary RS, Monteith DK, Matson J, Truong L, Fitchett J, Levin AA (2004) Tissue disposition of 2′-O-(2-methoxy) ethyl modified antisense oligonucleotides in monkeys. J Pharm Sci 93(1):48–59. https://doi.org/10.1002/jps.10473
16. Nair JK, Attarwala H, Sehgal A, Wang Q, Aluri K, Zhang X, Gao M, Liu J, Indrakanti R, Schofield S, Kretschmer P, Brown CR, Gupta S, Willoughby JLS, Boshar JA, Jadhav V, Charisse K, Zimmermann T, Fitzgerald K, Manoharan M, Rajeev KG, Akinc A, Hutabarat R, Maier MA (2017) Impact of enhanced metabolic stability on pharmacokinetics and pharmacodynamics of GalNAc–siRNA conjugates. Nucleic Acids Res 45(19):10969–10977. https://doi.org/10.1093/nar/gkx818
17. Baek M-S, Yu RZ, Gaus H, Grundy JS, Geary RS (2010) In vitro metabolic stabilities and metabolism of 2′-*O*-(methoxyethyl) partially modified Phosphorothioate antisense oligonucleotides in preincubated rat or human whole liver homogenates. Oligonucleotides 20(6): 309–316. https://doi.org/10.1089/oli.2010.0252
18. Fontaine SD, Reid R, Robinson L, Ashley GW, Santi DV (2015) Long-term stabilization of maleimide–thiol conjugates. Bioconjug Chem 26(1):145–152. https://doi.org/10.1021/bc5005262
19. Wei C, Zhang G, Clark T, Barletta F, Tumey LN, Rago B, Hansel S, Han X (2016) Where did the linker-payload go? A quantitative investigation on the destination of the released linker-payload from an antibody-drug conjugate with a maleimide linker in plasma. Anal Chem 88(9):4979–4986. https://doi.org/10.1021/acs.analchem.6b00976
20. Brinckerhoff LH, Kalashnikov VV, Thompson LW, Yamshchikov GV, Pierce RA, Galavotti HS, Engelhard VH, Slingluff CL Jr (1999) Terminal modifications inhibit proteolytic degradation of an immunogenic mart-127–35 peptide: implications for peptide vaccines. Int J Cancer 83(3):326–334. https://doi.org/10.

1002/(sici)1097-0215(19991029)83:3<326::Aid-ijc7>3.0.Co;2-x

21. Wolfrum C, Shi S, Jayaprakash KN, Jayaraman M, Wang G, Pandey RK, Rajeev KG, Nakayama T, Charrise K, Ndungo EM, Zimmermann T, Koteliansky V, Manoharan M, Stoffel M (2007) Mechanisms and optimization of in vivo delivery of lipophilic siRNAs. Nat Biotechnol 25(10): 1149–1157. http://www.nature.com/nbt/journal/v25/n10/suppinfo/nbt1339_S1.html

22. Biscans A, Coles A, Haraszti R, Echeverria D, Hassler M, Osborn M, Khvorova A (2019) Diverse lipid conjugates for functional extrahepatic siRNA delivery in vivo. Nucleic Acids Res 47(3):1082–1096. https://doi.org/10.1093/nar/gky1239

23. Nowakowski GS, Dooner MS, Valinski HM, Mihaliak AM, Quesenberry PJ, Becker PS (2004) A specific heptapeptide from a phage display peptide library homes to bone marrow and binds to primitive hematopoietic stem cells. Stem Cells 22(6):1030–1038. https://doi.org/10.1634/stemcells.22-6-1030

INDEX

Peter Scheiffele and Oriane Mauger (eds.), *Alternative Splicing: Methods and Protocols*,
Methods in Molecular Biology, vol. 2537, https://doi.org/10.1007/978-1-0716-2521-7,

MIX
Papier aus verantwortungsvollen Quellen
Paper from responsible sources
FSC® C105338

If you have any concerns about our products,
you can contact us on
ProductSafety@springernature.com

In case Publisher is established outside the EU,
the EU authorized representative is:
Springer Nature Customer Service Center GmbH
Europaplatz 3, 69115 Heidelberg, Germany

Printed by Libri Plureos GmbH
in Hamburg, Germany